Clean Energy Production Technologies

Series Editors

Neha Srivastava, Department of Chemical Engineering and Technology
IIT (BHU) Varanasi, Varanasi, India

P. K. Mishra, Department of Chemical Engineering and Technology
IIT (BHU) Varanasi, Varanasi, India

The consumption of fossil fuels has been continuously increasing around the globe and simultaneously becoming the primary cause of global warming as well as environmental pollution. Due to limited life span of fossil fuels and limited alternate energy options, energy crises is important concern faced by the world. Amidst these complex environmental and economic scenarios, renewable energy alternates such as biodiesel, hydrogen, wind, solar and bioenergy sources, which can produce energy with zero carbon residue are emerging as excellent clean energy source. For maximizing the efficiency and productivity of clean fuels via green & renewable methods, it's crucial to understand the configuration, sustainability and techno-economic feasibility of these promising energy alternates. The book series presents a comprehensive coverage combining the domains of exploring clean sources of energy and ensuring its production in an economical as well as ecologically feasible fashion. Series involves renowned experts and academicians as volume-editors and authors, from all the regions of the world. Series brings forth latest research, approaches and perspectives on clean energy production from both developed and developing parts of world under one umbrella. It is curated and developed by authoritative institutions and experts to serves global readership on this theme.

Akhilesh Kumar Yadav
Editor

Powering Future: Clean Energy Solutions for a Sustainable Planet

Springer

Editor
Akhilesh Kumar Yadav
Department of Environmental Engineering
and Management
Chaoyang University of Technology
Taichung, Taiwan

Department of Electronics and Communication Engineering
Manipal University Jaipur
Jaipur, India

Department of Civil Engineering
Manipal University Jaipur
Jaipur, India

ISSN 2662-6861 ISSN 2662-687X (electronic)
Clean Energy Production Technologies
ISBN 978-981-96-5054-5 ISBN 978-981-96-5055-2 (eBook)
https://doi.org/10.1007/978-981-96-5055-2

This Springer imprint is published by the registered company Springer Nature Singapore Pte Ltd.
The registered company address is: 152 Beach Road, #21-01/04 Gateway East, Singapore 189721, Singapore

If disposing of this product, please recycle the paper.

Dedicated to Environmental Engineers and Scientists Around the World

Foreword

I must congratulate the editor for an excellent review of clean energy requirement and identification of the gaps in the present knowledge in the area of clean energy need for sustainable development. Dr. Akhilesh Kumar Yadav's hard work to bring this volume is highly appreciated.

The contribution of eminent academicians is distinctive, appropriate, and judicious for the sustainable growth on the planet earth.

Our relationship with the world is significantly impacted by energy. We must acknowledge that the reliance on fossil fuels is unsustainable. Our future is at risk due to environmental degradation, climate change, and resource depletion. As you embark on this journey toward a brighter future, this book, *Powering the Future: Clean Energy Solutions for a Sustainable Planet* will engage you in a captivating examination of the technologies that are on the brink of revolutionizing our energy landscape. This book explores many renewable energy sources, including solar panels and geothermal electricity. The document contains details regarding nuclear energy and the remarkable wind turbines that harness wind power. The journey continues throughout each subsequent generation. Sustainability requires the utilization of two separate approaches and a more thorough examination. This book explains the fundamental concepts of energy efficiency and conservation, allowing us to decrease our use and maximize our resources. The book's writing style is both accessible and fascinating, making it a suitable choice for anyone with an interest in energy and environmental subjects. This comprehensive resource is intended for inquisitive individuals seeking to acquire knowledge, influential decision-makers who define our future, and ambitious business executives who are catalysts for change. Adhere to the guidelines outlined in this paragraph. Let it inspire you to promote and back the adoption of sustainable and eco-friendly energy options. The advent of renewable energy heralds a promising future that begins right now.

Director, Indian Institute of Technology Patna
Patna, India

Triloki Nath Singh

Preface

The motivation behind this volume, titled *Powering the Future: Clean Energy Solutions for a Sustainable Planet,* is to present innovative techniques for securing a sustainable future. This book serves as your guide to understanding the cutting-edge technologies driving you towards a cleaner, more sustainable world. It covers a wide range of topics, from harnessing the sun's energy through solar panels to tapping the earth's geothermal heat, from wind power captured by majestic turbines to the potential of nuclear energy. However, clean energy is not only about generation; it also involves using less energy. This contributed book highlights the critical role of energy efficiency and conservation in achieving sustainability. Written in a clear and engaging style, it empowers readers with a broad interest in energy and environment. The content is rich with informative text that is easy to follow, complemented by diagrams and photographs. Whether you're a curious citizen, policy-maker shaping the future, and business leader driving change, this contributed book is an essential resource for you. The book is organized into five sections. **Part I** provides an overview of wind energy and ecological sustainability with four chapters. **Part II**, which consists of four chapters, focuses on biomass and biogas, while **Part III** covers solar energy in three chapters. **Part IV** is dedicated to solar technology with three chapters, and the final section, **Part V**, consists of four chapters addressing clean energy solutions.

I would first like to thank the Almighty for His blessings throughout the completion of this project, enabling us to successfully explore this challenging research domain with a technical perspective. The authors are sincerely grateful to Springer for the opportunity to publish this work on such a contemporary topic of modeling and simulations. Special thanks go to all the contributors who made this volume a valuable source of knowledge, reporting the latest findings in their respective fields. Out of 76 abstracts submitted, 47 were selected for full chapter contributions, and 18 chapters were ultimately included, based on scope and quality of the submissions. The authors have shown tremendous effort throughout the process, from writing and revising based on comments to addressing Springer's evaluation reports and final proofreading. The review process adhered to a double-blind method, where both authors' and reviewers' identities were kept confidential to ensure unbiased decision-making. I also like to thank the Springer publication team for their timely responses and guidance. Lastly, I acknowledge Jaydeep Kumar (BCAS, New Delhi, India) and Advocate Dinesh Chandra Gupta (High Court, Allahabad, India) for their unwavering help and support during the execution of this project.

The editor welcomes any feedback, comments, or suggestions to improve future editions. Proposals for new chapters for subsequent editions are also encouraged and can be sent directly to the volume editor.

Taichung, Taiwan Akhilesh Kumar Yadav

Contents

About the Editor

Akhilesh Kumar Yadav, Ph.D., MIE, earned a Bachelor of Technology (B.Tech.) in Electronics and Communication Engineering from Chaudhary Charan Singh University in Meerut, India. He then obtained a Master of Technology (M.Tech.) in Environmental Engineering from Madan Mohan Malaviya Engineering College in Gorakhpur, India (affiliated with Dr. A.P.J. Abdul Kalam Technical University, Lucknow). His doctoral degree, with a research focus on air pollution, was awarded by the Indian Institute of Technology (BHU), Varanasi, India.

Dr. Yadav has served at various institutions, including Indian Institute of Technology (Banaras Hindu University), Varanasi, India; Indian Institute of Technology Bombay, Mumbai, India; Bhabha Atomic Research Centre, Mumbai, India (as a research collaborator); and Chaoyang University of Technology in Taichung, Taiwan. He has authored/co-authored several research and review articles, book chapters, and edited/authored books, and holds a patent.

Dr. Yadav received the Young Engineer Award from the Institution of Engineers (India) and the Young Scientist Award from VDGOOD Technology Factory, Kolkata. His research interests include air, water, and soil pollution; climate change; vulnerability; human health risk assessments; and GIS applications in environmental pollution and management. He is an active member of several reputed international professional bodies, including the IEI (Kolkata), ECI (New Delhi), AEACI (Mumbai), MSI (New Delhi), and IASTA (Mumbai).

Part I
Wind Energy and Ecological Sustainability

Chapter 1
Energy Efficiency and Conservation: The Silent Heroes of Sustainability

Mridu Kulwant, Saba Shirin, and Akhilesh Kumar Yadav

Abstract Electronic wastes are emerging pollution problems in nowadays global context. Countries such as the USA, Canada, Europe, and India receive thousands of tons of e-waste from the various countryside. The electronic industry (in developing countries) will be advancing with the rapid speed manufacturing industry in terms of supply and demand based on the lifespan of materials. E-waste consists of several issues that directly and indirectly harm human as well as ecological health. Mining is the only source of metals, which pollutes almost all natural resources. Recycling and recovery of metals from electronic waste can be second wide source of metals. Moreover, electronic waste recycling is attractive because of the presence of precious metals (PMs) economically and commercially. Recovery of valuable metals will accelerate the economy as well as a healthy environment.

This chapter offers an extensive summary of e-waste, issues involved, composition, sources, relevant regulations and laws from an Indian perspective, government initiatives, analysis of traditional and innovative alternative methods of metal recovery, and their importance in the management strategy. It focuses on tools and tech-

M. Kulwant (✉)
Department of Environmental Studies, The Maharaja Sayajirao University of Baroda, Vadodara, India
e-mail: mridu.k-envphd@msubaroda.ac.in

S. Shirin
Department of Environmental Science, School of Vocational Studies and Applied Sciences, Gautam Buddha University, Greater Noida, India
e-mail: sabashirin83@gmail.com

A. K. Yadav
Department of Environmental Engineering and Management, Chaoyang University of Technology, Taichung, Taiwan

Department of Electronics and Communication Engineering, Manipal University Jaipur, Jaipur, India

Department of Civil Engineering, Manipal University Jaipur, Jaipur, India
e-mail: yadavbasti@gmail.com

A. K. Yadav (ed.), *Powering Future: Clean Energy Solutions for a Sustainable Planet*, Clean Energy Production Technologies, https://doi.org/10.1007/978-981-96-5055-2_1

nological approaches for recycling and recovering metals from the following wastes. Mechanical techniques of metal recovery are followed by metallurgical and absorption processes. The chapter concerns the delimitation of the frontiers or boundaries of e-waste recycling. It describes the importance of global-level scientific initiatives to evaluate e-waste management quantitatively and qualitatively. The use of different commercial eradicators for metal recovery was examined thoroughly.

Keywords E-waste · Pollution · Natural resources · Precious metals

1.1 Introduction

Electronic waste is a major issue globally, but it is particularly problematic in developing nations like India. To resolve this problem, we focus on the basic principle of environmental, viz. 5 R techniques (Refuse, Reduce, Reuse, Repurpose, Reuse). The first challenge before e-waste management is to understand what will be adverse impact due to e-waste on a living being and the environment. The fast advancement of technology, upgrades in lifestyle and comfort. The evolution of the electrical and electronics industry contributes a considerable quantity of e-waste, mainly refrigerators, washing machines, computers and printers, televisions, mobile phones, iPods, and other items that often contain harmful compounds. Ferrous metals make up 50% of e-waste, polymers 21%, and nonferrous metals 13%. The Global E-waste Monitor 2020 reports that in 2019, users disposed of 53.6 million devices which is 20% more than they did in the preceding 5 years. India produced 3.2 million tons of e-waste last year, putting it in third place, after China (10.1 million tons) and the United States (6.9 million tons). The nation produces around 7.2 tons of commercial hazardous trash, 400,000 tons of electronic waste, 1.5 tons of plastic garbage, 1.7 tons of medical waste, and 48 tons of home waste annually, according to the Comptroller and Auditor General (CAG) assessment. Table 1.1 lists the numerous contaminants found in electronic devices. Approximately 95% of e-waste is recycled inefficiently in India's informal sector. An e-waste report released by the United Nations (UN) at the World Economic Forum on January 24 states that the waste stream reached 48.5 tons in 2018 and is expected to triple if no action is taken (E-waste UN 2019). Merely 20% of the global e-waste is gathered. According to a UN study, only 30% of the total amount of cobalt (a metal in high demand in laptops, smartphones, and electric vehicle batteries) from e-waste is recovered due to inefficient extraction methods. Chinese recycling companies already recycle more cobalt per year than the country produces itself. Furthermore, compared to metal made from raw ore, recycled metal has an energy efficiency of two to ten times. A resource that is in the wrong place at the wrong time is called waste. A wide definition of electronic trash can found in the European Parliament's WEEE Directive 2002/96/EC (European Union 2003).

> "**electrical and electronic equipment**" or "EEE" is equipment which is dependent on electric currents or electromagnetic fields to work properly and equipment for the generation, transfer, and measurement of such currents and fields falling under the categories set out in Annex IA and designed for use with a voltage rating not exceeding 1000 Volts for alternating current and 1500 Volts for direct current.

Table 1.1 Occurrence of different pollutants

S. no.	Pollutants	Occurrence
1.	Liquid crystal	Displays
2.	Lithium	Mobile telephones, photographic equipment, video equipment, batteries
3.	Mercury	A copper machine, a steam iron, a clock, a pocket calculator, switches, LCDs
4.	Nickel	Battery, relay, semiconductor, pigment, alloys
5.	PCBs (polychlorinated biphenyls)	Transformers, capacitors, softening agents for paints, glue, plastic
6.	Selenium	Photoelectric cells, pigments, photocopiers, fax machines
7.	Silver	Capacitors, switches (contacts) batteries, resistors
8.	Zinc	Steel, brass, alloys, disposable and rechargeable batteries, luminous substances
9.	Arsenic	Semiconductors, diodes, microwaves, LEDs (light emitting diodes), solar cells
10.	Barium	Electron tubes, filler of plastic and rubber, lubricant additives
11.	Brominated flame-proofing agent	Casings, circuit boards (plastic), cables, and PVC cables
12.	Cadmium	Batteries, pigments, solders, alloys, circuit boards, computer batteries, monitors, cathode ray tubes (CRTs)
13.	Chrome	Dyes/pigments, switches, solar
14.	Cobalt	Insulator
15.	Copper	Conductor cables, copper ribbons, coils, circuitry, pigments
16.	Lead	Circuit boards, lithium batteries, lead-acid batteries, transistors, solar panels, PVC (polyvinyl chloride), stabilizers, lasers, LEDs, thermoelectric elements, and solar panels

> The term **'e-waste'** is explained as any waste generated from appliances that use electric power and reach to End-of-Life (EoL) (Sharma et al. 2020).

As shown in Fig. 1.1, personal devices such as computers, monitors, phones, tablets, televisions, other major household appliances, and heating and cooling equipment account for 50% of all metals produced (according to a report by global recycling).

1.2 Characterization and Composition

Numerous valuable materials that can be recycled can be found in electronic garbage such as aluminum, ferrous metals, gold, tantalum, mercury, lead, chromium, arsenic, zinc, cadmium, copper, beryllium, and silver, encouraging the conservation of energy needed to create natural resources as well as their preservation. We require new electronics made with fresh materials. A typical ton of WMP comprises 128 kg of copper, 15 kg of nickel, 10 kg of tin, 6 kg of charges, 3.63 kg of silver, 1 kg of

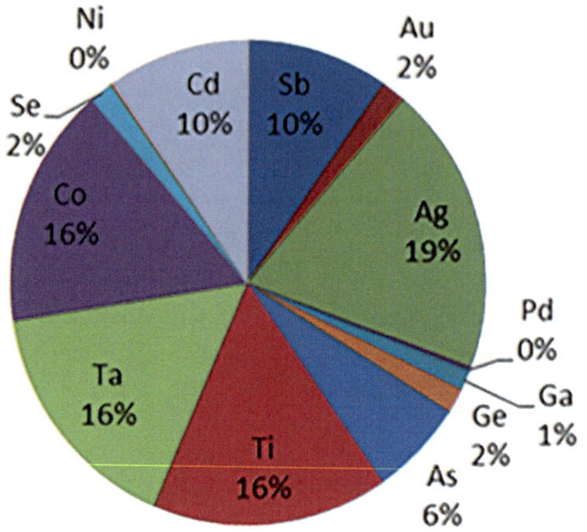

Fig. 1.1 Rare earth and valuable metal in PCB

antimony, 0.347 kg of gold, and 0.15 kg of palladium (Ari 2015). Material recovery from electrical and electronic waste, including mobile phones, tops the list of commonly discarded products. Mining the basic ores is far more expensive than recycling the circuit board waste from mobile phones. Zeng et al. published research results focusing on mobile phone recycling in 2018. With limited reserves and ever-growing demand, recycling these abundant minerals has enormous economic value (Li and Liu 2015; Zhang et al. 2017; Zeng et al. 2018). From an economic perspective, PCBs have the highest metal concentrations among all WMP components (Hira et al. 2017), and these metal contents are higher than computer PCBs, at 63 and 45 wt.%, respectively (Yamane et al. 2011). Valuable metal content is higher in PCBs than in PCBs from other electrical and electronic equipment (EEE) such as computers, refrigerators, and Internet routers (Priya and Hait 2017; Shah et al. 2014). Because e-waste is composed of several components, its composition is complicated. Nevertheless, it can be broken down into five components. Many materials such as polymers, glass, nonferrous and ferrous metals, etc. can be employed (Ongondo et al. 2011).

(a) ***Non-metals***: Waste PCBs have 70% of non-metallurgic materials, epoxy, resins, and glass.

- ***Metals***: The average composition of basic, rare earth, and valuable metals in PCBs of e-waste is shown in Table 1.2, which explains the metal content of about 30% (Debnath et al. 2018). Metallic gold increases the conductivity of electronic parts used in current refined electronic gadgets (mobile phones, calculators, sensors, global positioning system units, and so on). Gold is an extremely efficient and dependable conductor with excellent conductivity and anti-corrosion properties. All electronic gadget switches, connectors, connecting wires, relay contacts, and soldered joints are made from gold or a coating, and their construction is consistent.

Table 1.2 An illustration of the typical composition of basic materials, rare earth, and valuable metals in PCBs from e-waste

S. no.	Composition of precious metals, rare earth metals, and basic metals in PCBs from e-waste, by percentage	
1.	Lead (Pb)	1–5
2.	Nickel (Ni)	1–3
3.	Gold (Au)	0.03
4.	Copper (Cu)	~16
5	Palladium (Pd)	0.01
6.	Silver (Ag)	0.05
7.	Ferrite	~3
8.	Solder	~4
9.	**Basic metals in PCB**	**Conc. in wt.%**
10.	Copper (Cu)	6.928
11.	Aluminium (Al)	14.172
12.	Lead (Pb)	6.298
13.	Tin (Sn)	1.007
14.	Nickel (Ni)	0.850
15.	Zinc (Zn)	2.204
16.	Iron (Fe)	20.471
17.	Others [Bismuth (Bi), Antimony (Sb), Tantalum (Ta), etc.]	Less than 0.01

Reprinted from Debnath et al. (2018) with permission from Frontiers of Environmental Science & Engineering.

- The National Institute for Materials Science (NIMS) estimates that 16% of the world's gold reserves are initiated in electronic waste, which includes precious metals incorporating copper, silver, platinum, and gold. Above 6800 tons of gold have been discovered in urban mines (NIMS 2015).
- Toxic components include lead, mercury, cadmium, and halogenated flame retardants. E-waste is also considered a terrifying solid waste, rather than precious metals.
- By 2030, poorer countries produce twice the amount of e-waste as industrialized nations, according to Scientific Reports (Panda et al. 2020). There are 10 categories of electronic devices WEEE Directive (European Union 2003), following as:

    1. Large household appliances
    2. Small household appliances
    3. IT and telecommunications equipment
    4. Consumer equipment
    5. Lighting equipment
    6. Tools of electrical and electronic (excluding large-scale stationary industrial tools)

Table 1.3 Listing different kinds of plastics encompasses in e-waste

S. no.	Types of plastic	% Contribution in e-waste
1.	Acrylonitrile butadiene styrene (ABS) is the largest contributor	30
2.	High impact polystyrene (HIP)	25
3.	Polycarbonate (PC)	10

7. Toys, leisure, and sports equipment
8. Medical devices (except for all implanted and infected products)
9. Monitoring and control instruments
10. Automatic dispensers

- E-waste includes various types of plastic material, including Acrylonitrile Butadiene Styrene (ABS) which has a large contribution (30%) and polycarbonate (PC) with the same percentage contribution. Table 1.3 lists all kinds of plastic encompasses in e-waste. In addition to catalytic upgrading of ABS/PC, pyrolysis, PCB biochar, and char collected from electronic waste, these catalysts are considered alternative green catalysts employed in metal or oil separation (Leeabai et al. 2021). WPCBs comprise materials with high hardness and tenacity properties, multiple types of metals, and resin with a mixture of intertwined glass (Li and Xu 2010; Huang et al. 2009). Hence electronic waste covers 25% of copper (by mass) in accumulation to precious metals. WEEE leachates are generally composed of a high level of chloride ion which is hostile (Creamer et al. 2006).

1.3 Sources of Electronic Waste

A staggering amount of electrical and electronic equipment (EEE) waste has been created recently that has been introduced into the market every month. China is the first, and the United States has become the second leading fabricator of electronic waste around the globe due to the never-ending progression of e-waste expansions of electronic gadgets (Yang et al. 2019). Washing machines, vacuum cleaners, toasters, drying machines, refrigerators, irons, air conditioners, coffee machines, and other electric appliances contribute significantly to household e-waste. Mobile phones, personal computers, DVD players, television sets, laptops, MP_3 players, and other electronic devices are often used for personal purposes (Jitendra 2011). The four major contributor sectors are as follows:

1. IT sector
2. Institutions
3. Government buildings
4. Other

Table 1.4 Quantity of e-waste in different equipment

S. no.	Equipment	Quantity of e-waste (in million metric tons)
1.	Small equipment	16.8
2.	Large equipment	9.1
3.	Temperature exchange (freezing and cooling) equipment	7.6
4.	Screens and monitors	6.6
5.	Small IT	3.9
6.	Lamps	0.7

Reprinted from Baldé et al. (2017) with permission from United Nations University, CC-BY

Globally, 71 kilotons of BFR plastics and 50 tons of mercury from e-waste are found undocumented flows, harming the health of exposed workers and polluting the environment heavily. Table 1.4 comprises the quantity of waste from different equipment, as closely corresponds to their waste management characteristics, there are around 54 EEE product categories clustered into six wide-ranging categories (Baldé et al. 2017).

1.4 Indian Scenario

ASSOCHAM-EY joint report (Shenoy 2019) entitled, "Electronic waste management of India" highlights that India going to produce 5 million tons of principle waste computer equipment and mobile phones by 2021. In India, the creation of electronic trash varies by state based on the use and demand throughout 2019, and Maharashtra, Tamil Nadu, and Andhra Pradesh were the top three e-waste-producing regions whereas Uttar Pradesh, West Bengal, Delhi, Karnataka, Gujarat, Madhya Pradesh, and Punjab also followed. Metropolitan regions tend to be significant producers of garbage in India; 65 cities alone produced more than 60% of total rubbish created in India. Mumbai is the nation's biggest provider of e-waste, later Delhi, Bengaluru, Chennai, as well as Kolkata (Wikipedia 2020). Table 1.5 shows the portion of e-waste by different Indian states. India is expected to produce over 5 million tons of electrical waste next year (Shenoy 2019). According to Ashley Delaney (Founder of Group TenPlus, an e-waste management company, in Goa), "An ordinary circuit board from a mobile or laptop contains roughly 16 different metals," (Shenoy 2019).

1.5 Worldwide Scenario

The production of electronic garbage has surged by almost 21% over the last 5 years, according to Global E-waste Monitor 2020. Globally, 53.6 million tons of electronic garbage were produced in 2019, and by 2030, that digit is predicted to

Table 1.5 State contributes in e-waste generation in India 2019

Sr. no.	State	E-waste generation in %
1.	Maharashtra	19.8
2.	Tamil Nadu	13
3.	Uttar Pradesh	10.1
4.	West Bengal	9.8
5.	Delhi	9.5
6.	Karnataka	8.9
7.	Gujarat	8.8
8.	Madhya Pradesh	7.6

rise to 74 million tons. With 3.2 million tons, India is the third-largest contributor. In 2019, Asia generated 24.9 million tons of electronic garbage, more than any other region. This can be attributed to the expanding population in that region. Among the countries that generated electronic waste in 2019, China generated 10.1 million tons, followed by the United States with 13.1 million tons, Europe with 12 tons, and the United Kingdom with 6.9 million tons (United Nations University 2020). According to reports, only 17.4% of electronic garbage is recovered. Every year, the manufacturing, consumption, and e-waste disposal emit 94 million tons of CO_2 released into the atmosphere.

Customers disposed of 53.6 million tons of electronics globally in 2019, 20% more than in the previous 5 years, as reported by the Global E-waste Monitor 2020. Every year, America generates more e-waste ($55 billion) than many other countries' GDP (United Nations University 2020). Electronic waste is stored and recycled only 20% as per the document, although have high recoverable potential materials such as copper and gold (Forti et al. 2020). A versatile phone reusing can give over 35,000 pounds of copper, 33 pounds of palladium, 772 pounds of silver, and 75 pounds of gold. Humans can suffer many health issues from lead in e-waste, and it can damage kidneys and peripheral nervous systems too. If released into the environment, lead is an environmental pollutant (LeBlanc 2020). To remediate the e-waste issue in Nigeria, a $15 million collaborative movement was pronounced. The Nigerian Government and the Around the World Environment Office are partnering in 2019, and the UN Environment will work together (The Global Environmental Facility 2019). Globally, 8% of electronic waste is assumed for yearly growth (United Nations University January 10, 2020). In Asia, over 40% of e-waste is sent from several affluent nations such as United States, Canada, and Europe (Forti et al. 2020). World Financial Gathering, the Worldwide Environment Office, and the World Trade Chamber for economical improvement collaborated on the repairing of ancient electronic frameworks (UN Environment Programme 2019). To deal with growing e-waste, 67 countries collaborated to work on legislation (World Economic Forum, accessed January 11, 2020; Ryder and Houlin 2019). In Japan, around 2139 kilotons of e-waste are made but only 26% of it are appropriately collected. The USA and UK dumped junk roughly 19.4 and 24.9 kg per individual more noteworthy than Japan's 16.9 kg of e-waste and the Asian per capita normal of 4.2 kg (capita normal) (The Global E-Waste Statistics Partnership 2020). E-waste, which has 100 times more gold than comparable metals, makes up around 7% of the world's gold reserves.

1.6 E-Waste Management Rules

In accordance with the government's official declaration, 2011 witnessed the implementation of the e-waste (Management and Handling) Rules. Plan 1 of a Run the Show in 2011 administers the generation, deal, and acquirement, and and a component or piece of electrical or electronic hardware is prepared. E-waste (Management) Rules, 2016 (EWMR) came in to effect successful on October 1, 2016 (Ministry of Environment, Forest, Government of India 2011; Ministry of Environment, Forest and Climate Change, Government of India 2016).

1.6.1 MoEF Notification 2016

Obligatory necessities of the new 2016 Rule should comprise the collection system, registration of dismantlers and recyclers, reduction of the hazardous substances (RoHS), and extended producer responsibility (EPR) focus. Components and electrical and electronic equipment are included under Schedule I. The schedule specifically targets producers and consumers, or bulk buyers. Recycling of e-waste through assembly, destruction, and reuse. MoEF, Central Pollution Control Board, Delhi (CPCB), State Pollution Control Boards or Committees of Union Boards (SPCBs), Port Authority beneath the Indian Ports Act, 1908 (15 of 1908) and Customs Authority beneath the Customs Act, 1962 (52 of 1962) (PCCs), and Urban Local Bodies (Municipal Committee or Committee or Enterprise) (ULBs) (Urban Development and Urban Housing Department 2017). In Koko Port, Nigeria e-waste regulation came in 1988 when 4000 tons of harmful squandered from Italy. African nations boycott the moment of electronic waste beneath the Bamako Convention. An arrangement oversees the administration of unsafe squanders and their transboundary transport inside Africa (Department of Drinking Water and Sanitation, Ministry of Jal Shakti 2012). According to MoUD 2017, Fig. 1.2 shows the Swachh Bharat Mission's achievements under the Swachh Bharat Mission. Figure 1.3 shows the increasing demand for heavy metals in different sectors due to the vast applications of metals. There are many directives for the use of heavy metal in different appliances following standard directives and rules. Unused refrigerators, personal computers, televisions, copy machines, and other small home appliances are directly associated with 15 countries' e-waste generated per person (Akcil et al. 2015; Widmer et al. 2005). In the other 27 countries of the EU, it was calculated that 19.1 kg/year/person by 2012 and 10.5 million tons of electronic waste by 2014 (Akcil et al. 2015; Huisman 2010). Numerous nations, including the European

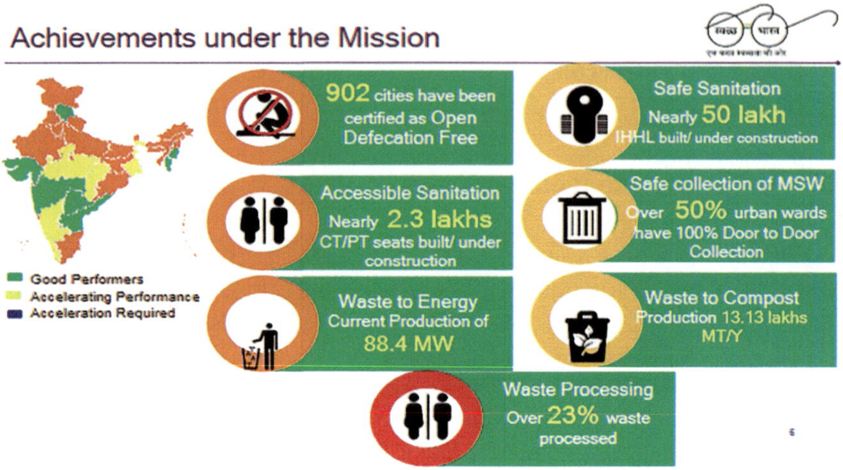

Fig. 1.2 Achievement under Swachh Bharat mission, SBM reporting MoUD 2017

Fig. 1.3 Application of heavy metals

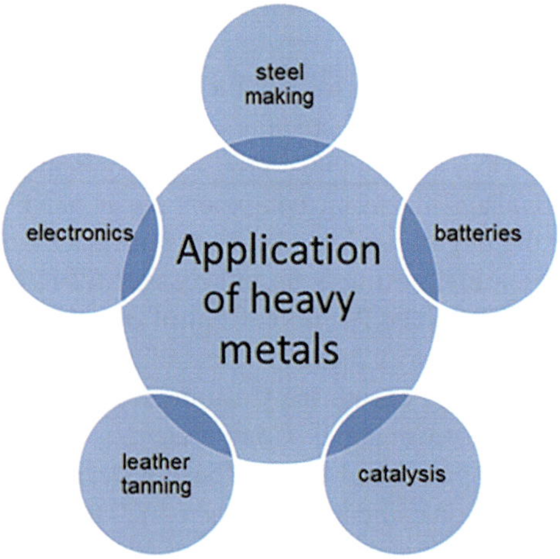

Union (EU), China, Japan, and South Korea, have taken the lead and implemented a variety of legislation and regulations to handle the expanding e-waste issues globally (Yang et al. 2017).

The 2002/96/EC rule suggests stricter regulations and limitations on disposal sites and businesses, along with the mandated recovery of metal values from e-waste. These are key thoughts for creating e-waste recycling plans. As per the Turkish Ministry of Environment and Forestry (European Union 2003), "Regulation of Controlling Waste of Electric and Electronic Equipments (WEEE)" (No. 2830) focuses on requirements for recycling and reusing electronic left-over and stringent policies, 2012 as follows:

- E-waste can be taken care of and managed conveniently by local authorities.
- The duty must be taken by the manufacturers and recycle the discarded product at the center itself, licensed centers of recycling, and separation recycling centers.
- The manufacturers should develop research platforms, projects, and enhanced recycling strategies for e-waste to evolve better technologies (Thomson Reuters IP Analytics 2013).

To enhance e-waste management, several methods are used, particularly in developed nations, likewise Extended Producer Responsibility (EPR), Material Flow Analysis (MFA), Multi-Criteria Analysis (MCA), Life Cycle Assessment (LCA), and so on. From 61 in 2014 to 67 in 2017, and 78 in 2019, nations have enacted national policies, laws, or regulations on e-waste.

1.7 Risk Assessment of E-Waste

Metals present in electrical and electronic wastes have many adverse effects and affect environmental health (Sharma et al. 2020). Similarly, Akcil et al. (2015) suggested that recovering metal content through discarded electronic items is worth it, as it is a potent source of metal content, and recycling will reduce related pollution. Figure 1.4 shows the different kinds of metals in e-waste. Heavy metal has various applications in such as steel making, electronics, batteries, leather tanning, catalysis, and so on and when discarded unsustainably causes direct or indirect discharges, and significant wastewater laden with heavy metals and solid waste has multiple environmental and health effects (Jin and Zhang 2020; Yadav 2021).

1.7.1 Effect on Environment

Numerous issues are related to the incineration and getting rid of electronic waste and the high chance of harming the environment, including aquatic and terrestrial ecosystems, as well as threats to living systems directly or indirectly are alarming

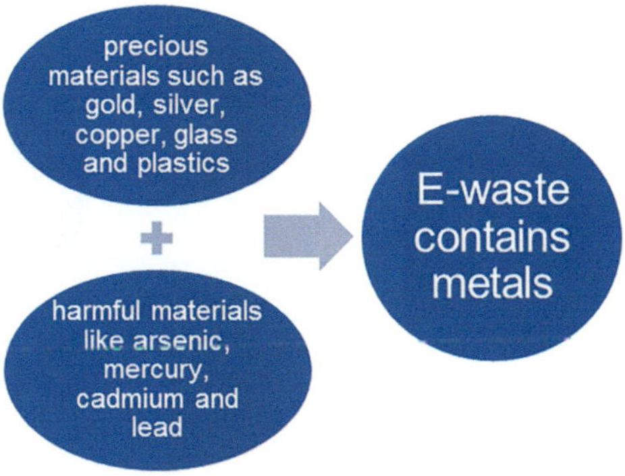

Fig. 1.4 E-waste contains different kinds of metals

signs. It is possible to recycle e-waste, secondary pollutants, and burning waste which produces dioxins, furans, polybrominated organic pollutants, and polycyclic aromatic hydrocarbons. These gases are harmful and cause serious diseases. Landfill sites are susceptible to polluting groundwater via the leaching of heavy metal dissolution. A great deal of ecologically harmful organic materials may also appear in e-waste, such as fire retardants and other organic components like Cl-Br (Akcil et al. 2015). As Jinglan and Hong (Hong et al. 2015) discussed in their work, there are several pollution problems associated with improper disposal of e-waste, such as

1. Climate change
2. Human toxicity
3. Air pollution
4. Photochemical oxidant formation
5. Particulate matter formation
6. Terrestrial acidification
7. Freshwater eutrophication
8. Marine eutrophication
9. Terrestrial eco-toxicity
10. Freshwater eco-toxicity
11. Marine eco-toxicity
12. Fossil depletion

The methods for recycling electronic trash are aggressive. These appliances typically contain hazardous elements that, when incorrectly disposed of, pollute the environment (Jitendra 2011). The common contaminants and factors contributing to climate change in the atmosphere are explained in Fig. 1.5. Arsenic and mercury emissions as air pollutants and selenium as water pollutants play vital roles in human toxicity; nitrogen oxide and phosphate support eutrophication as well as an important pollutant of photochemical oxidant formation (Yadav 2015; Yadav et al.

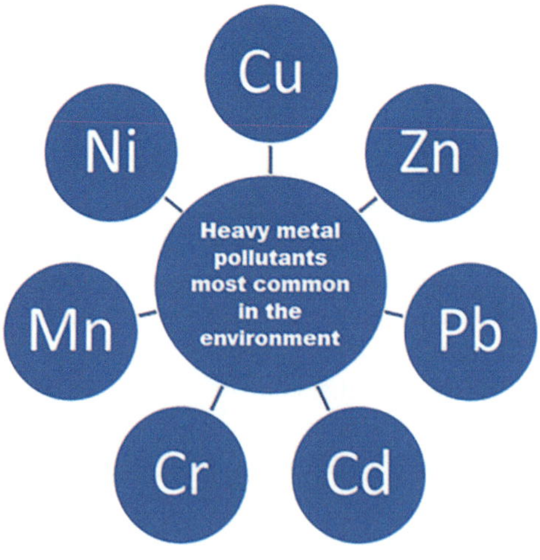

Fig. 1.5 Most common heavy metal pollutants in the environment

2022). Significant pollutants include ammonium and nitrate in water and volatile organic chemicals (non-methane) released into the atmosphere. Nitrogen oxides and sulfur dioxide emissions support terrestrial acidification. Various metals like vanadium, nickel, beryllium, and selenium play important roles in the eco-toxicity of marine and freshwater (Hong et al. 2015). As per Borthakur (2016), e-waste contains several metals that are potential persistent organic pollutants (POPs) that can reach to food chain through different ways such as bioaccumulation and biomagnification (Borthakur 2016).

1.7.2 Effect on Human

Overwhelming metal contamination can cause an assortment of well-being issues, such as lung fibrosis, dyspnea, frailty, solid and joint inconvenience, renal harm, and high blood weight (Ahmad et al. 2016). There are various effects of heavy and toxic metals on living beings. According to a recent report by Logical India released in July 2020, "E-waste contains toxic or hazardous extracts such as mercury, which damages the human brain and coordination system." The nervous system is affected by this condition, which reduces energy levels, alters blood composition, and later affects vital organs like the lungs, kidneys, and liver. Different types of environmental pollution can result from the buildup of heavy metals in nature, such as neurological and muscular degenerative disorders that are not biodegradable and permanent (Engwa et al. 2018).

1.8 Management Strategies

Recycling electronic waste has become essential in wealthy nations, whereas developing countries (e.g., Africa, Nigeria, Malaysia, China, and India) rely heavily on informal sector due to socio-technological challenges (Pandey and Govind 2014). E-waste is challenging to recycle because of its substantial metallic quantity and its thermally stable plastic, which makes harder for it to fall down naturally in a comparable shape (Patel et al. 2017). Adsorption is an essential tool in the metal extraction and recovery process (Table 1.6). Due to increased use and high content, copper is widely used among all metals around 30% in PCB. There are various processes like pyrometallurgy, hydrometallurgy, bioleaching, and plasma treatment to treat e-waste which have various positive and negative effects. *Aspergillus niger* and *Penicillium chrysogenum* are two fungus strains that have been utilized in biofiltration processes that contain valuable metals like gold, silver, and platinum as well as combinations of metals like copper, lead, nickel, and tin. Microorganisms produce a lixiviant pool (Senophiyah-Mary et al. 2018). E-waste recycling method has three major stages: dismantling, refining, and valorization (Ashiq et al. 2019). Metal

Table 1.6 Comprises some chapters with different goals and responsibilities

Chapter	E-waste management rules 2011 (Ministry of Environment, Forest, Government of India 2011)	E-waste management rules 2016 (Ministry of Environment, Forest and Climate Change, Government of India 2016)
I	Preliminary	Preliminary
II	Responsibilities	Responsibilities
III	Procedure for seeking authorization and registration for handling e-waste	Procedure for seeking and grant of authorization for management of e-waste
IV	Procedure for storage of e-waste	Procedure for storage of e-waste
V	Electric and electronic equipment can be manufactured with fewer hazardous substances	Reduce the use of hazardous materials in the manufacture of electrical and electronic equipment, components, consumables, parts, and spares
VI	Miscellaneous	Miscellaneous

recovery from e-waste offers a number of environmental and economic benefits (Zhang and Forssberg 1999) including:

- Recovery of non-metal materials such as plastic.
- Conservation of primary metal resources.
- Reduction in the generation of solid waste.
- Ferrous metals, nonferrous metals, and precious metals recovery.
- The use of secondary mines of metal will save the energy of primary metal resource mining.
- Preclusion of heavy metal pollution related to flame-retardant problems, plastics, and toxic gas pollution from the environment released from e-waste (Akcil et al. 2015).

Therefore, WEEE recycling provides enormous advantages to both the ecosystem along economic growth (Widmer et al. 2005). E-waste confines a diverse range of metals, incorporating precious metals, valuable metals, dangerous metals, poisonous metals, rare earth metals, and so on. Because of the abundance of metals, e-waste is regarded as the second mine of metals. Aside from the negative aspects of e-waste, there are some benefits. Instead of mining ore from the ground, metals may be collected from e-waste through recycling (Table 1.7).

1.8.1 Non-technical Approach

The procedure for extracting metals from e-waste could start with disassembling and destroying all electronic components (ECs) by re-soldering used PC boards (Yang et al. 2017). To reclaim or recover some metals from e-waste informal sector unknowingly works toward exposing hazardous metals and waste to the environment by dumping remains. Several chemicals are used to recover valuable metals

Table 1.7 Extended producer responsibility (EPR) goals

No.	Years of initial years of implementation of the rules	E-waste collection target (no./wt.) % of waste reported in the EPR plan
1.	First and second	30%
2.	Third and fourth	40%
3.	Fifth and sixth	50%
4.	Seventh and more	70%

that directly and indirectly deplete environmental health (Shenoy 2019). Various problems are addressed whether in dealing with precarious operation conditions or hazardous gas release, handling viscous and risky acids, poisonous smoke release, or polluted ashes faced by workers appealing WPCB recycling (Zhao et al. 2010).

There are very few nations in the globe that have a unified way to measure and recycle this type of waste. However, informal collection and dismantling processes play an important role and contribute significantly to electronic waste recycling. Unfortunately, most workers are unaware of the dangers of mishandling e-waste, including the possibility of dangerous waste after it has been recycled and disassembled. In connection with the purchase of the e-waste (Management and Handling) Rules, 2011 (Ministry of Environment, Forest, Government of India 2011) which went into force on May 1, 2012, the government body (Central and State Pollution Control Boards) registered several organizations interested in collecting and disassembling, with the nation's e-waste recycling. The informal sector, rather than registered recyclers/collectors, receives most of the e-waste in the nation. According to the latest CPCB statistics (2016), 178 e-waste recyclers are officially registered /dismantler companies nationwide, with an approximate yearly capacity of 438,086 tons for recycling (CPCB 2018). A study by SRM University, Tamil Nadu discloses information about contaminated lands in some metropolitan cities like Mumbai, Delhi, Kolkata, and Chennai India for the reason that informal metal recovery by electronic waste (Manish and Chakraborty 2019).

E-waste is high-risk prone waste. Its informal handling causes pollution to the air by direct burning and contaminates soil and water with gaseous or liquid toxins and its transportation contaminates the ocean and so on. Informal handling of e-waste is prone to cause pollution and extends sustainable limits. Non-technical collection, metal recovery processes, and landfilling can lead to toxic substances entering the soil, leaching of heavy metals leading to groundwater contamination, and greenhouse gas emissions leading to air pollution. Subsequently, many countries are challenged by the imperfectly managed Waste Electrical and Electronic Equipment (WEEE), widely threatening sources of substantial environmental and human health. Even nations with official e-waste management programs face dangers from a reasonably lower frequency of collection and low rates of recycling (Forti et al. 2020).

1.8.2 Technical Approach

Numerous investigations on metal recovery have been carried out by experts utilizing different inorganic acids and oxidants. The greatest recovery rates for Cu and Au were achieved with nitrate and chloride solutions. Additionally, some recommended pyrolysis step ahead with leaching to improve metal recovery (Akcil et al. 2015), and some suggested HNO_3/HCl leaching media for fast Au leaching from circuit boards (Sheng and Etsell 2007). Aqua regia is well known for being a reactive system that is used extensively to remove metals from electronic trash, but it is also costly and corrosive by nature. Printed circuit boards yield large (>95%) concentrations of Cu and Pb when dissolved in nitric acid. Using electrolytic method metals can be recovered from acid-leached solutions (Mecucci and Scott 2002). According to a Chinese patent work on metal recovery, metal can be recovered by burning e-wastes for 8–12 h at a temperature from 400 to 500 °C range for recovery, with HCl and H_2SO_4 leaching at 90 °C. Solid-liquid separation has been done along with analysis of leached pulp. The solid content was silver and was diluted into HNO_3 at 60 °C with a solid ratio. Resultant Au and Pd precious metals were recovered with 92% after HCl and $NaClO_3$ leaching (Zhou et al. 2004). A two-step method was developed for the leaching purpose of Cu, Fe, Zn, Ni, and Al from circuit boards; successful leaching of these elements was done using H_2SO_4 and H_2O_2, and 95% was recovered (Oh et al. 2003). Researchers have found that biosorption is the most effective and financially feasible method of reducing heavy metal pollution from wastewater originating from various sources. Both sugar bagasse ash and chitosan, two well-known biosorbents, have qualities that make them affordable, readily available, environmentally benign, and capable of biosorption (Nguyen and Nguyen 2019).

1.8.2.1 Metallurgical Processes in Metal Recycling

According to numerous experts, the following methods can be used to recover conservative metals from e-waste: through metallurgical processes, namely physical processes, biological/hydrometallurgical processes, and pyrometallurgical processes (Cui and Forssberg 2003; Cui et al. 2008; Hagelüken 2006); Hydro-metallurgical (Mecucci and Scott 2002; Oh et al. 2003; Quinet et al. 2005; Kim et al. 2011). Bio-hydro-metallurgical (Brandl et al. 2001; Choi et al. 2004; Ilyas et al. 2007; Yang et al. 2009; Priya and Hait 2017). Pyrometallurgical (Kim et al. 2004; Hagelüken 2006; Hall and Williams 2007). Recycling electronic waste encompasses disassembly, retrieval, preparatory treatment, and final processing. Metallic parts are purified and recovered during pre-treatment in hydrometallurgical, pyrometallurgical, electrometallurgical, and biometallurgical processes (Khaliq et al. 2014; Schluep et al. 2009). Pyrometallurgical processing is how it works by using heat to melt components in the metallurgical processing of e-waste. Hydrometallurgical processing involves dissolving metals in a suitable solvent or liquid, and then using different chemicals, currents, and voltages (e.g., electrolytic cells) to refine and sometimes precipitate the desired metals depending on their properties (Kaya 2016). In hydrometallurgy, the initial step for recovering metal is leaching, where the metal is dissolved from the waste into the solution by using acidic or alkaline media; a variety of methods may be used to selectively recover metals from a solution, including adsorption, precipitation, ion exchange, electrowinning, or solvent extraction (Birloaga et al. 2014), electrorefining, precipitation, cementation, adsorption, solvent extraction, and/or ion exchange (Abdelbasir et al. 2018; Tsydenova and Bengtsson 2011). To provide insights about gold (and other metals) recovery from garbage, several techniques have been employed to treat e-waste sources, including leaching, reducing agents, coagulants, cementing, peeling, adsorbents, agglomeration solvents, ion exchange resins, and biosorbents (Won et al. 2014).

1.8.2.2 Some Other Technical Approaches for e-Waste Management

- Solidification
- Stabilization
- Gravity
- Pyrolysis
- Pyro-, hydro-, and biometallurgical and thermo-metallurgic processes

Physical/Mechanical Recycling Techniques (Kaya 2016)

1. Dismantling (manual, automated, semi-automated)
2. Comminution for liberation of metallic and non-metallic fractions (shredding/crushing)
3. Sieving for classification
4. Chemical composition determination
5. Gravity separation
6. Magnetic separation
7. Electrostatic separation
8. Froth flotation

Metal Recovery by Adsorption

The adsorption approach has numerous benefits over other metal recovery methods, including being simple to use and manage, sludge-free, potentially rejuvenating, and having a high metal ion efficiency (Ahmauzzaman 2011; Ahmad et al. 2016; Wang et al. 2017; Wu et al. 2017). There are many materials used as an adsorbent for metal recovery, considered as cheaper and greener substitutes such as chitin and chitosan (CTS), agricultural wastes, such as nutshells, can accomplished in activation bagasse pith, maize cob, coconut shells, rice husks, and wheat straw (Corapcioglu and Huang 1987; Chuang et al. 2005), carbon nanotubes, lignin, seaweed, wool waste, fly ash, silica gel, and clay materials, such as diatomaceous earth, montmorillonite, bentonite, and kaolinite (Wu et al. 2010).

Some other adsorbents are listed below:

- Brown algae (Davis et al. 2003)
- The petiolar felt sheath of palm-PFP (Iqbal et al. 2002)
- Stimulated alumina (Lin et al. 2001; Singh and Pant 2004)
- Basic yttrium carbonate, silica gel (Yu et al. 2015)
- Coral limestone loaded with metal (Khaliq et al. 2014)
- Hematite (Lata et al. 2015)
- Feldspar (Singh et al. 2002)
- Hydrous zirconium oxide (Hang et al. 2011)
- Using 3-aminopropyl triethoxysilane (APTEOS) as an inorganic precursor, phosphorylated cellulose triacetate (CTA)/silica composite adsorbents have a very high adsorption capacity for Ni(II). Diethylene thiamine pentaacetic acid (DTPA) based method is used to extract copper, iron, manganese, and zinc from e-waste (da Fonseca et al. 2010)

Photocatalysis

PW12O403-, SiW12O404-, and P2Mo18O626- have been employed as photocatalysts to recover copper and produce fine metal particles. It was discovered that a revolving cathode photo electrocatalytic reactor had exceptional efficiency in recovering metallic ions and dissolving metal-EDTA groups (Chen et al. 2017).

Pyrolysis Process

Pyrolysis involves pyrolysis of a PCB at 6000 °C under vacuum, solder separation by vacuum centrifugation and subsequent remelting. A solid residue of 72 wt.% is produced, along with a gaseous remnant of 6.35 wt.% and an oil fraction with lead and tin (Zhao et al. 2010).

Biological Process

Removing and recovering certain metals from aqueous solutions (with different pollutants) biologically is a promising, eco-friendly, and effective process. A biological adsorbent, brown algae was used to recover metals Cd(II), Cu(II), Zn(II), Pb(II), Cr(III), and Hg(II) (Davis et al. 2003). Researchers Iqbal and Saeed employed a biodegradable felt sheath made from palm-PFP (bio-adsorbent) to adsorb, desorb, and filter polluted water for heavy metals (Iqbal et al. 2002). Similar works performed using different biological strains and processes:

- Bioextraction (using fungal microorganism, *Aspergillus niger*)
- The formerly extracted heavy metal with microbes and proceeded with the extraction by gluconate and later used a microbe-free culture
- Metal recovery by microbial electrometallurgy
- Bio-mineralization
- Phytoremediation
- Bioremediation (Patel et al. 2024)
- Biohydrometallurgy
  1. Biosorption
  2. Bioleaching

Alternative Methods

- Ozone oxidation
- Bioremediation
- Electrochemical treatments

They are also used to remove arsenic (Ahmed 2001; Ravenscroft et al. 2005; Mandal and Suzuki 2002; Choong et al. 2007).

Other Processes for Recovery of Metals

Expanded technologies of metal recovery which have been practical in industries and mining locations function by different principles of membrane filtration, ion exchange, and adsorption. Membrane filtration offers high-efficiency metal recovery in the range of 82–100% of ultrafiltration (UF), 47–99.5% nanofiltration (NF), and 70–99.5% of reverse osmosis. Table 1.8 showing below some more recovery processes of metals (Seelam et al. 2018).

1.9 Removal Techniques

1.9.1 Heavy Metal(s) and Metalloid (Arsenic) Removal Techniques (Table 1.9)

- Coagulation and flocculation
- Precipitation
- Adsorption
- Ion-exchange
- Membrane filtration

Table 1.8 Management processes

S. no.	Management process	About	Reference
1.	Ion exchange	It is possible to convert thermosetting epoxy resins (TEPRs) into ion-exchange resins by treating them with ion exchange capacities (IECs) measured with sulfuric acid The greatest adsorption capacities were 23.30 mg/g Cu(II) and 56.27 mg/g Ca(II)	Zhang et al. (2015)
2.	Ultrasonically enhanced two-stage acid leaching	By stabilizing the two-stage leaching system, Cu 97.42% is recovered from electroplating sludge, Ni 98.46% is recovered from nickel 98.46%, Zn 98.63% is recovered from zinc 98.63%, and Cr 98.32% is recovered from iron 98.32%	Li and Xu (2010)
3.	Chemical leaching	Maximum removal efficiency of several metals was accomplished during acid leaching operations—72% and 79% Zn, 70% and 75% Cu, 69% and 73% Ni, 57% and 70% Pb, 55% and 65% Cd, and 11% and 22% Cr	Bayat et al. (2009)
4.	Adsorption on commercial-activated carbon (CAC) banana peel (BAC)	Adsorption is an important extraction process tool. This process helps to reduce waste and transforms the waste into activated carbon. The copper adsorption capacity resulted from more than 80% BAC	Senophiyahmary and Thirumoorthy (2018)
5.	Various electrochemical metal recovery techniques have been developed, including electrodeposition, electrosorption, electrodialysis, electro-deionization, bioelectrochemistry, and photochemistry	These technologies have sparked widespread interest in metal pollution due to environmental sustainability, great productivity/selectivity/practicality, and low budget Overcoming the restrictions on concentration polarization of weak metal ions, resulting in a spongy and dendritic deposits, and overcoming the side effects of snail dynamics of ion transport, oxygen reduction process, and hydrogen evolution in agriculture	Jin and Zhang (2020)
6.	Reverse osmosis solvent extraction	In the electrochemical-osmotic process (EOS), gold, silver, and mercury can be recovered from wastewater by using electrical energy and water produced at a coordinated rate	Sun et al. (2020)
7.	Membrane adsorption techniques	Membranes for membrane adsorption (MA) serve two functions: adsorption and membrane filtration, allowing them to remove tiny contaminants such as nitrates, anionic phosphates, and cationic heavy metals with high efficiency	Khulbe and Matsuura (2018)

(continued)

Table 1.8 (continued)

S. no.	Management process	About	Reference
8.	Electrodialysis (ED)	ED processes can be used to reduce, dilute, desalinate, rejuvenate, recover, and eliminate organic debris, acids and bases, heavy metal ions, including nutrition. The elimination of charge energy in wastewater is facilitated by water treatment	Luigi Gurreri (2020)
9.	Photocatalysis	Using photocatalytic chemistry can extract seven precious metal elements from scrap blanks: Ag, Au, Pd Pt, Ru, and Rh with dissolution rates of 99%, respectively; auto catastases Ir and ores. These substances are also valuable as organic acids in the presence of radiocarbon ore	Yao et al. (2020)
10.	Titration and DTPA methods	Copper, iron, aluminum, and other heavy metals can be recovered from e-waste using DTPA and titration techniques	Ruchi and Kanjan (2016)
11.	Microbial catalyzed leaching	Bioleaching tests have shown that the recovery of copper with Zn, Al, and Ni is 98–99%. The optimum copper yield in this process justifies the possibility of increasing the density of high-strength pulp. It is an inexpensive, environmentally friendly device	Garg et al. (2019)
12.	Photocatalysis with TiO_2	The (Cu(2)O/TiO(2), has greater deterioration ability than commercial P25 and appears as an effective catalyst. The stocking of Cu(2)O on the surface of TiO(2) improves photocatalytic effect of the catalyst	Yang et al. (2009)
13.	Photo electrocatalytic oxidation of metal-EDTA	Cu, Co^{2+}, Ni^{2+}, and Zn^{2+} obtained experimentally	Chen et al. (2017)
14.	Mesoporous silica	It was discovered that sulfonated mesoporous silica is a potentially useful adsorbent for uranium (U) extraction from granitic liquids	Abd El-Magied et al. (2018)
15.	Chitosan/SiO_2 coating on magnetic iron sand material	The investigation centered on metal-specific adsorption and desorption (Au, Cu, Ni)	Harumi et al. (2019)

Table 1.9 Methods of metal recovery

S. no.	Methods of metal recovery	References
1.	Magnetic separation	Wei et al. (2020)
2.	Eddy current separation	Jujun et al. (2014)
3.	Density separation	Kang and Schoenung (2005)
4.	Combination of electrostatic and magnetic separation	Veit et al. (2005)
5.	Corona discharge method	Shunli and Forssberg (1998) and Li et al. (2007)
6.	Density-based separation	Mou et al. (2004)
7.	Milling	Koyanaka et al. (2006)
8.	Froth floatation	Ogunniyi and Vermaak (2009)
9.	Co-pyrolysis (with biomass)	Liu et al. (2013) and Abnisa and Daud (2014)
10.	Vacuum pyrolysis	Zhou and Qiu (2009)
11.	Microwave-induced pyrolysis	Sun et al. (2003) and Zhang et al. (2021)
12.	Catalytic pyrolysis	Guo et al. (2020)

1.9.2 Corona Electrostatic Separation

In growing recovery work, corona electrostatic separation (CES) is considered an encouraging way to recover metals from WPCBs (pulverized), eco-friendly, economical, and highly efficient. These processes are a bit low efficient in recovering precious metals because of the high concentration (which continues to be a combination of several metals, including Cu, Al, Pb, Au, and Ag) of metallic materials obtained after the above processes (Huang et al. 2009).

1.10 Conclusion

Because of their inconsistent collection methods, emerging nations import a significant amount of electronic garbage worldwide. Household waste in every country is the biggest waste to consider, with its worries and problems. Why not select e-waste over all other rubbish, given it makes up over 78% of electronic waste in underdeveloped regions? Due to a deficit of technological advancements, many developing countries continue to be concerned about electronic waste, skills, labor, poor management, etc. In addition to damaging the environment, global warming and climate change also harm human health. Electronic waste processing contains several harmful components and in the informal sector, e-waste is mainly recycled, through large subsidies, it receives from developed countries where products can be reused and recycled. However, metropolitan industries and businesses do not realize how important it is to dispose of e-waste rather than just generate and store it. This paper discusses several strategies and challenges to reduce e-waste inequality in developing countries by integrating life cycle assessment and integrated waste management approaches with the leading along with prosperous unauthorized e-waste recycling industry. This is dependent on creative approaches that permit the unauthorized sector to continue engaging in safe recycling procedures, despite some companies that recycle within the formal sector. The E-waste Rules, 2016, must be implemented in India by the government to allow consumers, dismantlers, recyclers, and bulk consumers to bear the responsibility of producing, storing, collecting, transporting, or sustainably disposing of electronic waste, rather than being delegated onto SPCBs. Poorer countries should avoid this approach and instead build their EPR systems based on their capacity to enforce such standards.

Acknowledgments The authors express their gratitude to the reviewers and the editor for dedicating their valuable time to enhancing this work.

References

Abdelbasir S, El-Sheltawy C, Abdo M (2018) Green processes for electronic waste recycling: a review. J Sustain Metall 4:295. https://doi.org/10.1007/s40831-018-0175-3

Abd El-Magied MO, Dhmees AS, Abd Allah M, Eldesouky EM (2018) Uranium extraction by sulfonated mesoporous silica derived from blast furnace slag. J Nuc Mat 509:295–304

Ahmad A, Khatoon A, Mohd-Setapar S-H, Kumar R, Rafatullah M (2016) Chemically oxidized pineapple fruit peel for the biosorption of heavy metals from aqueous solutions. Desalin Water Treat 57(14):6432–6442. https://doi.org/10.1080/19443994.2015.1005150

Ahmauzzaman M (2011) Industrial wastes as low-cost potential adsorbents for the treatment of wastewater laden with heavy metals. Adv Colloid Interf Sci 166(1–2):36–59

Ahmed MF (2001) An overview of arsenic removal technologies in Bangladesh and India. In: Technologies for arsenic removal from drinking water, BUET-UNU international workshop, Dhaka, Bangladesh

Abnisa F, Daud W (2014) A review on co-pyrolysis of biomass: An optional technique to obtain a high-grade pyrolysis oil. Energy Convers Manag 87:71–85

Akcil A, Erust C, Gahan CS, Ozgun M, Sahin M, Tuncuk A (2015) Precious metal recovery from waste printed circuit boards using cyanide and non-cyanide lixiviants—a review. Waste Manag 45:258–271. https://doi.org/10.1016/j.wasman.2015.01.017

Ari V (2015) A review of technology of metal recovery from electronic waste. https://doi.org/10.5772/61569

Ari V, Das A (2012) Kinetics and efficacy of froth flotation for the recovery of metal values from pulverized printed circuit boards

Ashiq A, Kulkarni J, Vithanage M (2019) Chapter 10: Hydrometallurgical recovery of metals from e-waste. In: Prasad MNV, Vithanage M (eds) Electronic waste management and treatment technology. Butterworth-Heinemann. https://doi.org/10.1016/B978-0-12-816190-6.00010-8

Baldé CP, Forti V, Gray V, Kuehr R, Stegmann P (2017) The global E-waste monitor—2017. United Nations University (UNU), International Telecommunication Union (ITU) & International Solid Waste Association (ISWA), Bonn

Bayat O, Sever E, Bayat B, Arslan V, Poole C (2009) Bioleaching of Zinc and Iron from Steel Plant Waste using Acidithiobacillus Ferrooxidans. Appl Biochem Biotechnol 152:117–126

Birloaga I, Coman V, Kopacek B, Vegliò F (2014) An advanced study on the hydrometallurgical processing of waste computer printed circuit boards to extract their valuable content of metals. Waste Manag 34(12):2581–2586. https://doi.org/10.1016/j.wasman.2014.08.028

Borthakur A (2016) International perspectives/special report: Health and environmental hazards of electronic waste in India. J Environ Health 78:18–23. https://doi.org/10.2307/26330439

Brandl H, Bosshard R, Wegmann M (2001) Computer-munching microbes: metal leaching from electronic scrap by bacteria and fungi. Hydrometallurgy 59(2):319–326. https://doi.org/10.1016/S0304-386X(00)00188-2

Chen Y, Zhao X, Guan W et al (2017) Photoelectrocatalytic oxidation of metal-EDTA and recovery of metals by electrodeposition with a rotating cathode. Chem Eng J 324:74–82. https://doi.org/10.1016/j.cej.2017.05.031

Choi M-S, Cho K-S, Kim D-S, Kim D-J (2004) Microbial recovery of copper from printed circuit boards of waste computer by Acidithiobacillus ferrooxidans. J Environ Sci Health A 39(11–12):2973–2982. https://doi.org/10.1081/LESA-200034763

Choong T, Chuah AL, Yunus R, Koay G, Idris A (2007) Arsenic toxicity, health hazards and removal techniques from water: an overview. Desalination 127:139–166. https://doi.org/10.1016/j.desal.2007.01.015

Chuang C, Fan M, Xu M et al (2005) Adsorption of arsenic(V) by activated carbon prepared from oat hulls. Chemosphere 61:478–483. https://doi.org/10.1016/j.chemosphere.2005.03.012

Corapcioglu MO, Huang CP (1987) The adsorption of heavy metals onto hydrous activated carbon. Water Res 21(9):1031–1044. https://doi.org/10.1016/0043-1354(87)90024-8

CPCB (2018) Attention: Authorized recyclers of e-waste documents to be retained for verification for compliance of collection targets

Creamer NJ, Baxter-Plant VS, Henderson J, Potter M, Macaskie LE (2006) Palladium and gold removal and recovery from precious metal solutions and electronic scrap leachates by Desulfovibrio desulfuricans. Biotechnol Lett 28(18):1475–1484. https://doi.org/10.1007/s10529-006-9120-9

Cui J, Forssberg E (2003) Mechanical recycling of waste electric and electronic equipment: a review. J Hazard Mater 99(3):243–263. https://doi.org/10.1016/S0304-3894(03)00061-X

Cui P, Yang H, Fang Z (2008) Numerical analysis and experimental validation of heat transfer in ground heat exchangers in alternative operation modes. Energy Build 40(6):1060–1066. https://doi.org/10.1016/j.enbuild.2007.10.005

da Fonseca AF, Caires EF, Barth G (2010) Extraction methods and availability of micronutrients for wheatunder a no-till system with a surface application of lime. Sci Agric 67:60–70

Davis T, Volesky B, Mucci A (2003) A review on biochemistry of heavy metal biosorption by brown algae. Water Res 37:4311–4330. https://doi.org/10.1016/S0043-1354(03)00293-8

Debnath B, Chowdhury R, Ghosh SK (2018) Sustainability of metal recovery from E-waste. Front Environ Sci Eng 12(6):2. https://doi.org/10.1007/s11783-018-1044-9

Department of Drinking Water and Sanitation, Ministry of Jal Shakti (2012) Target vs achievement of BLS Swachh Bharat Mission—Gramin report

Engwa GA, Ferdinand PU, Nwalo FN, Unachukwu MN (2018) Mechanism and health effects of heavy metal toxicity in humans. https://doi.org/10.5772/intechopen.82511

European Union (2003) Directive 2002/96/EC of the European Parliament and of the Council. Official Journal of the European Union L 37/24

E-waste UN (2019) A new circular vision for electronics time for a global reboot. Platform for accelerating the circular economy. https://www.weforum.org/publications/a-new-circular-vision-for-electronics-time-for-a-globalreboot/

Forti V, Baldé CP, Kuehr R, Bel G (2020) The global e-waste monitor

Garg H, Nagar N, Ellamparuthy G, Angadi S, Gahan C (2019) Bench scale microbial catalysed leaching of mobile phone PCBs with an increasing pulp density. Heliyon 5:e02883

Guo D, Hu M, Chen Z, Cui B, Zhang Q, Liu Y, Luo S, Ruan R, Liu Y (2020) Catalytic pyrolysis of rain tree biomass with nano nickel oxide synthetized from nickel plating slag: A green path for treating waste by waste. Bioresour Technol 315:123831

Gurreri L, Tamburini A, Cipollina A, Micale G (2020) Electrodialysis Applications in Wastewater Treatment for Environmental Protection and Resources Recovery: A Systematic Review on Progress and Perspectives. Membranes 10(7):146. https://doi.org/10.3390/membranes10070146

Hagelüken C (2006) Recycling of electronic scrap at umicore. Precious metals refining. Acta Metall Slovaca 12:111

Hall WJ, Williams PT (2007) Separation and recovery of materials from scrap printed circuit boards. Resour Conserv Recycl 51(3):691–709. https://doi.org/10.1016/j.resconrec.2006.11.010

Hang C, Li Q, Gao S, Shang J (2011) As(III) and As(V) adsorption by hydrous zirconium oxide nanoparticles synthesized by a hydrothermal process followed with heat treatment. Ind Eng Chem Res 51:353–361. https://doi.org/10.1021/ie202260g

Harumi M, Santosa SJ, Nuryono (2019) Recovery of Au(III) From Printed Circuit Board Waste by Chitosan/SiO2 Coated On Iron Sand Magnetic Material. Materials Today: Proceedings 19:1101–1110

Hira M, Yadav S, Morthekai P, Linda A, Kumar S, Sharma A (2017) Mobile phones—an asset or a liability: a study based on characterization and assessment of metals in waste mobile phone components using leaching tests. J Hazard Mater 342:29. https://doi.org/10.1016/j.jhazmat.2017.08.008

Hong J, Shi W, Wang Y, Chen W, Li X (2015) Life cycle assessment of electronic waste treatment. Waste Manag 38:357–365. https://doi.org/10.1016/j.wasman.2014.12.022

Huang K, Guo J, Zhenming X (2009) Recycling of waste printed circuit boards: a review of current technologies and treatment status in China. J Hazard Mater 164(2):399–408. https://doi.org/10.1016/j.jhazmat.2008.08.051

Huisman J (2010) WEEE recast. The International Solid Waste Association

Ilyas S, Anwar MA, Niazi SB, Ghauri MA (2007) Bioleaching of metals from electronic scrap by moderately thermophilic acidophilic bacteria. Hydrometallurgy 88:180. https://doi.org/10.1016/j.hydromet.2007.04.007

Iqbal M, Saeed A, Akhtar N (2002) Petiolar felt-sheath of palm: a new biosorbent for the removal of heavy metals from contaminated water. Bioresour Technol 81(2):151–153. https://doi.org/10.1016/s0960-8524(01)00126-2

Jin W, Zhang Y (2020) Sustainable electrochemical extraction of metal resources from waste streams: from removal to recovery. ACS Sustain Chem Eng 8(12):4693–4707. https://doi.org/10.1021/acssuschemeng.9b07007

Jitendra (2011) Source of E-waste. Waste management

Jujun R, Yiming Q, Zhenming X (2014) Environment-friendly technology for recovering nonferrous metals from e-waste: Eddy current separation. Resour Conserv Recycl 87:109–116

Kang H-Y, Schoenung J (2005) Electronic Waste Recycling: a Review of U.S. Infrastructure and Technology Options. Resour Conserv Recycl 45:368-400

Kaya M (2016) Recovery of metals and nonmetals from electronic waste by physical and chemical recycling processes. Waste Manag 57:64–90. https://doi.org/10.1016/j.wasman.2016.08.004

Khaliq A, Rhamdhani MA, Brooks G, Masood S (2014) Metal extraction processes for electronic waste and existing industrial routes: a review and Australian perspective. Resources 3:152–179. https://doi.org/10.3390/resources3010152

Khulbe KC, Matsuura T (2018) Removal of heavy metals and pollutants by membrane adsorption techniques. Appl Water Sci 8:19

Kim B-S, Lee J-C, Seo S-P, Park Y-K, Sohn HY (2004) A process for extracting precious metals from spent printed circuit boards and automobile catalysts. JOM 56:55–58. https://doi.org/10.1007/s11837-004-0237-9

Kim EY, Kim MS, Lee JC, Pandey BD (2011) Selective recovery of gold from waste mobile phone PCBs by hydrometallurgical process. J Hazard Mater 198:206–215. https://doi.org/10.1016/j.jhazmat.2011.10.034

Koyanaka S, Endoh S, Ohya H (2006) Effect of impact velocity control on selective grinding of waste printed circuit boards. Adv Powder Technol 17:113–126

Lata S, Singh PK, Samadder SR (2015) Regeneration of adsorbents and recovery of heavy metals: a review. Int J Environ Sci Technol 12(4):1461–1478. https://doi.org/10.1007/s13762-014-0714-9

LeBlanc R (2020) E-waste recycling facts and figures

Leeabai N, Areeprasert C, Khaobang C et al (2021) The effects of color preference and noticeability of trash bins on waste collection performance and waste-sorting behaviors. Waste Manag 121:153–163. https://doi.org/10.1016/j.wasman.2020.12.010

Li J, Liu L (2015) Solving spent lithium-ion battery problems in China: opportunities and challenges. Renew Sustain Energy Rev 52:1759–1767. https://doi.org/10.1016/j.rser.2015.08.014

Li J, Xu Z (2010) Environmental friendly automatic line for recovering metal from waste printed circuit boards. Environ Sci Technol 44(4):1418–1423. https://doi.org/10.1021/es903242t

Li J, Xu Z, Zhou Y (2007) Application of corona discharge and electrostatic force to separate metals and nonmetals from crushed particles of waste printed circuit boards. J Electrostat 65:233–238

Lin J, Jihuai W, Yang Z, Minli P (2001) Synthesis and properties of poly(acrylic acid)/mica superabsorbent nanocomposite. Macromol Rapid Commun 22:422. https://doi.org/10.1002/1521-3927(20010301)
Liu W-J, Tian K, Jiang H, Zhang X-S, Yang G-X (2013) Preparation of liquid chemical feedstocks by co-pyrolysis of electronic waste and biomass without formation of polybrominated dibenzo-p-dioxins. Bioresour Technol 128:1–7
Mandal BK, Suzuki KT (2002) Arsenic round the world: a review. Talanta 58(1):201–235. https://doi.org/10.1016/S0039-9140(02)00268-0
Manish A, Chakraborty P (2019) E-waste management in India: challenges and opportunities
Mecucci A, Scott K (2002) Leaching and electrochemical recovery of copper, lead and tin from scrap printed circuit boards. J Chem Technol Biotechnol 77(4):449–457. https://doi.org/10.1002/jctb.575
Ministry of Environment, Forest and Climate Change, Government of India (2016) EWM-Rules-2016
Ministry of Environment, Forest, Government of India (2011) Notification
Mou P, Wa L, Xiang D, Gao J, Duan G (2004) A physical process for recycling and reusing waste printed circuit boards. IEEE International Symposium on Electronics and the Environment, 2004. Conference Record. 2004, pp. 237–242
Nguyen L, Nguyen T (2019) Enhanced heavy metals biosorption using chemically modified chitosan coated microwave activated sugarcane baggage ash composite biosorbents. SN Appl Sci 1. https://doi.org/10.1007/s42452-019-1607-9
NIMS (2015) E-waste management and recycling options for MSMEs
Oh CJ, Lee SO, Yang HS, Ha TJ, Kim MJ (2003) Selective leaching of valuable metals from waste printed circuit boards. J Air Waste Manage Assoc 53(7):897–902. https://doi.org/10.1080/10473289.2003.10466230
Ogunniyi IO, Vermaak MKG (2009) Investigation of froth flotation for beneficiation of printed circuit board comminution fines. Miner Eng 22:378–385
Ongondo FO, Williams ID, Cherrett TJ (2011) How are WEEE doing? A global review of the management of electrical and electronic wastes. Waste Manag 31(4):714–730. https://doi.org/10.1016/j.wasman.2010.10.023
Panda R, Dinkar OS, Jha MK, Pathak DD (2020) Recycling of gold from waste electronic components of devices. Korean J Chem Eng 37:111–119. https://doi.org/10.1007/s11814-019-0412-x
Pandey P, Govind M (2014) Social repercussions of e-waste management in India: a study of three informal recycling sites in Delhi. Int J Environ Stud 71. https://doi.org/10.1080/00207233.2014.926160
Patel D, Kulwant M (2024) Bioremediation of Xenobiotics Contamination in Soil, Bioremediation of Emerging Contaminants from Soils, Elsevier. ISBN9780443139949
Patel S, Patel R, Gautam A, Gautam S (2017) Influence of oxidizing agent on recovery of metals including gold and silver from printed circuit boards. Int Res J Eng Technol 4:830
Priya A, Hait S (2017) Comparative assessment of metallurgical recovery of metals from electronic waste with special emphasis on bioleaching. Environ Sci Pollut Res 24(8):6989–7008. https://doi.org/10.1007/s11356-016-8313-6
Quinet P, Proost J, Van Lierde A (2005) Recovery of precious metals from electronic scrap by hydrometallurgical processing routes. Mining Metall Explor 22(1):17–22. https://doi.org/10.1007/BF03403191
Ravenscroft P, Burgess WG, Ahmed KM, Burren M, Perrin J (2005) Arsenic in groundwater of the Bengal Basin, Bangladesh: distribution, field relations, and hydrogeological setting. Hydrogeol J 13(5):727–751. https://doi.org/10.1007/s10040-003-0314-0
Ruchi C, Kanjan U (2016) Recovery of heavy metal from electronic waste. Inter J App Res (IJAR) 2:417–419
Ryder G, Houlin Z (2019) The world's e-waste is a huge problem. It's also a golden opportunity
Schluep M, Hagelüken C, Kuehr R et al (2009) Recycling—from e-waste to resources
Seelam JS, Maesara SA, Mohanakrishna G, Patil SA, ter Heijne A, Pant D (2018) Chapter 18: Resource recovery from wastes and wastewaters using bioelectrochemical systems. In: Bhaskar T, Pandey A, Mohan SV, Lee D-J, Khanal SK (eds) Waste biorefinery. Elsevier. https://doi.org/10.1016/B978-0-444-63992-9.00018-5

Senophiyah-Mary J, Loganath R, Meenambal T (2018) A novel method for the removal of epoxy coating from waste printed circuit board. Waste Manag Res 36(7):645–652. https://doi.org/10.1177/0734242X18782392

Senophiyahmary J, Thirumoorthy M (2018) Removal of Copper from Bioleachate of Electronic Waste Using Banana-Activated Carbon (BAC) and Comparison with Commercial-Activated Carbon (CAC). pp. 233–242

Shah M, Tipre D, Dave S (2014) Chemical and biological processes for multi-metal extraction from waste printed circuit boards of computers and mobile phones. Waste Manag Res 32:1134. https://doi.org/10.1177/0734242X14550021

Sharma M, Joshi S, Kumar A (2020) Assessing enablers of e-waste management in circular economy using DEMATEL method: an Indian perspective. Environ Sci Pollut Res 27(12):13325–13338. https://doi.org/10.1007/s11356-020-07765-w

Sheng PP, Etsell TH (2007) Recovery of gold from computer circuit board scrap using aqua regia. Waste Manag Res 25(4):380–383. https://doi.org/10.1177/0734242X07076946

Shenoy J (2019) India's e-waste to touch 5.2 MMT by 2020: ASSOCHAM-EY study

Shunli Z, Forssberg E (1998) Mechanical recycling of electronics scrap - the current status and prospects. Waste Management & Research 16:119–128

Singh T, Pant K (2004) Equilibrium, kinetics and thermodynamic studies for adsorption of As(III) on activated alumina. Sep Purif Technol 36:139–147. https://doi.org/10.1016/S1383-5866(03)00209-0

Singh D, Tiwary R, Tewary B (2002) Feldspar for the treatment of As(III) contaminated water. Water Qual Res J Can 37:757. https://doi.org/10.2166/wqrj.2002.051

Sun L, Lu J, Wang S, Zhang J, Zhou H (2003) Experimental research on pyrolysis characteristics of printed circuit board wastes. J Chem Ind & Eng 54:408–412

Sun M, Qin M, Wang C, Weng G, Huo M, Taylor AD Qu J, Elimelech M (2020) Electrochemical-Osmotic Process for Simultaneous Recovery of Electric Energy, Water, and Metals from Wastewater. Environ Sci & Technol 54:8430–8442

The Global Environmental Facility (2019) Nigeria turns the tide on electronic waste

The Global E-Waste Statistics Partnership (2020) In-depth review of the WEEE collection rates and targets

Thomson Reuters IP Analytics (2013) Patent landscape report on e-waste recycling technologies

Tsydenova O, Bengtsson M (2011) ChemInform abstract: Chemical hazards associated with treatment of waste electrical and electronic equipment. Waste Manag (New York) 31:45–58. https://doi.org/10.1016/j.wasman.2010.08.014

UN Environment Programme (2019) UN report: time to seize opportunity, tackle challenge of e-waste

United Nations University (2020) Global e-waste surging: up 21% in 5 years

Urban Development and Urban Housing Department (2017) An overview of the functioning, accountability mechanism and financial reporting issues of urban local bodies

Veit H, Diehl T, Salami AP, Rodrigues J, Bernardes A, Tenório J (2005) Utilization of magnetic and electrostatic separation in the recycling of printed circuit boards scrap. Waste Management 25(1):67–74

Wang Z, Li X, Liang H, Ning J, Zhou Z, Li G (2017) Equilibrium, kinetics and mechanism of Au3+, Pd2+ and Ag+ ions adsorption from aqueous solutions by graphene oxide functionalized persimmon tannin. Mater Sci Eng C 79:227–236. https://doi.org/10.1016/j.msec.2017.05.038

Wei X-X, Nie C-C, Yu Y-X, Wang J-X, Lyu X-J, Wu P, Zhu X-N (2020) Environment-friendly recycling of resin in waste printed circuit boards. Process Saf Environ Prot 146:694–701

Widmer R, Oswald-Krapf H, Sinha-Khetriwal D, Schnellmann M, Böni H (2005) Global perspectives on e-waste. Environ Impact Assess Rev 25(5):436–458. https://doi.org/10.1016/j.eiar.2005.04.001

Wikipedia (2020) Electronic waste in India

Won SW, Kotte P, Wei W, Lim A, Yun Y-S (2014) Biosorbents for recovery of precious metals. Bioresour Technol 160:203. https://doi.org/10.1016/j.biortech.2014.01.121

Wu F-C, Tseng R-L, Juang R-S (2010) A review and experimental verification of using chitosan and its derivatives as adsorbents for selected heavy metals. J Environ Manag 91(4):798–806. https://doi.org/10.1016/j.jenvman.2009.10.018
Wu Y, Fang Q, Yi X, Liu G, Li R-W (2017) Recovery of gold from hydrometallurgical leaching solution of electronic waste via spontaneous reduction by polyaniline. Progr Nat Sci Mater Int 27(4):514–519. https://doi.org/10.1016/j.pnsc.2017.06.009
Yadav AK (2015) Elemental composition and source apportionment of suspended particulate matters and health risk assessment in mining and nonmining areas of Odisha, India. J Hazard Toxic Radioact Waste 19(3):04014037. https://doi.org/10.1061/(asce)hz.2153-5515.0000248
Yadav AK (2021) Human health risk assessment in opencast coal mines and coal-fired thermal power plants surrounding area due to inhalation. Environ Challenges 3:100074. https://doi.org/10.1016/j.envc.2021.100074
Yadav AK, Sahoo SK, Patra AC et al (2022) Enrichment factor and chemical composition of size separated airborne particulate matter of Singrauli coalfield, India. Aerosol Sci Eng 6(4):414–436. https://doi.org/10.1007/s41810-022-00155-4
Yamane L, Moraes V, Espinosa D, Tenório J (2011) Recycling of WEEE: Characterization of spent printed circuit boards from mobile phones and computers. Waste Manag (New York) 31:2553–2558. https://doi.org/10.1016/j.wasman.2011.07.006
Yang T, Zheng X, Wen J, Yang L (2009) Factors influencing bioleaching copper from waste printed circuit boards by Acidithiobacillus ferrooxidans. Hydrometallurgy 97(1):29–32. https://doi.org/10.1016/j.hydromet.2008.12.011
Yang C, Tan Q, Liu L, Dong Q, Li J (2017) Recycling tin from electronic waste: a problem that needs more attention. ACS Sustain Chem Eng 5(11):9586–9598. https://doi.org/10.1021/acssuschemeng.7b02903
Yang J, Wang H, Zhang G, Bai X, Zhao X, He Y (2019) Recycling organics from non-metallic fraction of waste printed circuit boards by a novel conical surface triboelectric separator. Resour Conserv Recycl 146:264–269. https://doi.org/10.1016/j.resconrec.2019.03.008
Yao Y, Gao X, Li Z, Meng X (2020) Photocatalytic reforming for hydrogen evolution: A review. Catalysts 10(3):335
Yu Y, Ling Y, Chen J (2015) Introduction of an yttrium–manganese binary composite that has extremely high adsorption capacity for arsenate uptake in different water conditions. Ind Eng Chem Res 54:150209161327009. https://doi.org/10.1021/ie5037098
Zeng X, Ming X, Li J (2018) Examining the sustainability of China's nickel supply: 1950–2050. Resour Conserv Recycl 139:188–193. https://doi.org/10.1016/j.resconrec.2018.08.011
Zhang S, Forssberg E (1999) Intelligent liberation and classification of electronic scrap. Powder Technol 105(1):295–301. https://doi.org/10.1016/S0032-5910(99)00151-5
Zhang J, Liu S, Chen J, Zu J, Wang Y (2015) Preparation of anion exchange resin by recycling of waste printed circuit boards. RSC Advances 5(129):106680-106687
Zhang S, Ding Y, Liu B, Chang C-c (2017) Supply and demand of some critical metals and present status of their recycling in WEEE. Waste Manag 65:113. https://doi.org/10.1016/j.wasman.2017.04.003
Zhang T, Mao X, Qu J, Liu Y, Siyal AA, Ao W et al. (2021). Microwave-assisted catalytic pyrolysis of waste printed circuit boards, and migration and distribution of bromine. J Hazard Mater 402:123749
Zhou Y, Qiu K (2009) A new technology for recycling solder from waste printed circuit boards. J Hazardous Materials 175:823–828
Zhao G, Zhou H, Wang Z (2010) Concentrations of selected heavy metals in food from four e-waste disassembly localities and daily intake by local residents. J Environ Sci Health A Toxic Hazard Subst Environ Eng 45(7):824–835. https://doi.org/10.1080/10934521003709016
Zhou P, Zheng Z, Tie J (2004) Technological process for extracting gold, silver and palladium from electronic industry waste

Chapter 2
Harnessing the Power of Air for a Sustainable Tomorrow

Prathamesh S. Utturkar, Rutuja Mokal, Sachin V. Jadhav, and Haresh G. Manyar

Abstract Harnessing the power of air has long been a tradition, predominantly through the use of wind turbines to extract wind energy for electricity generation. Despite reduced attention over time, it is gaining importance due to its reliability, sustainability, and environmental friendliness. In a world of diminishing reserves of non-renewable resources, emphasis is placed on the role of air-based power generation in contributing to future energy security. Although wind energy's worldwide application is well known, recent technological advancements like cross axis, vertical axis, and magnetic-based wind turbines are expanding the possibilities for efficient energy capture. Also, the integration of wind energy with solar and hydropower shows opportunities to improve sustainable electricity generation. Atmospheric water harvesting represents a novel application of air utilization, particularly beneficial for regions with water scarcity. This technology offers a sustainable solution for producing clean drinking water, with the potential to integrate with power generation systems through biomimetic materials and innovative turbine designs. Additionally, the use of wind energy for desalination provides another way to harness the power of air for water generation, further strengthening the sustainable connection between air and water. Air pollution remains a challenge driven by industrialization and conventional power generation practices. Hence, utilizing air for energy production is insufficient without addressing air quality concerns. Strategies such as carbon capture and utilization and reducing emissions from fossil fuel combustion are crucial for ensuring the prolonged and effective sustainability. Therefore, integrating the application of innovative technologies and improving air quality can harness the potential of air resources for sustainable power generation in the future.

P. S. Utturkar · R. Mokal · S. V. Jadhav (✉)
Department of Chemical Engineering, Institute of Chemical Technology, Mumbai, Maharashtra, India
e-mail: sv.jadhav@ictmumbai.edu.in

H. G. Manyar
School of Chemistry and Chemical Engineering, Queen's University Belfast, Belfast, UK

A. K. Yadav (ed.), *Powering Future: Clean Energy Solutions for a Sustainable Planet*, Clean Energy Production Technologies, https://doi.org/10.1007/978-981-96-5055-2_2

Keywords Wind energy · Turbine technologies · Renewable energy · Future power generation

2.1 Introduction

Air is an indivisible part of the Earth's environment. The wind is an element that has a significant amount of energy that can be easily extracted. Wind is simply the flow of air from one location to another due to the presence of differential pressure between these regions. Various natural activities in the environment generate wind flows. Harnessing wind power offers significant advantages. It has grown substantially over the past 10 years. In India, a notable percentage of electricity production comes from renewable sources, with wind energy contributing 37% of this total (Ministry of New and Renewable Energy 2022) and has the potential to generate up to 1800 GW from wind energy alone (Ministry of New and Renewable Energy 2021). Globally, wind energy is widely used, with China, the US, and Germany generating over 50 GW each (Wind Power by Country 2024). As of March 2023, India had an installed wind energy capacity of 42,633 MW; this contribution significantly contributed to the nation's renewable energy portfolio. Renewable energy sources, other than large hydroelectric power, account for 30% (125,160 MW) of India's total installed power capacity of 416,059 MW as of March 31, 2023. Wind energy comprising 34% of this renewable capacity, making it the leading supplier of clean energy (Indian Wind Turbine Manufacturers Association 2023). The Government of India aims to achieve 500 GW of renewable energy by 2030, with 140 GW expected from wind energy. The National Institute of Wind Energy (NIWE) initially estimated India's wind potential at 49 GW at a 50 m hub height, which increased to 102 GW at 80 m and 302 GW at 100 m. A new study at 120 m hub height estimates a potential of 695 GW. This shows great potential for harnessing wind power with the application of new advances in this field.

The use of wind energy dates back thousands of years, and it dates back to ancient civilizations. Persia and China used windmills to grind grain into flour and pump water. Later, they were primarily used for agricultural tasks, such as milling grain and irrigation. Wind energy technology continued to evolve over the centuries. In the nineteenth century, wind turbines began to be used for generating electricity, with the first wind turbine built in Scotland in 1887 by Professor James Blyth (The Scottish Engineering Hall of Fame 2012). This technology has evolved significantly in the modern era, becoming a sophisticated and vital component of the global energy mix. The focus on sustainable energy solutions has grown due to increasing energy consumption, urbanization, geopolitical conflicts, depleting fossil fuels, and global warming. The transition to renewable energy is crucial because reliance on fossil fuels is damaging the environment. The Paris Agreement sets a goal to limit the global temperature rise to 1.5–2.0 °C. Countries such as the USA, Australia, and EU aim for carbon neutrality by 2050, while India aims for 2070. From 1989 to 2021, the average annual temperature has increased by 0.34 °C per decade. To limit the rise in temperature to 1.5 °C, renewable energy sources like wind and solar energy must grow by 20% annually until 2030 (Díaz and Moya 2024). According to a report on the "Evaluation of Wind Energy in India" published on August 2, 2022, to achieve energy goals with net zero emissions, the Committee recommended replacing old and less efficient turbines with advanced turbines and formulating a policy for re-powering old turbines and issuing guidelines for recycling old turbines (Singh 2022).

Today, wind energy is a key player in the global transition to renewable energy, as it offers a sustainable and clean source of power that reduces greenhouse gas emissions and dependence on fossil fuels. As the wind is ubiquitous, wind energy can be utilized efficiently, even in remote places such as hot deserts, lonely islands, or high-altitude places where survival is challenging. Wind power is now a major contributor to electricity generation in many countries, and on-going technological innovations continue to enhance its efficiency and cost-effectiveness. The continuous development of wind energy technologies underscores its important role in creating a sustainable future.

2.2 Wind Energy for Power Generation: Conventional Approach

The primary aim of harnessing wind energy for electricity production was to transform kinetic energy into rotational mechanical energy, which could then be used to generate electricity, so wind energy is considered an energy supplement in rural regions. However, history also provides evidence that ancient humans utilized wind energy to ease their work. Since time immemorial, boats have been sailing in lakes, rivers, and seas because of the wind; transportation has become easy and quick. This transportation connected the diverse civilizations of the ancient world and allowed the exchange of goods, ideas, and cultures. Windmills were used to pump water, sawmilling of timber, mill grain, extract oil extraction from seeds, etc., ultimately converting the kinetic energy of the wind into mechanical energy (Jain 2010). To understand wind power generation, it is essential to understand the factors that cause wind formation. Earth's rotation around the sun leads to uneven heating across different regions of the planet. Tropical regions receive a greater amount of solar radiation compared to the poles. This heating decreases the air density, causing it to rise and creating a low-pressure area. Consequently, air from high-pressure regions flows toward low-pressure areas, resulting in airflow. As these generated winds possess a significant amount of kinetic energy and such energy can be utilized for electricity generation with mechanical arrangements (Jain 2010).

During the nineteenth century, wind turbines were created to provide electricity in rural regions. A wind turbine is a device that converts the kinetic energy of the wind into electrical power. Technologies for generating usable wind energy have evolved since the 1970s, with significant advancements occurring since the 1990s.

In particular, in the last decade, significant improvement has been made in this area. This rapid development was observed due to the high growth rate in the wind energy market. Recent research and technologies focus on reducing costs and improving efficiency and reliability (Cheng and Zhu 2014). The conventional approach to wind energy-based electricity generation is widely applied with the integration of new advances, which has enhanced its efficiency. The Horizontal-Axis Wind Turbine (HAWT) became commercially applicable after detailed research of the aerodynamics of turbine blades. HAWTs have a horizontal rotation axis and usually feature two or three blades positioned up or downwind of the vertical support tower. Modern HAWTs installed particularly for commercial energy production have a capacity ranging from 1 to 5 MW (Leishman 2011).

Early wind energy systems had a constant speed configuration with a multistage gearbox and a squirrel-cage induction generator connected directly to the grid. With the advancement of power electronics, variable-speed systems incorporating a doubly fed induction generator and a partial-scale converter were introduced to improve efficiency and expand the operational wind speed range. Because gearboxes are prone to issues, gearless direct drive systems with synchronous generators and full-scale converters were developed. To further enhance efficiency and reliability, permanent magnet synchronous generators have replaced traditional synchronous generators (Cheng and Zhu 2014). A wind turbine blade cross-section is an airfoil. A symmetric airfoil with uniform wind speed and pressure does not have a lift because there are equal pressures on both surfaces. Tilting the airfoil or having a more curved upper surface creates a pressure difference, with higher speed and lower pressure on the upper surface compared to the lower surface. This generates lift perpendicular to the wind direction and drag parallel to it, causing the turbine blade to rotate and harness wind energy (Leishman 2011).

Initially, small turbine blades were used for easier installation. However, after conducting detailed aerodynamic studies to capture more wind energy economically, modern horizontal-axis wind turbines (HAWTs) have significantly increased in size, often exceeding 50 m in diameter. These design trends have led to the development of longer and more flexible turbine blades that can elastically bend and twist in response to aerodynamic forces (Leishman 2011). HAWT is commercially used for electricity generation because of its high efficiency up to 70%. These are typically used to optimize wind conditions, ensuring a steady flow and direction of wind to capture maximum wind energy (Shah Irshad et al. 2024). As they are ineffective in turbulent wind conditions; therefore, they are placed in areas with consistent and directional airflow, such as open fields. Due to their size and the need to avoid the turbulence caused by the spinning blades, HAWTs must be placed with sufficient space between each turbine. Therefore, the harvesting of wind energy through HAWTs in a limited space is a challenge.

Another type of wind turbine is the vertical axis wind turbine (VAWT) (Soraghan et al. 2012). The axis of VAWT is parallel to the earth's surface. A small VAWT is omnidirectional, which means it rotates in the direction of the wind without needing a yaw mechanism. A yaw system is required in HAWT to properly position the turbine toward the wind. VAWT turbines do not require high wind speeds or constant regulation and are most effective in areas with turbulent winds, such as rooftops, beaches, streets, and similar locations. Smaller VAWTs compete directly with smaller horizontal-axis wind turbines (HAWTs) on the market, unlike their larger counterparts, which have not attracted much investment. Due to the high demand for low-power applications, many producers and users favor small VAWTs. In addition to their omnidirectional capability, small VAWTs have lower start wind speeds compared to HAWTs and can be installed in areas where tall structures are prohibited. They also produce less noise and are less dependent on large tower structures. VAWTs offer simpler and more predictable features, enhancing turbine durability and performance (Shah Irshad et al. 2024).

Wind energy has emerged as a vital contributor to electricity generation, offering sustainable solutions to energy needs. The evolution of turbine size, aerodynamics, and power electronics has propelled the growth of wind energy, making it a competitive and reliable source of renewable power. As research and development continues, wind energy is expected to play an increasingly important role in global energy transitions.

2.3 Recent Advancement in Wind Energy for Power Generation

Historically, the conversion of wind energy into the desired form was carried out using sailboats, windmills, and wind pumps. But today, wind energy is mainly harvested to produce electricity. Professor James Blyth of Anderson's College, Glasgow, constructed Scotland's first windmill for electric power production in July 1887 (Price 2005). Since then, scientists and engineers have come up with many modifications in the design and construction of these turbines. Numerous inventions and innovations related to these turbines have increased their efficiency and made the energy harvesting from remote places or using low-speed winds feasible. As discussed in the previous section, there are situations where low wind speed, high turbulence, and frequent changes in wind direction can adversely affect the working of HAWTs. In this case, VAWTs are expected to work efficiently, but generally have a low-power coefficient. To address these issues, a novel cross-axis wind turbine (CAWT) is designed to harness wind energy from both horizontal and vertical directions of incoming wind, maximizing power generation from wind.

This innovative CAWT design aims to maximize wind power utilization regardless of the direction of the wind, eliminating the need for an orientation mechanism and improving self-starting capabilities. The CAWT occupies the same space as a VAWT but includes additional surfaces for generating lift force. The cross-axis configuration of this turbine is created using three main vertical blades connected to six horizontal blades through specially designed connectors. A significant advantage of the CAWT is its ability to operate with omnidirectional airflow from the sides of the vertical axis wind turbine and from the bottom through the horizontal-axis blades (Chong et al. 2017) (Fig. 2.1).

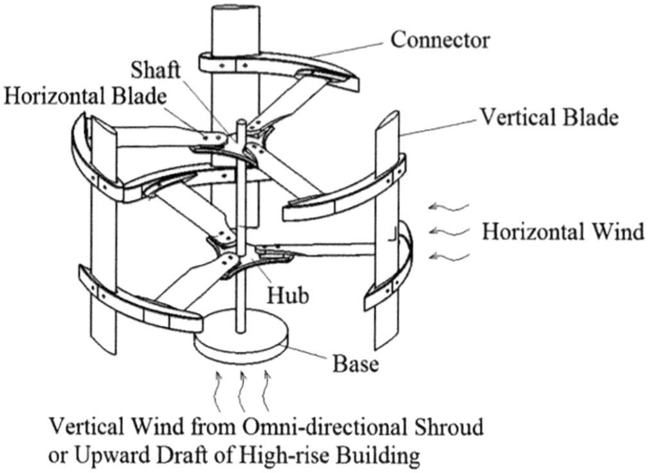

Fig. 2.1 Arrangement of the CAWT (wind direction is indicated with arrows). (Reprinted from Chong et al. 2017 with permission from Energy Procedia. Copyright 2017, Elsevier Ltd.)

A detailed computational fluid dynamics study of the CAWT by Chong et al. confirmed that CAWT provides 166% more rotational speed as compared to VAWT. The increase in maximum rotational speed indicates that the horizontal blades in CAWT contribute to additional power generation by capturing the redirected wind energy (Chong et al. 2017). However, the performance of wind turbines is mainly dependent on their location. Along with it, the speed of the wind, a hindrance to flow, and the degree of seasonal variation are some of the many parameters that can affect the generation of electricity. Therefore, high mountains, open fields, grasslands, etc. are well-suited sites for harvesting wind energy. But urban areas are recognized as the primary anthropogenic contributors to energy demand and global warming.

As a result, building an integrated wind turbine energy system is an effective way to utilize urban winds. In this system, wind energy is harnessed through Small Wind Turbines (SWTs), which can be integrated into building roofs, highways, railways, and various urban sites, and even incorporated into building designs from the outset (Díaz and Moya 2024). Here, the energy produced by the system can be used directly at the installation site; therefore, transportation losses are minimal, and the investor saves on the costs due to high-voltage transmission lines and control devices. However, this type of installation requires a thorough understanding of the environmental conditions (height above the mean sea level, heavy rain, storm, etc.). Also, the disadvantages, such as noise, shadow flickers, etc., should be studied. The above limitations (noise and visual pollution) can be efficiently tackled by using triboelectric nanogenerators (TENG), which can harvest energy from low wind speeds, even below 3 m/s. If harvested and stored efficiently, such energy could become a highly desirable and adequate source to meet the microscale energy needs of smart cities, such as lighting pathways, tunnels, and traffic lights (Bian et al.

2018). The cylindrical electrode TENG (CE-TENG) has been found to be effective in capturing wind energy even at low speeds of 1.5 m/s. This innovative design includes a lightweight and flexible aluminum foil for the hollow cylindrical top electrode. The friction layers are made up of aluminum foil and a drop-cast composite film made from polystyrene (PS) and graphene oxide (GO). This technology allows wind energy from various sources, including moving vehicles, air conditioning vents, and natural breezes (Mudgal et al. 2024).

Due to such advancements, wind generators are the most commonly used type of intermittent renewable energy harvester at present, but in the future, such growth will be limited by the availability of windy areas that are suitable for installations. Thus, the scientific community came up with the concept of airborne wind energy (AWE). This energy is harvested using high-altitude winds that blow at a height of 0.2–10 km above ground level. Airborne Wind Energy Systems (AWES) generally include two main components: a ground system and one or more aircrafts connected by ropes, which may also have electrical connections. The ground generation system is one of the AWES, and the energy harnessed from the wind is converted into electricity. This ground system can be either a moving vehicle or a stationary (fixed) system. Another concept is Fly-Gen systems, where electrical energy is generated on the aircraft and transmitted to the ground through a specialized tether carrying electrical cables. In such systems, wind turbines are typically used for electrical energy conversion (Cherubini et al. 2015). Finding suitable sites for installation may present complexities due to height and permitting requirements. However, the relative ease of installing airborne wind energy systems (AWE) should help in addressing such challenges. Furthermore, AWE exhibits a 40% lower carbon intensity than traditional wind technology due to its reduced material usage. Achieving even 90% lower carbon intensity with advanced AWES (Kristian et al. 2022). Therefore, it can be easily concluded that technological advancements will pave the way to a better future by increasing the efficiency and reliability of wind turbines. These inventions and innovations can harness wind energy sustainably in adverse conditions, remote places, and urbanized locations. Along with this, the cost-effectiveness has also made wind energy, the fastest-growing energy source in the world. However, there are several challenges to be overcome in order to further increase profitability and long-term sustainability.

2.4 Wind-Solar Integrated Approach for Power Generation

In recent years, research on sustainable energy has driven significant advances in renewable energy technologies, particularly in wind and solar power. Hybrid renewable energy systems are emerging as promising alternatives, and they are being used to make energy more reliable and efficient while reducing dependence on traditional energy sources. Solar power is known for its high efficiency during daylight hours, whereas wind power can function effectively both at night and in various weather conditions. Integrating these two energy sources can create a more reliable energy supply and significantly reduce the possibility of shortages during adverse weather conditions. Combining solar and wind energy ensures a more consistent power supply by balancing their peak production times. This minimizes the risk of shortage during times with less sunlight or wind and better manages seasonal changes (Hassan et al. 2023). One of the key advantages of hybrid systems is the integration of energy storage technology. This technology allows us to store surplus energy produced during peak times, which can be used when production is lower. When there is excess energy, it is possible to increase the efficiency of the entire system and reduce waste. Moreover, these systems help stabilize the grid by providing power during peak demand periods, reducing congestion, and additionally improving reliability. This makes renewable energy sources more reliable and efficient throughout the year (Hassan et al. 2023). Several countries have proven the successful implementation of hybrid renewable energy systems. For example, King Island in Australia has managed to generate 65% of its electricity from renewables by using a combination of wind and solar power. Similarly, Kodiak Island in Alaska has combined wind and hydropower to completely eliminate the use of diesel for heating purposes, resulting in significant savings in energy costs (Hassan et al. 2023). Furthermore, in 2018, the government of India introduced the National Wind-Solar Hybrid Policy to promote and standardize the wind-solar hybrid sector. This policy offers special benefits and incentives to developers in India (Kumar et al. 2022). However, the implementation of new hybrid power generation systems is not without its challenges. Economic feasibility is a major concern, as the initial costs of setting up these systems can be quite high. Additionally, the development of appropriate policy frameworks is essential to support the growth and sustainability of hybrid energy systems. Technological advances are essential to improve the efficiency and reliability of these systems (Bogdanov et al. 2021).

Despite these challenges, the benefits of hybrid renewable energy systems are undeniable. Combining solar and wind power can create a more balanced and reliable energy supply. This not only helps reduce the risk of energy shortages, but also makes better use of the available natural resources. The integration of energy storage technology further enhances the efficiency of these systems by ensuring that surplus energy is not wasted, but is stored for future use.

2.5 Wind Energy-Powered Desalination

Water is one of the most important resources required for the survival of life. The earth contains only 3% of freshwater, whereas the remaining 97% is saline water. The requirement for freshwater has been increasing since the onset of industrialization, and pollution has increased it manifolds. In the present scenario, saline water can be proved to be an effective solution to address the potable water crisis. Methods such as thermal-based desalination, ion exchange separation, and chemical and electrical methods have been developed in recent times for operation, but among them, membrane-based (RO) desalination is the most widely used process (Abdelkareem et al. 2024).

However, being an energy-intensive operation, people worldwide strive to use renewable energy for this process. Studies show that photovoltaic RO plants are the most prevalent and have a share of 32% of the worldwide renewable energy-powered desalination plants. However, in coastal regions where wind energy can be harvested efficiently, wind energy-based desalination facilities are more advantageous. Among all, stand-alone wind desalination plants have great potential in remote coastal areas with limited freshwater resources. Although such regions have ample amounts of wind that blows at sufficient speed, the intermittent nature of the wind is troublesome. To achieve an unhindered supply, storage devices such as batteries, fuel cells, etc., are coupled with wind energy systems (Abdelkareem et al. 2024). However, in the case of RO systems powered by wind energy, it is crucial to design a reliable control system that can improve or maintain yield composition by studying the variations of the slow wind (Rashidi et al. 2022).

A study carried out by Dahioui and Loudiyi on the performance of wind-powered RO systems showed that wind fluctuations resulted in an inconsistent supply of electricity. As a result, a variation in the pressure of the feed water was observed. However, to lessen the effect of intermittency in wind flow, pressure stabilizers or similar devices can be used (Dahioui and Loudiyi 2013). One can avoid external systems or devices to deal with wind fluctuations by using dynamic regulation systems, which use a supercapacitor bank as proposed by Carta et al. The system alters the flow rates and pressure to adapt the wind-generated power so that the desalination system works uninterruptedly (Rashidi et al. 2022). The installation of renewable energy-based desalination plants has significantly reduced environmental degradation. For example, Dehmas et al. found that the utilization of a wind farm with 10 MW capacity for the RO plant based in Algeria can reduce greenhouse gas emissions by 8500 tons or even more. Along with environmental benefits, wind energy also has economic benefits depending on the type of system and site of installation (Dehmas et al. 2011). According to Brendel et al., wind-powered RO systems have the lowest levelized water cost (Brendel et al. 2020). In recent times, due to automation of desalination processes, continuous supply of electricity is necessary, and therefore complete dependence on renewable energy, especially wind energy, is not feasible. In these cases, integration of wind turbines with non-renewable technologies can be helpful. These hybrid systems commonly use diesel generators to power desalination as needed. Efficient control systems (real-time analysis) can minimize the use of non-renewable energy by analyzing the power which wind turbines can generate. However, installation and operation costs increase, thereby increasing the levelized cost of water produced (Kershman et al. 2005).

2.6 Harnessing Water from Air

Water is integral to power generation within the interconnected framework of the energy-water-food nexus. It is essential for power generation directly or indirectly. Water availability is vital for energy production for multiple reasons. Conventional power generation methods like coal-based and nuclear energy-based thermal power plants utilize steam to rotate the turbine, which requires freshwater sources. After 2017, the government restricted water withdrawal to up to 2.5 m^3/MWh of power generated (Srinivasan et al. 2018). Considering future power sources, solar photovoltaic systems have an operational water withdrawal intensity of approximately 0.08 m^3/MWh, primarily due to panel cleaning. The conclusion can be drawn that water is essential for power generation, even for renewable energy sources such as solar (Utturkar et al. 2024). Currently, the electricity generation sector has a substantial water footprint. Additionally, as the global population increases, the per capita water footprint will also increase drastically. This may create a water crisis in the coming decades and it is necessary to identify the water source from which drinkable water can be obtained.

As most of the water footprint involves the vaporization of water as a greenhouse gas, it is possible to recover water from the humid air using various atmospheric water-harvesting methods. Atmospheric water generation involves extracting water from humidity in the air. This method does not adversely affect the environment, as the removal of humidity is part of the natural hydrological cycle (Raveesh et al. 2021). However, the question arises about the absence of minerals in the generated water, which makes it ineffective for human consumption. Instead, this water can be readily used for power generation purposes, such as thermal power plants, where the least total suspended solids are desired in the water.

Several techniques for air-water generation include active refrigeration, passive cooling, vapor concentration, and hybrid generation. The active refrigeration method involves cooling the humid air below its dew point. This required cooling for condensation is achieved primarily using a conventional vapor compression refrigeration (VCR) system. In this method, ambient temperature, relative humidity, and dew point are crucial parameters that define the system's efficiency (Bagheri 2018). Higher relative humidity, lower ambient temperature, and higher dew point temperature result in better system efficiency. Since these systems involve the compression of refrigerant, they are inherently energy-intensive. However, the use of renewable energy sources, such as wind, solar, and hydropower, can make the system more sustainable. Additionally, employing non-freon-based refrigerants can further enhance the system's environmental sustainability (Raveesh et al. 2021; Patel et al. 2023). Furthermore, the passive cooling method involves extracting water through the harvesting of dew and fog. When water vapor condenses on a surface by releasing heat via the passive radiant cooling method, which needs the temperature of the dew point, then it is dew harvesting (Zhao et al. 2019). Fog harvesting is a method for AWG that collects fog droplets using a mesh structure and allows coalescence, followed by water collection under gravity action. This method is simple, cost-effective and doesn't require an external energy source. For example, a dew harvesting plant in Gujarat, India, produces 500 L of water daily and claims to be more economical than reverse osmosis, with no environmental impact (Raveesh et al. 2021).

Sorption methods can be employed to reduce significant energy losses associated with air cooling in the active refrigeration method to increase the air humidity. This increase in relative humidity makes the refrigeration process much more efficient, leading to considerable energy savings. Another method involves using sorbents to capture atmospheric water vapor until saturation. Once saturated, it is heated to release the captured water vapor, which is then condensed at ambient temperature to produce water. Therefore, the active refrigeration method is found to be most effective in increasing the desalination process and making it a commercially scalable unit (Raveesh et al. 2021). Choosing the optimal technique for efficient operation depends on various factors such as climatic conditions, economic feasibility, and commercial scalability. Currently, there is no universally applicable, scalable, energy-efficient technique that can operate across a wide range of conditions. Despite significant advances in this field over the past decade, no technique has yet proven to be effective as most are still in the early stages. Future research and development should focus primarily on the parameters influencing energy to create opportunities for scalable operation.

2.7 Carbon Capture and Utilization for Sustainable Power Generation

Air is composed of a mixture of different gases, and the current level of carbon dioxide (CO_2) in air has alarmingly increased to 429 ppm in 2023 (Pro Oxygen 2024). The exponential rise in the CO_2 levels since the Industrial Revolution in 1750, resulting from various manufacturing activities, has led to climate emergency. Carbon dioxide (CO_2) sequestration and storage (CSS) processes have gained significant attention to mitigate greenhouse gas emissions. However, CSS increases the availability of captured CO_2 as a zero-cost feedstock. Consequently, the carbon capture and recycling (CCR) process, which involves the conversion of CO_2 into fuels and chemicals, has become a widely adopted solution. This approach transforms CO_2 into valuable products and helps reduce emissions, making it a more favorable option than CSS (Al-Saydeh and Zaidi 2018). This process aims to create a cyclical system in which CO_2 generated from fuel combustion is recaptured and converted back into fuel. This approach not only helps to manage CO_2 emissions but also provides a sustainable method for the production of circular fuels. Therefore, CO_2 fuel production can be an option for harnessing power from the air, due to emerging technologies and extensive research in the same field (Ishaq and Crawford 2023). The carbon capture and utilization process begins with a concentration of the CO_2 stream. The combustion of carbon-rich fuel produces a combustion gas that is concentrated with CO_2. Researchers have developed several technologies to capture the CO_2. Various methods are used for carbon capture, including direct air capture, post-combustion capture, precombustion capture, and the combustion of oxygen-rich fuels (Prajapati et al. 2024). Chemical scrubbing using basic amines is one of the most developed technologies in the case of post-combustion carbon capture. It is found to be applicable and effective at the commercial level (Sieborg et al. 2024).

With chemical conversions from CO_2 as feedstock, it is possible to produce fuels and chemicals such as methanol, formic acid, and syngas. During the 1920s and 1930s, the first methanol production plant was established in the USA, using CO_2 and H_2 for the synthesis of methanol (Al-Saydeh and Zaidi 2018). Methanol can be utilized as a feedstock for various chemical products such as formaldehyde, gasoline, and dimethyl ether. In addition, syngas can be used to produce methanol or other hydrocarbons and gasoline. As described in chemical processes for a sustainable future, the anthropogenic carbon cycle for methanol demonstrates that recycling carbon dioxide followed by hydrogenation produces methanol (Letcher et al. 2014). This methanol can be further upgraded to create other chemicals or fuels or can be used to generate syngas. When these fuels are combusted, they produce CO_2-rich combustion gases. Capturing this CO_2 provides a feedstock for the next cycle of hydrogenation, thereby continuing the CO_2 recycling process (Fig. 2.2).

Recent advances in CO_2 utilization technologies have led to the development of methods that couple CO_2 capture with thermo-catalytic conversion, biofixation using microalgae, electrochemical catalytic conversion, and photoelectrocatalysis. These methods aim to utilize CO_2 in the air while reducing the associated energy costs. However, the thermodynamic stability of CO_2 creates a high activation barrier, which requires high temperatures and otherwise costly catalysts to have sufficient reactivity. The thermocatalytic process needs agents that can capture CO_2.

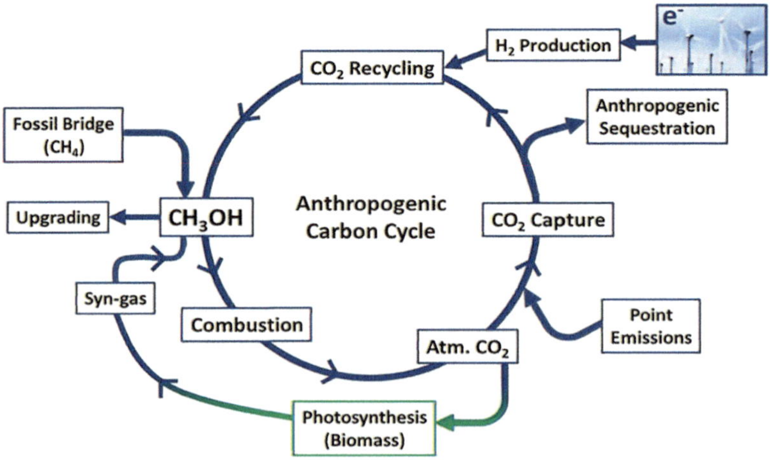

Fig. 2.2 Anthropogenic carbon cycle for methanol production. (Reprinted from Letcher et al. 2014 with permission from RSC. Copyright 2017, Royal Society of Chemistry)

These agents can withstand high temperatures to convert CO_2 with reducing agents such as light alkanes or hydrogen. However, electrochemical catalysis employs expensive metal-based catalysts to lower temperature requirements. However, impurities in the flue gas such as SO_x and NO_x can poison the catalyst. To prevent poisoning, extensive flue gas is essential (Sieborg et al. 2024). In addition to the challenges mentioned above, the energy-intensive nature and scalability concerns of these technologies pose significant barriers to their wide applicability. However, continued research and advancement in chemical conversion processes have the potential to offer effective solutions to mitigate excessive CO_2 emissions. As these technologies advance, they could significantly help reduce greenhouse gas emissions and support a sustainable future for future generations.

2.8 Conclusions and Recommendations

Wind, a significant source of energy, is the result of air movement driven by pressure differences. Wind energy has historically been used to convert kinetic energy into mechanical energy for transportation, which, over centuries, facilitated global connectivity through trade and cultural exchange. Along with that, the wind energy was used to mill, grind, and pumping water.

Developing since ancient times, wind turbines now play a crucial role in global energy transition efforts, with the aim of achieving sustainability and reducing fossil fuel dependence amidst rising energy demands and climate concerns. Modern wind turbines harness this energy to generate electricity, initially for rural areas. Technological advances have significantly enhanced efficiency and reliability, particularly in horizontal-axis wind turbines (HAWTs) since the nineteenth century. These turbines, characterized by their aerodynamic blades and advanced electronics, dominate commercial wind energy due to their high efficiency in consistent airflow. Despite challenges in turbulent environments, vertical axis wind turbines (VAWTs) offer simpler, quieter, and more flexible options for low-power applications. The Cross-Axis Wind Turbine (CAWT) represents a novel approach that integrates vertical and horizontal blades to maximize power generation regardless of wind direction, outperforming Vertical Axis Wind Turbines (VAWT).

Urban areas are increasingly adopting small wind turbines (SWTs) integrated into buildings and infrastructure to meet local energy demands efficiently, while triboelectric nanogenerators (TENGs) offer microscale energy solutions by harvesting wind energy from low-speed environments. In addition, airborne wind energy systems (AWES) present a promising frontier by harnessing high-altitude winds with a lower carbon intensity compared to traditional wind technologies.

On a commercial scale, regardless of such advancements, shortages due to adverse weather and seasonal variations affect the working of wind energy systems; thus, using hybrid renewable energy systems combining solar efficiency during daylight with wind's round-the-clock operation, enhances reliability and reduces reliance on traditional energy sources. Energy storage integration further boosts efficiency by storing surplus energy for later use, stabilizing grids, and enhancing overall reliability. Despite challenges such as high costs and policy frameworks, hybrid systems offer substantial benefits in sustainable energy production worldwide.

Today, the world has progressed a lot in terms of innovations and inventions, but water remains critical in power generation within the energy-water-food nexus. India's increasing electricity demand is closely related to water availability, which is crucial for conventional steam-powered thermal power plants. Despite water-intensive processes, renewables such as solar energy also require water for panel cleaning. As global water footprints increase, methods such as atmospheric water generation offer sustainable solutions, including extracting humidity from air using various techniques, each with unique efficiencies and environmental impacts, supporting diverse power generation needs. Similarly, desalination of seawater is crucial due to freshwater scarcity. Increasingly, renewable energy, particularly from photovoltaic and wind sources, powers desalination plants to mitigate energy intensity. Wind energy systems are prevalent in coastal regions with robust wind resources. Challenges include wind intermittency, addressed by storage systems, and the need for robust control systems to stabilize operations. Despite costs, wind-powered desalination offers environmental benefits and economic advantages in remote areas.

It is evident from the above-mentioned points that wind energy continues to evolve as a pivotal renewable energy source, driving global sustainability efforts. But in the process of harnessing the maximum energy from the wind, attention should be paid to improving the quality of air for a sustainable and healthy future.

In that sense, carbon dioxide sequestration and storage (CSS) methods aim to mitigate emissions but also make captured CO_2 available as a zero-cost feedstock. Carbon capture and recycling (CCR) processes have emerged as an alternative by converting CO_2 into fuels and chemicals, promoting the circular economy and effectively reducing greenhouse gas emissions effectively; however, scalability and energy efficiency challenges still remain to be addressed.

To ensure the sustainability of our planet, it is imperative to focus on the commercialization of recent technological advancements. The use of energy from the air facilitates electricity production and offers solutions to water scarcity and air quality improvement. In addition to research, investments will be crucial in advancing these technologies and ensuring their widespread adoption. By fostering innovation and supporting the development of sustainable technologies, we can pave the way for a better and more resilient India. Through these collective efforts, we can protect the environment and secure a sustainable future for all.

Acknowledgments The authors declare that they have no conflict of interest. This work did not receive any specific grants from public, commercial, or not-for-profit funding agencies. The authors acknowledge the Institute of Chemical Technology, Mumbai, and Queen's University Belfast, UK, for the resources provided in writing this chapter.

References

Abdelkareem MA, Al Radi M, Mahmoud M, Sayed ET, Salameh T, Alqadi R, Kais E-CA, Olabi AG (2024) Recent progress in wind energy-powered desalination. Therm Sci Eng Prog 47:102286. https://doi.org/10.1016/j.tsep.2023.102286

Al-Saydeh SA, Zaidi SJ (2018) Carbon dioxide conversion to methanol: opportunities and fundamental challenges. In: Carbon dioxide chemistry, capture and oil recovery. InTech. https://doi.org/10.5772/intechopen.74779

Bagheri F (2018) Performance investigation of atmospheric water harvesting systems. Water Resour Ind 20:23–28. https://doi.org/10.1016/j.wri.2018.08.001

Bian Y, Jiang T, Xiao T, Gong W, Cao X, Wang Z, Wang ZL (2018) Triboelectric nanogenerator tree for harvesting wind energy and illuminating in subway tunnel. Adv Mater Technol 3(3). https://doi.org/10.1002/admt.201700317

Bogdanov D, Ram M, Aghahosseini A, Gulagi A, Oyewo AS, Child M, Caldera U, Sadovskaia K, Farfan J, De Souza Noel Simas Barbosa L, Fasihi M, Khalili S, Traber T, Breyer C (2021) Low-cost renewable electricity as the key driver of the global energy transition towards sustainability. Energy 227:120467. https://doi.org/10.1016/j.energy.2021.120467

Brendel LPM, Shah VM, Groll EA, Braun JE (2020) A methodology for analyzing renewable energy opportunities for desalination and its application to Aruba. Desalination 493:114613. https://doi.org/10.1016/j.desal.2020.114613

Cheng M, Zhu Y (2014) The state of the art of wind energy conversion systems and technologies: a review. Energy Convers Manag 88:332–347. https://doi.org/10.1016/j.enconman.2014.08.037

Cherubini A, Papini A, Vertechy R, Fontana M (2015) Airborne wind energy systems: a review of the technologies. Renew Sustain Energy Rev 51:1461–1476. https://doi.org/10.1016/j.rser.2015.07.053

Chong W-T, Wong K-H, Wang C-T, Gwani M, Chu Y-J, Chia W-C, Poh S-C (2017) Cross-axis-wind-turbine: a complementary design to push the limit of wind turbine technology. Energy Procedia 105:973–979. https://doi.org/10.1016/j.egypro.2017.03.430

Dahioui Y, Loudiyi K (2013) Wind powered water desalination. In: 2013 International renewable and sustainable energy conference (IRSEC), pp 257–262. https://doi.org/10.1109/IRSEC.2013.6529659

Dehmas DA, Kherba N, Hacene FB, Merzouk NK, Merzouk M, Mahmoudi H, Goosen MFA (2011) On the use of wind energy to power reverse osmosis desalination plant: a case study from Ténès (Algeria). Renew Sustain Energy Rev 15(2):956–963. https://doi.org/10.1016/j.rser.2010.11.004

Díaz AV, Moya IH (2024) Urban wind energy with resilience approach for sustainable cities in tropical regions: a review. Renew Sustain Energy Rev 199:114525. https://doi.org/10.1016/j.rser.2024.114525

Hassan Q, Algburi S, Sameen AZ, Salman HM, Jaszczur M (2023) A review of hybrid renewable energy systems: solar and wind-powered solutions: challenges, opportunities, and policy implications. Results Eng 20:101621. https://doi.org/10.1016/j.rineng.2023.101621

Indian Wind Turbine Manufacturers Association (2023) Indian wind energy. https://www.indianwindpower.com/

Ishaq H, Crawford C (2023) CO2-based alternative fuel production to support development of CO2 capture, utilization and storage. Fuel 331:125684. https://doi.org/10.1016/j.fuel.2022.125684

Jain P (2010) Wind energy engineering. McGraw Hill Professional

Kershman SA, Rheinländer J, Neumann T, Goebel O (2005) Hybrid wind/PV and conventional power for desalination in Libya—GECOL's facility for medium and small scale research at Ras Ejder. Desalination 183(1–3):1–12. https://doi.org/10.1016/j.desal.2005.04.021

Kristian P, Alexander B, Jon G (2022) Getting airborne—the need to realise the benefits of airborne wind energy for net zero

Kumar A, Pal D, Kar SK, Mishra SK, Bansal R (2022) An overview of wind energy development and policy initiatives in India. Clean Technol Environ Policy 24(5):1337–1358. https://doi.org/10.1007/s10098-021-02248-z

Leishman GJ (2011) In: Sathyajith M, Philip GS (eds) Advances in wind energy conversion technology: aerodynamics of horizontal axis wind turbines. Springer, Berlin. https://doi.org/10.1007/978-3-540-88258-9

Letcher T, Scott J, Patterson D (2014) In: Letcher T, Scott J, Patterson D (eds) Chemical processes for a sustainable future. The Royal Society of Chemistry. https://doi.org/10.1039/9781839168642

Ministry of New and Renewable Energy (2021) Wind overview, Jun 14. Ministry of New and Renewable Energy. https://mnre.gov.in/wind-overview/

Ministry of New and Renewable Energy (2022) Renewable energy in India, Sept 9. https://pib.gov.in/FeaturesDeatils.aspx?NoteId=151141&ModuleId%20=%202

Mudgal T, Tiwari M, Bharti D (2024) Cylindrical-electrode triboelectric nanogenerator for low-speed wind energy harvesting. Nano Energy 123:109388. https://doi.org/10.1016/j.nanoen.2024.109388

Patel D, Kulwant M, Shirin S, Varshney R, Pandey G, Yadav AK (2023) Fluoride removal from aqueous solution using iron-based materials: preparation, characterization, and applications. In: Yadav AK, Saba S, Singh VP (eds) Advanced treatment technologies for fluoride removal in water. Springer Nature, Cham. https://doi.org/10.1007/978-3-031-38845-3_4

Prajapati M, Thesia D, Thesia V, Rakholia R, Tailor J, Patel A, Pardiwala JM, Shah M (2024) Carbon capture, utilization, and storage (CCUS): a critical review towards carbon neutrality in India. Case Stud Chem Environ Eng 10:100770. https://doi.org/10.1016/j.cscee.2024.100770

Price TJ (2005) James Blyth—Britain's first modern wind power pioneer. Wind Eng 29(3):191–200. https://doi.org/10.1260/030952405774354921

Pro Oxygen (2024) Daily CO_2, Jun 16. https://www.co2.earth/daily-co2

Rashidi MM, Mahariq I, Murshid N, Wongwises S, Mahian O, Alhuyi Nazari M (2022) Applying wind energy as a clean source for reverse osmosis desalination: a comprehensive review. Alex Eng J 61(12):12977–12989. https://doi.org/10.1016/j.aej.2022.06.056

Raveesh G, Goyal R, Tyagi SK (2021) Advances in atmospheric water generation technologies. Energy Convers Manag 239:114226. https://doi.org/10.1016/j.enconman.2021.114226

Shah Irshad A, Kargar N, Elkholy MH, Ahmad Ludin G, Elias S, Hilali A, Senjyu T, Gamil M, Pinter G (2024) Techno-economic evaluation and comparison of the optimal PV/wind and grid hybrid system with horizontal and vertical axis wind turbines. Energy Convers Manag X 23:100638. https://doi.org/10.1016/j.ecmx.2024.100638

Sieborg MU, Oliveira JMS, Ottosen LDM, Kofoed MVW (2024) Flue-to-fuel: bio-integrated carbon capture and utilization of dilute carbon dioxide gas streams to renewable methane. Energy Convers Manag 302:118090. https://doi.org/10.1016/j.enconman.2024.118090

Singh R (2022) Evaluation of wind energy in India

Soraghan CE, Leithead WE, Yue, H (2012) Control based power smoothing for aggregated vertical axis wind turbines. https://api.semanticscholar.org/CorpusID:59046162

Srinivasan S, Kanudia A, Roshna N, Dharmala N (2018) Impact of power sector growth on water resources

The Scottish Engineering Hall of Fame (2012) James Blyth. The Institution of Engineers in Scotland. https://engineeringhalloffame.org/profile/james-blyth

Utturkar PS, Prajapat VR, Jadhav SV, Manyar HG (2024) Exploring water's role in sustainable electricity generation for power in future. In: Integrated management of water resources in India: a computational approach. Springer, pp 449–471. https://doi.org/10.1007/978-3-031-62079-9_26

Wind Power by Country (2024) World population review. https://worldpopulationreview.com/country-rankings/wind-power-by-country

Zhao D, Aili A, Zhai Y, Lu J, Kidd D, Tan G, Yin X, Yang R (2019) Subambient cooling of water: toward real-world applications of daytime radiative cooling. Joule 3(1):111–123. https://doi.org/10.1016/j.joule.2018.10.006

Chapter 3
Harnessing Wind Energy for Future Sustainability

S. C. Vetrivel, V. Sabareeshwari, K. C. Sowmiya, and V. P. Arun

Abstract Using the power of wind appears to be a viable answer as the world struggles to find sustainable energy sources to combat climate change. In this chapter, we examine the potential of wind energy and other aerial power sources to create a sustainable future. One of the most well-known renewable energy sources, wind energy has grown significantly in the last few years. Due to developments in turbine technology, wind farms can now compete on a broad scale with traditional fossil fuels because of their increased efficiency and cost-effectiveness. Furthermore, the scalability and versatility of wind power make it adaptable to various geographical locations, from coastal regions to inland areas with strong wind resources. Beyond wind turbines, other innovative technologies are emerging to harness the power of the wind. For example, kite and drone systems have the ability to harvest wind energy at higher altitudes with more regular and powerful winds. Tethered flying devices and wind turbines are examples of airborne wind energy technologies that offer innovative methods of producing electricity from the atmosphere. For sustainability, switching to wind-based renewable energy sources has several advantages. These technologies help mitigate greenhouse gas emissions and combat climate change by decreasing dependence on fossil fuels. Furthermore, decentralization of energy generation by distributed wind farms improves energy security and resilience, especially in isolated or off-grid areas. To fully realize the potential of wind energy, there are

S. C. Vetrivel (✉)
Department of Management Studies, Kongu Engineering College,
Perundurai, Tamil Nadu, India

V. Sabareeshwari
Department of Soil Science and Agricultural Chemistry, Amrita Vishwa Vidyapeetham,
Coimbatore, Tamil Nadu, India

K. C. Sowmiya
Research Department of Physics, Sri Vasavi College, Erode, Tamil Nadu, India

V. P. Arun
Department of Business Administration, JKKN College of Engineering and Technology,
Kumarapalayam, Tamil Nadu, India

A. K. Yadav (ed.), *Powering Future: Clean Energy Solutions for a Sustainable Planet*,
Clean Energy Production Technologies, https://doi.org/10.1007/978-981-96-5055-2_3

still obstacles to overcome. Concerns about wildlife impacts, visual aesthetics, and noise pollution require careful planning and community participation in the construction of wind farms. Technological advances and regulatory frameworks are essential to overcome barriers to the widespread adoption of wind energy solutions.

Keywords Sustainability · Renewable energy · Mitigation of climate change mitigation · Turbine technology · Innovation · Distributed energy · Decentralization

3.1 Introduction

3.1.1 History and Evolution of Wind Energy

The utilization of wind energy has a rich history that spans several millennia. Early applications can be traced back to around 500–900 AD when the Persians are believed to have constructed the first windmills. These primitive machines, with their vertical-axis design, were primarily used to grind grains and pumping water, tasks essential for agrarian societies. The design and utility of windmills evolved significantly in the twelfth century in Europe, particularly in the Netherlands, where horizontal-axis windmills became prevalent. These windmills not only supplied ground grain, but also played a crucial role in the drainage of wetlands and the reclamation of land. The advent of modern wind turbines can be traced back to the late nineteenth century with the work of the Danish scientist Poul-la-Cour. His experiments in the 1890s laid the groundwork for electricity generation from wind, marking the transition from mechanical to electrical applications of wind power. Charles F. Brush constructed the first wind turbine to produce energy in Cleveland, Ohio, in 1888. This turbine was built with spruce wood blades and had a capacity of 12 kW. Its rotor diameter measured 17 m. The 1970s oil crises sparked a resurgence of interest in wind energy and led to significant funding for research and development from both the public and private sectors. Significant technological advances were achieved during this time, such as innovations in materials, aerodynamics, and blade design, which improved efficiency and dependability. Large-scale wind farms became commercially viable in the 1990s and early 2000s, especially in China, Europe, and the United States, because of the encouragement of regulations and advances in technology. The incredible advancement in wind energy technology is demonstrated by the fact that contemporary wind turbines have capacities more than 10 MW, and offshore wind farms are becoming more and more feasible and attractive from an economic standpoint.

3.1.2 Value of Sustainable Energy Resources

The global energy landscape is characterized by a growing demand for energy coupled with the urgent need to mitigate the adverse effects of fossil fuel consumption. Fossil fuels, which have powered industrialization and economic growth for more than a century, are finite resources that pose significant environmental challenges (Ahmed et al. 2016). The main cause of greenhouse gas emissions, which in turn lead to air pollution and climate change with dire consequences for ecosystems and human health, is the burning of coal, oil, and natural gas. A sustainable substitute that can deliver a steady stream of clean energy is provided by renewable energy sources (Wiser 2009). Renewable energy sources such as wind, solar, and hydropower have less impact on the environment when they are used than fossil fuels because they do not diminish over time. Due to its widespread availability and advanced technology, wind energy, in particular, is one of the renewable energy sources with the fastest rate of growth (Akpinar and Akpinar 2005). Wind energy has major environmental benefits, such as a notable decrease in carbon emissions and other pollutants, which contributes to the fight against climate change and improves air quality. Geopolitical and economic factors are also driving the transition to renewable energy. Although dependence on imported fossil fuels can cause vulnerabilities and compromise national security, the use of renewable energy sources can improve resilience and energy independence. Furthermore, the generation of jobs and economic growth is now significantly influenced by the renewable energy sector (Allouhi et al. 2017). More than 11 million people were employed globally in the renewable energy sector in 2018, with wind energy contributing significantly to this number, according to the International Renewable Energy Agency (IRENA). A growing number of governments and international organizations are realizing these advantages, resulting in agreements and policies designed to accelerate the adoption of renewable energy sources and meet long-term sustainability objectives.

3.1.3 Overview of Wind Energy's Role in Sustainability

To achieve a sustainable energy future globally, wind energy is essential. Wind power is one of the most developed and widely used renewable energy sources and plays a major role in both diversifying the energy mix and reducing greenhouse gas emissions. Approximately 6% of the world's electricity was generated by wind energy in 2023, and installed capacity is still expanding rapidly. Countries like Denmark, Germany, and China have been at the forefront of this expansion, demonstrating the viability and scalability of wind energy on a large scale (Archer and Jacobson 2003). Wind energy is very beneficial economically. It creates jobs in several industries, including wind turbine production, installation, operation, and maintenance. For example, the wind energy industry alone employs approximately

120,000 people in the United States, and as additional projects are online, this figure is expected to increase. Wind farms often boost local economies and give landowners additional revenue streams, especially in rural and coastal locations, where there may not be as many other economic options. Wind energy is among the cleanest energy sources in terms of the environment. Its operation does not produce greenhouse gas emissions or air pollutants, hence lowering the carbon footprint of the energy production process. Furthermore, wind farms can coexist with agricultural practices and have a low environmental impact, allowing for the numerous uses of one piece of land. Wind energy is now a more sustainable option due to improvements in turbine technology and site planning, which have also reduced the impact on wildlife (Wind Europe 2017). In the future, stability and reliability should be improved by integrating wind energy with other renewable energy sources and energy storage technologies (Ayodele and Ogunjuyigbe 2016; Bousla et al. 2023). New technologies have the potential to further enhance the capacity and efficiency of wind energy systems. Examples of these include enhanced aerodynamics and floating offshore wind turbines. Wind energy is expected to be essential in achieving global sustainability goals and guaranteeing a cleaner and more resilient energy future as nations work to meet their climate commitments and shift to low-carbon economies.

3.2 Fundamentals of Wind Power

3.2.1 Principles of Wind Energy Conversion

Converting wind kinetic energy to mechanical energy, which is then converted into electricity, is the process of harnessing wind energy (Wind Europe 2019). This process occurs within the framework of a wind turbine, a device specifically designed to capture and utilize wind energy. The basic principle behind the conversion of wind energy involves the rotation of turbine blades when exposed to the wind. As the blades rotate, they spin a shaft connected to a generator, which produces electricity (Chandel et al. 2014). Wind speed, turbine size and design, and conversion system efficiency are some of the variables that affect how much energy is produced.

3.2.2 Types of Wind Turbines

Different designs and combinations of wind turbines are available to meet different environmental requirements and energy needs. Horizontal- and vertical-axis wind turbines are the two main types of wind turbine (Chang 2011). The most popular types of turbine have blades positioned on a horizontal shaft and are called horizontal-axis turbines. These turbines typically have a tall tower to capture the higher wind speeds found at elevated altitudes. Horizontal-axis turbines are further classified into two main types:

Onshore turbines: These turbines are installed on land and are most commonly used because of their accessibility and cost-effectiveness. Small home and large utility-scale turbines are among the sizes offered.

Offshore turbines: According to the World Wind Energy Association (2015), offshore wind farms or coastal areas are the usual locations for offshore wind turbine installations. Compared to onshore turbines, they benefit from faster and more steady wind speeds, which increase energy production. Offshore turbines require specialized designs to withstand harsh marine conditions and are typically larger in size.

On the contrary, the blades of vertical-axis turbines are oriented vertically around a central axis (Dabbaghiyan et al. 2016). Vertical-axis turbines are less common than horizontal-axis turbines, although they have several benefits, such as the ability to capture wind from any direction without the need for yaw devices. However, they are mostly used in experimental projects or specialized applications because of their lower efficiency.

3.2.3 Key Components of Wind Turbines

Together, these essential parts of wind turbines enable them to capture, transform, and transfer wind energy into electrical power: (a) *Blades*: The aerodynamic surfaces of the blades collect wind energy and transform it into rotational motion. They are designed for performance and efficiency and come in a variety of sizes and forms; (b) *Rotor*: The rotor is the assembly comprising the blades and the hub to which they are attached. It is responsible for capturing wind energy and transferring it to the drivetrain; (c) *Drivetrain*: The drivetrain consists of components such as the gearbox and generator, which convert the rotational motion of the rotor into electricity. To power the generator, which generates electricity, the gearbox increases the rotor's rotating speed; (d) *Tower*: The tower supports the rotor and nacelle at an elevated height, allowing the turbine to capture higher wind speeds and minimize ground-level turbulence. Towers vary in height depending on the size of the turbine and site characteristics; (e) *Nacelle*: The gearbox, generator, and control systems for the drive train are housed in the nacelle. For the best energy capture, it is usually placed atop the tower and rotates to face the wind; (f) *Yaw system*: The yaw system allows the turbine to align itself with the direction of the wind. It consists of motors and sensors that adjust the orientation of the nacelle and rotor to maximize energy production; and (g) *Foundation*: The foundation secures the turbine to the ground or seabed, providing stability and support against wind and environmental loads (Youness et al. 2019). The type of foundation used for an offshore installation may vary depending on the type of turbine and site conditions. Some alternatives include jackets, floating platforms, and monopiles. Together, these parts allow wind turbines to effectively capture wind energy and transform it into useable electricity for a variety of purposes, including providing power to the grid or powering homes and businesses.

3.3 Technological Advancements in Wind Energy

3.3.1 Innovations in Turbine Design

Recent years have seen significant advances in wind turbine design aimed at improving efficiency, reliability, and cost-effectiveness. One notable innovation is the development of larger and more efficient turbine blades. For example, at higher altitudes where wind speeds are generally stronger and more consistent, longer blades allow turbines to harness more wind energy (Dogan and Seker 2016). In addition, advances in aerodynamics, materials science, and manufacturing techniques have led to more streamlined blade designs that reduce drag and increase energy capture. Generator and drivetrain systems represent another area of innovation. Because they are simpler, more reliable, and require less maintenance than gearboxes, direct drive turbines have become more and more popular (Zamzoum et al. 2019). Permanent magnet generators are also becoming more prevalent, offering higher efficiency and power density compared to traditional induction generators (El Mourabit et al. 2020a). Furthermore, advancements in control systems and sensor technology have enabled turbines to operate more intelligently and adaptively. Algorithms for predictive maintenance employ real-time data to foresee component failures and enhance maintenance plans, hence decreasing operating expenses and downtime (El Mourabit et al. 2020b). Furthermore, under varied wind conditions, turbine performance is optimized and energy production is maximized through the application of machine learning and artificial intelligence technology.

3.3.2 Offshore vs. Onshore Wind Farms

The decision between offshore and offshore wind farms is influenced by several variables, such as access to wind resources, environmental concerns, legal limitations, and economic viability. Because they have the potential to produce more energy and cause fewer problems with land use, offshore wind farms have become a more attractive option (Ellabban et al. 2014). Higher capacity factors and energy output are the result of offshore wind speeds, which are often stronger and more consistent than onshore wind speeds. Furthermore, because offshore wind farms are frequently located closer to populous coastal areas, transmission losses and grid congestion are reduced. However, offshore wind farms pose unique challenges compared to their offshore counterparts. Installation and maintenance costs are typically higher due to the harsh marine environment, deeper water depths, and logistical challenges associated with offshore operations. Furthermore, grid connection and permitting processes for offshore projects can be more complex and time consuming (Zhao et al. 2012). Despite these obstacles, economies of scale and technological developments are bringing offshore wind energy closer to parity with traditional energy sources by lowering its cost. Additionally, the popularity of floating offshore wind technology, creating new options for offshore wind development in areas with little bottom infrastructure by enabling the installation of turbines in deeper waters away from shore.

3.3.3 *Smart Grid Integration*

The integration of wind energy into contemporary electrical grids equipped with cutting-edge automation, control, and communication technology is referred to as "smart grid integration." Smart grids make it possible to monitor and control energy flows in real time, giving grid managers the ability to better balance supply and demand and take into account variable renewable energy sources such as wind. The creation of grid-scale energy storage technologies is a crucial component of smart grid integration (Energy and Philibert 2013). Batteries, pumped hydro storage, and thermal energy storage are examples of energy storage devices that help reduce the erratic nature of wind energy by storing extra electricity produced during high wind availability and releasing it during low wind or high demand periods. Increased grid stability, less renewable energy, and increased grid integration are all possible by energy storage. Furthermore, the accuracy of wind energy projections is increased through the use of predictive analytics and sophisticated forecasting techniques (Equinor 2019). Grid managers can optimize resource allocation and reduce dependence on backup generation by anticipating variations in wind power output and modifying grid operations accordingly through improved wind condition forecasting. Demand response programs, which encourage customers to modify their electricity consumption in response to market signals and grid conditions, are made possible by smart grid technologies. By adjusting their electricity consumption to periods of abundant wind energy or low demand, customers can contribute to grid balance and reduce the dependency on expensive peak power plants.

3.4 Economic and Environmental Benefits

3.4.1 *Cost-Effectiveness of Wind Power*

In many parts of the world, wind power is now a competitive alternative to produce electricity because of its increasing cost-effectiveness. Economies of scale, legislative support, and technological breakthroughs are some of the reasons for the falling cost of wind energy. The levelized cost of electricity (LCOE) from wind generation has decreased due to technological developments in turbine design, manufacture, operation, and maintenance procedures, which have improved efficiency and dependability. Larger turbines with longer blades and manufacturing and operating costs are reduced by advances in materials and aerodynamics. The cost of wind

Table 3.1 Cost-effectiveness of wind power (in INR)

Factor	Description	Value range
Capital costs	Initial investment costs, including turbines, installation, and grid connection	₹90,000–₹127,500 per kW
Operational costs	Annual costs for operation and maintenance	₹2250–₹4500 per kW per year
Capacity factor	Ratio of actual output over a period of time to the maximum possible output	30–50%
Levelized cost of electricity	Average cost per MWh over the lifetime of the plant	₹2250–₹4500 per MWh
Lifetime	Expected operational lifetime of wind turbines	20–25 years
Subsidies/incentives	Government incentives and subsidies to reduce costs	Varies by region and policy
Environmental impact	Reduction in CO_2 emissions and other environmental benefits	Significant reduction
Energy payback time	Time taken for the energy produced to equal the energy invested in the construction and operation	6–8 months

Notes: (a) *Capital costs*: These are the upfront costs required to set up a wind power plant, including the costs of wind turbines, installation, and connection to the power grid; (b) *Operational costs*: These are the annual costs for running and maintaining the wind power plant; (c) *Capacity factor*: This is the ratio of the actual energy output over a given period to the maximum possible output, indicating the efficiency of the plant; (d) *Levelized cost of electricity*: This represents the per-unit cost (per MWh) of building and operating a generating plant over an assumed financial life and duty cycle; (e) *Lifetime*: The expected operational period of wind turbines; (f) *Subsidies/incentives*: Government policies that provide financial support to make wind power more competitive; (g) *Environmental impact*: Wind power significantly reduces greenhouse gas emissions compared to fossil fuels; and (h) *Energy payback time*: The duration it takes for the energy produced by the wind turbine to offset the energy used in its construction and operation

energy has also been significantly decreased by economies of scale (EW Association 2015). The cost per kilowatt hour of power generated has reduced as wind turbines have expanded in size and installed capacity (Yadav et al. 2024). Furthermore, as the sector develops and investors become more confident in the long-term sustainability of wind energy projects, project developers reap the benefits of reduced financing costs. The regulatory support in the form of tax breaks, renewable energy goals, and auction procedures has encouraged investment in wind power and contributed to cost reduction (Table 3.1). Governments everywhere are putting laws in place to encourage the use of renewable energy as they become more aware of its advantages for the economy and the environment.

3.4.2 *Job Creation and Economic Growth*

Through various stages of project development, manufacture, installation, operation, and maintenance, wind power supports economic growth and employment creation. The wind energy industry supports a wide variety of occupations, including engineering, technology, manufacturing, construction, and administration. Jobs are created in factories and supply chains by manufacturing components of wind turbines, such as blades, towers, and nacelles. For wind turbine installation, specialized manpower is needed for site preparation, transportation, and assembly. After being online, wind farms hire experts to perform regular maintenance and monitoring tasks, which guarantees maximum efficiency and dependability (Fried 2016). Additionally, wind energy projects boost local economies, especially in rural and coastal areas, where wind farms are often found. Investments in wind energy infrastructure contribute to local tax revenues, land lease payments to landowners, and the procurement of goods and services from nearby businesses. Studies have shown that the wind energy sector creates jobs at a faster rate than many other industries, providing stable employment opportunities and career paths for workers across the skill spectrum (Froese 2018). Furthermore, the economic benefits of wind power extend beyond direct employment to include indirect and induced impacts on the broader economy, such as increased consumer spending and investment in local infrastructure.

3.4.3 Environmental Impact and Carbon Footprint Reduction

Wind power utilization presents notable ecological advantages in contrast to traditional fossil fuel-derived electricity production. Through the utilization of wind energy, wind turbines provide power with low-carbon emissions and minimal influence on the environment. Reducing greenhouse gas emissions and preventing climate change is one of the main environmental advantages of wind power (Garcia et al. 1998). Since wind energy generation does not emit carbon dioxide or other air pollutants when it operates, it contributes to the decarburization of the electrical industry and reduces the carbon footprint of all human activity. Research has shown that switching from fossil fuel-based to wind-powered energy generation can greatly reduce greenhouse gases, sulfur dioxide, nitrogen oxide, and other dangerous pollutants, improving public health and air quality (Get-invest.eu 2019). Additionally, wind power helps conserve natural resources and protect ecosystems by reducing the need for fossil fuel extraction, transportation, and combustion. By displacing coal-, oil-, and natural gas-fired power plants, wind energy helps conserve water resources and minimize habitat destruction, biodiversity loss, and ecosystem degradation associated with fossil fuel extraction and combustion (Shirin and Yadav 2021; Shirin et al. 2023; Kulwant and Yadav 2024). In addition, wind energy is a low-impact and land-efficient method of producing electricity. Multiple benefits can be obtained from the same land area by allowing wind farms to coexist with various land uses and agricultural operations (Hossain 2015). Wind turbines can be installed in a variety of geographical areas because they require less land for each unit of power they produce, compared to other renewable energy sources such as solar photovoltaic panels.

3.5 Challenges and Solutions in Wind Energy

3.5.1 Intermittency and Reliability Issues

The intrinsic unpredictability and intermittency of wind energy is one of its main problems. Although conventional nuclear, natural gas, or coal-fired conventional power plants can operate nonstop to meet demand, wind turbines can only produce electricity when the wind blows within specific speed limits (IEA 2014). There may be problems with grid stability and dependability as a result of these variations in wind speed that could cause variations in electricity output. Several ideas have been proposed and put into practice to overcome difficulties with intermittency and reliability: (a) *Energy storage*: Batteries, pumped hydro storage, thermal storage, and other grid-scale energy storage devices can store the extra electricity produced during times of high wind availability and release it during times of low wind or high demand. Energy storage increases the dependability of renewable energy sources, stabilizes the grid, and reduces variations in wind power output; (b) *Hybrid systems*: One way to diversify the renewable energy mix and reduce the effects of intermittency is to integrate wind energy with other renewable energy sources, such as hydropower or solar photovoltaic panels (IRENA 2018). The advantages of hybrid renewable energy systems can be combined with the disadvantages of separate technologies to provide electricity that is more reliable and steady; (c) *Demand response*: Demand response systems encourage customers to modify their electricity use habits in response to changes in market signals and grid conditions. Customers can contribute to grid balancing and reduce the need for backup production by timing their electricity use during periods of abundant wind energy or low demand. This increases resilience and reliability; and (d) *Advanced forecasting*: Grid operators can better anticipate variations in wind power production and allocate resources and operations by using machine learning algorithms, improved meteorological modeling, and remote sensing technologies to improve the accuracy of wind energy forecasts.

3.5.2 Environmental and Wildlife Concerns

Although wind energy is typically seen as more environmentally beneficial than power generated from fossil fuels, there may still be effects on the environment and wildlife that need to be taken into account. Among the main worries are the following: (a) *Birds and bats*: Due to habitat damage and bird and bat accidents with turbine blades, wind turbines can be dangerous. Certain species of birds and bats are more vulnerable to these impacts, particularly those that migrate through or inhabit areas with high potential for wind energy (Islam et al. 2013). Mitigation measures, such as placing wind farms away from important bird and bat habitats, implementing radar and acoustic monitoring systems to detect wildlife, and employing deterrent technologies, can help minimize these risks. (b) *Visual and aesthetic impacts*: In particular, in rural and coastal locations where wind farms are frequently located, wind turbines can change the aesthetic environment and picturesque views. Depending on personal preferences and cultural norms, visual effects can be subjective and change over time (Jaramillo and Borja 2004). Proper location and design of wind projects, community involvement, and landscape planning can help minimize visual impacts and ensure acceptance of wind energy development by local communities. (c) *Noise and shadow flicker*: Wind turbines can generate noise and shadow flicker, which can affect nearby residents and wildlife. Noise impacts can be mitigated through proper turbine design, setback distances, and sound barriers (Jiang et al. 2017). Shadow flicker can be minimized by adjusting the turbine position and rotor orientation, as well as implementing mitigation measures such as turbine shutdown during sensitive times.

3.5.3 Policy and Regulatory Challenges

Policy and regulatory challenges can also present barriers to the widespread deployment of wind energy. Addressing these policy and regulatory challenges requires collaboration between government agencies, industry stakeholders, and civil society to develop clear, transparent, and consistent policies that promote sustainable deployment of wind energy while protecting environmental and social interests. Some of the key challenges include the following: (a) *Grid connection and interconnection*: Connecting wind farms to the electrical grid and obtaining permits for grid interconnection can be complex and time-consuming processes. Upgrades to the grid infrastructure may be required to accommodate large-scale wind energy integration, and the regulatory frameworks governing grid connection and interconnection of the grid may vary between regions; (b) *Land use and siting*: Siting wind projects requires careful consideration of land use conflicts, environmental impacts, and community concerns. Balancing competing interests and ensuring stakeholder participation in the siting process can be challenging, particularly in densely populated or environmentally sensitive areas; (c) *Market design and incentives*: Market design and incentives are essential to encourage wind energy investment and to provide project developers with steady income streams. The market dynamics and economics of wind projects can be shaped by regulatory frameworks such as feed-in tariffs, renewable portfolio standards, tax incentives, and auction processes; and (d) *Policy stability and long-term planning*: The wind energy industry needs stable policies and long-term planning to attract investment and promote innovation. The uncertainty surrounding government policies, regulatory changes, and market conditions can deter investment and undermine confidence in the industry. Long-term planning frameworks, such as national energy strategies and climate action plans, provide a roadmap for transitioning to a low-carbon economy and achieving renewable energy goals.

3.6 Case Studies and Global Perspectives

3.6.1 Leading Countries in Wind Energy Adoption

Several nations have become leaders in the use of wind energy, propelling the expansion of the wind energy sector worldwide and making a major contribution to the deployment of renewable energy. Several of the top nations in the world for wind energy adoption are as follows:

- **China**: China is the world's largest market for wind energy, with a rapidly expanding installed capacity and ambitious renewable energy targets (Justus et al. 1978). The country's robust manufacturing capabilities and supportive government policies have propelled its wind industry to the forefront of the global market. Due to investments in both offshore and onshore wind projects, China's wind energy industry has grown exponentially in recent years.
- **United States**: The United States has long been a key player in the wind energy market, with abundant wind resources and a mature supply chain. States like Texas, Iowa, and California lead the nation in wind power capacity, supported by favorable wind conditions, policy incentives, and technological advancements. The US wind industry continues to grow, driven by federal tax incentives, state mandates for renewable energy, and corporate demand for clean energy.
- **Germany**: Germany has led the way in the use of renewable energy and is adamant about moving away from nuclear and fossil fuels. By 2050, a significant portion of the nation's electricity mix is supposed to come from renewable sources, according to the Energiewende (energy transition) program. Germany's energy transition is heavily dependent on wind energy, especially in the north, where wind conditions are ideal. Additionally, offshore wind development is accelerating in the North and Baltic Seas.
- **India**: Due to its favorable legislative environment and ambitious renewable energy goals, India has become a major player in the global wind energy market. With an emphasis on increasing wind energy deployment in areas like Tamil Nadu, Gujarat, and Rajasthan, the nation's National Wind Energy Mission seeks to reach 60 GW of installed wind power capacity by 2022 (Kang et al. 2021). India's wind industry benefits from strong government incentives, competitive auction mechanisms, and growing investor interest in renewable energy.
- **Denmark**: Since the 1970s, when the first wind turbine technology was developed, Denmark has been at the forefront of wind energy innovation. With wind turbines providing more than 40% of Denmark's electricity needs in recent years, the nation has one of the highest percentages of wind power in its electrical mix.

Denmark's wind industry has thrived due to supportive government policies, strong public–private partnerships, and a culture of innovation and collaboration.

3.6.2 Globally Successful Wind Projects

The feasibility and scalability of wind energy deployment in various geographic regions and market conditions have been proven by a multitude of wind projects worldwide (Kannan and Vakeesan 2016). These successful wind projects highlight the growing momentum behind the deployment of wind energy around the world and demonstrate the diverse range of opportunities and challenges associated with harnessing wind power for sustainable development. Some successful wind projects include the following:

- **Hornsea Offshore Wind Farm, United Kingdom**: One of the largest offshore wind farms in the world is Hornsea Offshore Wind Farm, which is located off the coast of Yorkshire, England. When Hornsea is fully operational, its overall capacity of more than 6 GW will produce enough clean electricity to power over 4.5 million homes. The project is evidence of offshore wind energy's ability to reduce carbon emissions and fulfill the world's expanding electricity demand.
- **Gansu Wind Farm, China**: One of the largest offshore wind farms in the world is the Gansu Wind Farm, which is located in the Chinese province of Gansu. With more than 6 GW of installed capacity in general, the Gansu Wind Farm reduces China's dependence on coal-fired power plants while contributing to the country's energy demands. The project uses the abundant wind resources in the region and showcases China's commitment to the development of renewable energy.
- **Alta Wind Energy Center, United States**: One of the largest wind farms in the country is the Alta Wind Energy Center, located in California's Mojave Desert. With a total capacity of over 1.5 GW, the Alta Wind Energy Center supplies clean electricity to millions of homes and businesses in California. The project demonstrates the potential of wind energy to contribute to grid reliability and decarbonization efforts in the United States.
- **Jaisalmer Wind Park, India**: One of the largest wind energy projects in Asia is the Jaisalmer Wind Park, which is located in the Indian state of Rajasthan. The Jaisalmer Wind Park, which has an installed capacity of more than 1 GW, contributes to the region's economic growth and helps India reach its renewable energy goals (Lund and Boyd 2016). The project highlights India's commitment to increasing the deployment of wind energy and takes advantage of Rajasthan's ideal wind conditions.
- **Netherlands' Borssele Offshore Wind Farm**: One of the premier offshore wind projects in Europe is the Borssele Offshore Wind Farm, which is situated in the North Sea off the country's coast. Borssele, with a total capacity of 1.5 GW,

will help The Netherlands meet its renewable energy targets and supply millions of Dutch consumers with clean electricity (Mahmood et al. 2020). The project is a prime example of how offshore wind energy can contribute significantly to Europe's energy transformation.

3.7 Future of Wind Energy

3.7.1 New Trends and Technologies

The future of wind energy is defined by ongoing technological innovation, which will reduce costs, improve efficiency, and increase the possibilities for wind power deployment. The following are some new developments and trends that will influence wind energy in the future.

- **Floating offshore wind**: Large stretches of the ocean are now available for the development of wind energy due to floating offshore wind technology, which allows wind turbines to be installed in deeper waters. Floating turbines can access stronger and more consistent wind resources farther offshore, reducing visual and environmental impacts compared to fixed bottom offshore wind farms.
- **Next-generation turbine designs**: Larger and more reliable wind turbines are being developed as a result of developments in manufacturing processes, materials research, and turbine design. Next-generation turbines feature innovative blade designs, advanced control systems, and integrated sensors for optimized performance and reduced operational costs.
- **Digitalization and data analytics**: Real-time monitoring, predictive maintenance, and optimization of the performance of wind turbines and wind farms are made possible by digitalization and data analytics technologies that are transforming the wind sector. Machine learning algorithms, artificial intelligence, and remote sensing technologies are being used to analyze vast amounts of data and improve decision-making in wind energy operations.
- **Hybrid wind and solar projects**: Hybrid renewable energy projects that combine wind and solar photovoltaic (PV) technologies offer synergies in resource complementarity, grid integration, and project economics. Hybrid projects can optimize land use, balance electricity generation profiles, and improve grid stability, making them attractive options for meeting diverse energy needs.

3.7.2 Integration with Other Renewable Energy Sources

In order to create a diversified and resilient energy system, wind energy must be integrated with other renewable energy sources, including solar, hydroelectric, and geothermal power. Integrated renewable energy systems have various advantages, such as:

- **Grid stability and reliability**: Combining wind energy with other renewable energy sources helps balance the variability of wind and solar generation, ensuring a more stable and reliable electricity supply (Mathew 2006). By diversifying the renewable energy mix, integrated systems can reduce the need for backup generation and improve grid resilience to fluctuations in weather conditions.
- **Optimized resource utilization**: Integrated renewable energy projects can take advantage of the strengths of different technologies to maximize resource utilization and energy output. For example, wind and solar energy generation profiles often complement each other, with wind turbines producing more electricity during periods of high wind speed and solar panels generating more electricity during daylight hours.
- **Synergies in infrastructure and operations**: Integrating wind energy with other renewable energy sources can result in synergies in infrastructure development, project financing, and operations and maintenance (Rukikaire 2019). Shared transmission lines, grid connections, and support infrastructure can reduce project costs and improve overall project economics.

3.7.3 The Role of Wind Energy in Achieving Global Sustainability Goals

The achievement of global sustainability objectives, such as reducing climate change, improving energy security, and stimulating economic growth, is highly dependent on wind energy. Wind energy provides a clean, abundant, and affordable way to fulfill rising energy while lowering dependence on fossil fuels as nation's transition to low-carbon economies and reduce greenhouse gas emissions. In numerous ways, wind energy supports the objectives of global sustainability:

- **Decarbonizing the electricity sector**: Wind energy produces electricity with minimal greenhouse gas emissions, helping to decarbonize the electricity sector and reduce the carbon footprint of energy production (Shata and Hanitsch 2006). Wind energy is vital in reducing carbon dioxide and other pollutants that contribute to climate change and air pollution, as it replaces coal, oil, and natural gas power facilities.
- **Promoting energy independence**: By broadening the energy mix and reducing reliance on imported fossil fuels, wind energy improves energy security (Liu et al. 2015). Countries can reduce their susceptibility to supply disruptions, price volatility, and geopolitical risks related to the import of fossil fuels using their own wind resources.
- **Creating jobs and economic opportunities**: Throughout the value chain, from installation and manufacturing to operation and maintenance, wind energy creates jobs and boosts the economy (Werapun et al. 2015). Countries can boost local economies, generate employment opportunities, and attract investment in

the renewable energy sector by allocating resources to wind energy infrastructure and technology.
- **Fostering innovation and technological advancement**: Innovation and technical progresses in fields such as materials science, grid integration, and turbine design are driven by wind energy (Weston 2016). Countries can accelerate the adoption of wind energy technologies and reduce prices by investing in research and development, increasing the affordability and accessibility of all.

3.8 Conclusion

A key component of the global shift toward a low-carbon, sustainable future is wind energy. Wind power has evolved over time from basic mechanical windmills to complex and effective turbines that can produce enormous amounts of clean electricity. Significant turning points in the history of wind energy have included advances in technology, legislative support, and public understanding of the financial and environmental benefits of renewable energy sources. These days, wind energy is essential to reduce greenhouse gas emissions, encourage energy independence, and stimulate local economies and job development around the world. With new trends and technologies that are poised to substantially improve wind power affordability, dependability, and efficiency, the future of wind energy looks quite promising. A more resilient and diverse energy system that can meet the growing demand for energy while reducing climate change and preserving the environment for future generations will be made possible by integration with other renewable energy sources, such as solar and energy storage. Wind energy will remain a vital facilitator of this shift as nations work to achieve their goals for global sustainability and make the transition to clean energy. We can create a more just, wealthy and sustainable world for everyone by using wind energy.

Acknowledgments We are thankful to Dr. Akhilesh Kumar Yadav for allowing us to be a part of this project. Also, we express our gratitude to him for his support and assistance throughout the creation of this chapter.

References

Ahmed A, Uddin GS, Sohag K (2016) Biomass energy, technological progress and the environmental Kuznets curve: evidence from selected European countries. Biomass Bioenergy 90:202–208

Akpinar EK, Akpinar S (2005) An assessment on seasonal analysis of wind energy characteristics and wind turbine characteristics. Energy Convers Manag 46:1848–1867

Allouhi A, Zamzoum O, Islam M, Saidur R, Kousksou T, Jamil A, Derouich A (2017) Evaluation of wind energy potential in Morocco's coastal regions. Renew Sustain Energy Rev 72:311–324

Archer CL, Jacobson MZ (2003) Spatial and temporal distributions of US winds and wind power at 80 m derived from measurements. J Geophys Res Atmos 108. https://doi.org/10.1029/2002JD002076

Ayodele T, Ogunjuyigbe A (2016) Wind energy potential of Vesleskarvet and the feasibility of meeting the South African's SANA IV energy demand. Renew Sustain Energy Rev 56:226–234

Bousla M, Haddi A, El Mourabit Y, Sadki A, Mouradi A, El Kharrim A (2023) Detection and prevention of repetitive major faults of a WTG by analysis of alarms through SCADA. In: Motahhir S, Bossoufi B (eds) Digital technologies and applications. ICDTA 2023, Lecture notes in networks and systems, vol 668. Springer, Cham

Chandel SS, Murthy KSR, Ramasamy P (2014) Wind resource assessment for decentralised power generation: case study of a complex hilly terrain in western Himalayan region. Sustain Energy Technol Assess 8:18–33

Chang TP (2011) Estimation of wind energy potential using different probability density functions. Appl Energy 88:1848–1856

Dabbaghiyan A, Fazelpour F, Abnavi MD, Rosen MA (2016) Evaluation of wind energy potential in province of Bushehr, Iran. Renew Sustain Energy Rev 55:455–466

Dogan E, Seker F (2016) The influence of real output, renewable and non-renewable energy, trade and financial development on carbon emissions in the top renewable energy countries. Renew Sustain Energy Rev 60:1074–1085

E.W. Association (2015) Wind energy scenarios for 2030. EWEA. https://www.ewea.org/fileadmin/files/library/publications/reports/EWEA-Wind-energy-scenarios-2030.pdf

El Mourabit Y, Derouich A, Allouhi A, El Ghzizal A, El Ouanjli N, Zamzoumyes O (2020a) Sustainable production of wind energy in the main Morocco's sites using permanent magnet synchronous generators. Int Trans Electric Energy Syst 30:e12390

El Mourabit Y, Derouich A, El Ghzizal A, El Ouanjli N, Zamzoum O (2020b) Nonlinear backstepping control for PMSG wind turbine used on the real wind profile of the Dakhla-Morocco city. Int Trans Electric Energy Syst 30:e12297

Ellabban O, Abu-Rub H, Blaabjerg F (2014) Renewable energy resources: current status, future prospects and their enabling technology. Renew Sustain Energy Rev 39:748–764

Energy FS, Philibert C (2013) Energy and climate change analyst IEA (atoms for the future 2013). https://www.slideshare.net/SFEN/cdric-philibert-energy-and-climate-change-analyst-iea-atoms-for-the-future?from_action=save

Equinor (2019) Equinor—the world's leading floating offshore wind developer. https://www.equinor.com/en/what-we-do/hywind-where-the-wind-takes-us.html

Fried L (2016) Wind power to dominate power sector growth. https://gwec.net/wind-power-to-dominate-power-sector-growth/

Froese M (2018) World's first floating wind farm delivers promising results. Windpower Engineering & Development. https://www.windpowerengineering.com/worlds-first-floating-wind-farm-delivers-promising-results/

Garcia A, Torres J, Prieto E, de Francisco A (1998) Fitting wind speed distributions: a case study. Sol Energy 62:139–144

Get-invest.eu (2019) Global market outlook 2019–2023. http://www.solarpowereurope.org/wp-content/uploads/2019/05/SolarPower-Europe-Global-Market-Outlook-2019-2023.pdf

Hossain J (2015) Wind energy 2050 on the shape of near-100% RE grid. World Wind Energy Association

IEA (2014) Energy technology perspectives. International Energy Agency

IRENA (2018) Global energy transformation: a roadmap to 2050. International Renewable Energy Agency, Abu Dhabi. https://www.irena.org/-/media/Files/IRENA/Agency/Publication/2018/Apr/IRENA_Report_GET_2018.pdf

Islam M, Mekhilef S, Saidur R (2013) Progress and recent trends of wind energy technology. Renew Sustain Energy Rev 21:456–468
Jaramillo OA, Borja MA (2004) Wind speed analysis in La Ventosa, Mexico: a bimodal probability distribution case. Renew Energy 29:1613–1630
Jiang H, Wang J, Wu J, Geng W (2017) Comparison of numerical methods and metaheuristic optimization algorithms for estimating parameters for wind energy potential assessment in low wind regions. Renew Sustain Energy Rev 69:1199–1217
Justus CG, Hargraves WR, Mikhail A, Graber D (1978) Methods for estimating wind speed frequency distributions. J Appl Meteorol 17:350–353
Kang S, Khanjari A, You S, Lee J-H (2021) Comparison of different statistical methods used to estimate Weibull parameters for wind speed contribution in nearby an offshore site, Republic of Korea. Energy Rep 7:7358–7373
Kannan N, Vakeesan D (2016) Solar energy for future world: a review. Renew Sustain Energy Rev 62:1092–1105
Kulwant M, Yadav AK (2024) Enhancing water conservation and management in India: a holistic approach. In: Yadav AK, Yadav K, Singh VP (eds) Integrated management of water resources in india: a computational approach. Springer Nature, Cham, pp 121–137. https://doi.org/10.1007/978-3-031-62079-9_7
Lund JW, Boyd TL (2016) Direct utilization of geothermal energy 2015 worldwide review. Geothermics 60:66–93
Mahmood FH, Resen AK, Khamees AB (2020) Wind characteristic analysis based on Weibull distribution of Al-Salman site, Iraq. Energy Rep 6:79–87
Mathew S (2006) Wind energy: fundamentals, resource analysis and economics. Springer, Berlin/Heidelberg, pp 45–88
Rukikaire KP (2019) Renewable energy investment. https://www.unenvironment.org/news-and-stories/press-release/renewable-energy-investment-2018-hit-usd-2889-billion-far-exceeding
Shata ASA, Hanitsch R (2006) Evaluation of wind energy potential and electricity generation on the coast of Mediterranean Sea in Egypt. Renew Energy 31:1183–1202
Shirin S, Yadav AK (2021) Assessment and suitability for the dumping of flyash in an abandoned mine. Environ Challenges 4:100111. https://doi.org/10.1016/j.envc.2021.100111
Shirin S, Jamal A, Emmanouil C, Singh VP, Yadav AK (2023) Assessment and characterization of waste material used as backfilling in an abandoned mine. Int J Coal Prep Util 43(8):1402–1410. https://doi.org/10.1080/19392699.2022.2118259
Werapun W, Tirawanichakul Y, Waewsak A (2015) Comparative study of five methods to estimate Weibull parameters for wind speed on Phangan Island, Thailand. Energy Procedia 79:976–981
Weston D (2016) GWEC envisages 5.8TW of wind by 2050. https://www.windpowermonthly.com/article/1412717/gwec-envisages-58tw-wind-2050
Wind Europe (2017) Wind energy in Europe, scenarios for 2030. https://windeurope.org/about-wind/reports/wind-energy-in-europe-scenarios-for-2030/
Wind Europe (2019) Europe installs 4.9 GW of new wind energy capacity in first half of 2019. https://windeurope.org/newsroom/press-releases/europe-installs-4-9-gw-of-new-wind-energy-capacity-in-first-half-of-2019/
Wiser R (2009) Wind power 2008 annual report. Wind Stats 22:2–7
World Wind Energy Association (2015) Wind energy 2050 on the shape of near-100% RE grid
Youness EM, Aziz D, Abdelaziz EG, Jamal B, Najib EO, Othmane Z, Khalid M (2019) Implementation and validation of backstepping control for PMSG wind turbine using dSPACE controller board. Energy Rep 5:807–821
Zamzoum O, El Y, Errouha M, Derouich A, El A (2019) Active and reactive power control of wind turbine based on doubly fed induction generator using adaptive sliding mode approach. Int J Adv Comput Sci Appl 10:397–406
Zhao P, Wang J, Xia J, Dai Y, Sheng Y, Yue J (2012) Performance evaluation and accuracy enhancement of a day-ahead wind power forecasting system in China. Renew Energy 43:234–241

Chapter 4
Ecological Health Evaluation for Sustainable Practices

Rahul Raj, Rashmi Raghav, Muskan Mayank, and A. R. Laiju

Abstract Environmental impact and sustainability assessment methodologies are indispensable tools for assessing the effects of human activity on the ecosystem and promoting sustainable development. This article provides an overview of the key concepts, methods, and implications of these assessments. Before decisions are made, proposed projects, programs, or plans are subjected to an environmental impact assessment (EIA), a methodical procedure that looks at potential negative consequences on the environment. It helps identify and mitigate adverse impacts on air, water, soil, biodiversity, and human health, promoting environmentally responsible decision-making. Beyond an EIA, a sustainability assessment considers social and economic factors in addition to environmental factors. It evaluates the overall sustainability of projects, products, or policies by assessing their impacts on ecosystems, communities, and future generations. Life cycle assessment (LCA) and social life cycle assessment (SLCA) are two examples of sustainability assessment frameworks that offer comprehensive insight into the social, economic, and environmental effects of human activity. Integrating sustainability principles into decision-making processes is essential to achieve long-term environmental and social goals. Businesses, governments, and other stakeholders are increasingly recognizing the importance of incorporating sustainability considerations into their strategies and operations. Regulatory frameworks and international agreements are crucial in incentivizing sustainable practices and ensuring accountability. This paper also discusses future directions in environmental impact and sustainability assessment, emphasizing the imperative to embrace innovation and technological advancement to address emerging environmental challenges. It advocates a paradigm shift toward

R. Raj (✉) · M. Mayank
Department of Civil Engineering, NIT Uttarakhand, Srinagar, Garhwal, India
e-mail: rahul.rajphd2021@nituk.ac.in

R. Raghav
Department of Chemistry, NIT Uttarakhand, Srinagar, Garhwal, India

A. R. Laiju
Department of Civil Engineering, NIT Trichy, Trichy, Tamil Nadu, India

A. K. Yadav (ed.), *Powering Future: Clean Energy Solutions for a Sustainable Planet*, Clean Energy Production Technologies, https://doi.org/10.1007/978-981-96-5055-2_4

a circular economy model, where resources are conserved, waste is minimized, and ecosystems regenerated. In the end, the study emphasizes how critical environmental impact and sustainability assessments are to navigating the complexity of today's world, encouraging adaptability, and preserving the integrity of the planet for coming generations.

Keywords LCA · SLCA · Biodiversity · Ecosystems · Resilience

Environmental Significance
Ecological health evaluation for sustainable practices is vital to the environment as it embodies a proactive approach to balancing human activities with ecological integrity. Guide decisions toward sustainable outcomes by assessing the health of ecosystems and gauging the impact of human practices. This process emphasizes the preservation of biodiversity, the maintenance of water and air quality, and the protection of soil health. By promoting sustainable practices, it encourages resource efficiency, waste reduction, and mitigation of pollution. These evaluations inform policies, businesses, and communities about the environmental consequences of their actions, fostering responsible stewardship of natural resources. Ultimately, this framework contributes to the preservation of ecosystems, the protection of species, and resilience of the planet in face of environmental challenges.

4.1 Introduction

In recent decades, sustainability has gained paramount importance in various fields, including ecology, agriculture, industry, and urban planning. The urgency to address environmental degradation and mitigate its impacts on ecosystems has led to a growing emphasis on sustainable practices. Central to this paradigm shift is the recognition that human activities are intimately interconnected with ecological health and that the well-being of both is inherently linked. Against this backdrop, this study encapsulates a critical area of inquiry that seeks to assess the ecological vitality of our planet and develop strategies to foster sustainability in human endeavors (Parkes and Horwitz 2009). Ecological health, in its essence, refers to the overall well-being and resilience of ecosystems. It encompasses the intricate relationships between living organisms, their habitats, and the physical environment. Healthy ecosystems exhibit robust biodiversity, efficient nutrient cycling, and resilience to disturbances. However, rampant exploitation of natural resources, pollution, habitat destruction, and climate change have severely compromised the health of many ecosystems around the world (Hinze et al. 2013). Thus, for the sake of present and the future generations, it is imperative to assess ecological health as it stands today and come up with plans to preserve and improve it. On the contrary, sustainable practices refer to human behaviors that meet present needs without jeopardizing the ability of future generations to meet their own needs (Nicholls et al. 2020). These techniques aim to improve social fairness, preserve natural resources, reduce their detrimental impacts on the environment, and ensure economic viability. From the adoption of renewable energy to waste reduction initiatives, sustainable practices encompass a wide range of strategies in different sectors. However, the effectiveness of these practices in promoting ecological health requires a rigorous evaluation and assessment. It underscores the interconnectedness (Mariani and Vastola 2015) between ecological health and sustainable practices. It implies a systematic and comprehensive assessment framework aimed at gauging impact of human activities on ecological systems and identifying strategies to promote sustainability. At its core, this title embodies a call to action for interdisciplinary research and collaboration between ecologists, environmental scientists, policymakers, and stakeholders from various sectors. Evaluation of ecological health for sustainable practices involves multifaceted approaches and methodologies (Yang et al. 2020). It may include field studies, laboratory experiments, remote sensing techniques, and modeling exercises to gather data on ecosystem dynamics, biodiversity trends, pollution levels, and other relevant indicators.

4.2 Environmental Impact Assessment

The EIA emerged in response to increasing environmental concerns of development projects. Originating in mid-twentieth century amidst rising industrialization and urbanization, it aimed to address widespread environmental degradation and pollution. EIA's modern practice took shape in the 1960s and 1970s, fueled by landmark legislation like the United States National Environmental Policy Act (NEPA) of 1969 and the UN Conference on the Human Environment in Stockholm (1972), advocating for global environmental planning. Since then, EIA has become a global practice, mandated by many countries for proposed projects, and expanded to include policies and programs. It is seen as vital for sustainable development, integrating environmental concerns into decision-making (MacLaren et al. 2020). Challenges persist, including ensuring process effectiveness and transparency and addressing emergent problems such as biodiversity loss and climate change. Continued research, innovation, and collaboration are crucial for strengthening practice and its role in protecting the environment while promoting sustainable development (Vandana et al. 2024).

Scoping: Determining the elements of the environment and any effects that must be taken into account during the assessment. This might involve the quality of the land, water, air as well as biodiversity, habitat integrity, and cultural legacy (Of et al. 2005).

Baseline data collection: Collecting data on current environmental parameters to create a baseline to evaluate any possible effects. Field surveys, data analysis, and stakeholder interaction may be necessary for this (Arunachalam et al. 2013).

Impact prediction: Evaluating the project or activity's possible effects on the environment. This entails determining and projecting the effects, both direct and indirect, while taking into account variables such as impacts' size, duration, scope, and reversibility (Batterman et al. 2009).

Mitigation and management measures: Formulating plans to prevent, reduce, or mitigate the effects on the environment that have been detected. This could involve making adjustments to the project's design, putting best management practices into action, restoring the harmed ecosystems, or implementing compensatory measures (Borja et al. 2016).

Alternatives analysis: Evaluating alternative options or project designs to identify the most environmentally sustainable solution. This may involve comparing different scenarios based on their environmental, social, and economic impacts (Lattimore et al. 2009).

Public consultation and participation: Include local communities, native Americans, and other relevant groups as stakeholders in the EIA process. By doing this, you can be sure that their opinions and concerns are heard and that a consensus is formed about the suggested project or activity (Chen et al. 2015).

Decision-making: Delivering the EIA's conclusions to decision-makers, together with information on probable environmental effects, mitigating strategies, and alternative analyses. Decision-makers can decide whether to accept, amend, or reject the suggested project or activity using knowledge based on this information (Campion and Essel 2013).

Monitoring and follow-up: Implement monitoring programs to track the actual environmental impacts once the project or activity is underway. This helps ensure that mitigation measures are effective and allows for adaptive management if unexpected impacts arise (Hunsberger et al. 2005).

Through minimizing negative effects on ecosystems and communities, encouraging long-term environmental stewardship and ensuring that environmental concerns are integrated into decision-making processes, EIAs play a key role in promoting sustainable development. The main steps involved in EIA are shown in Fig. 4.1.

4.3 Assessment of the Social Life Cycle

SLCA, which stems from the evolution of sustainability assessment methods, notably within LCA, addresses social aspects alongside environmental impacts. Its origins lie in movements highlighting social dimensions of sustainability, emerging in the twentieth century due to rising awareness of globalized supply chains' social impacts (Mathe 2014). The ISO 14040 standards laid the foundational groundwork, focusing on environmental aspects but enabling the expansion of the SLCA. Further efforts of researchers led to SLCA methodologies, helping various sectors in assessing social impacts, improving accountability, and promoting social responsibility

Fig. 4.1 Major steps involved in EIA

(Santos et al. 2019). As sustainability challenges persist, SLCA is crucial to guide equitable and sustainable development. Ongoing research and collaboration are vital for advancing SLCA's theory and practice and integrating social considerations into decision-making globally (Arcese et al. 2013), the objectives of SLCA are mentioned here.

Identify social hotspots: Within an item or service's life cycle, SLCA assists in locating social "hotspots" or areas of concern where major social consequences take place. This requires awareness of possible dangers and weaknesses related to various stages of the life cycle, extraction of raw materials, and final disposal of materials (De Vreese et al. 2016).

Assess social impacts: SLCA assesses the possible social impacts of a product or service using indicators and metrics relevant to social sustainability. These may include factors such as worker wages and working conditions, access to healthcare and education, community participation and empowerment, human rights violations, and impacts on vulnerable populations (Chardine-Baumann and Botta-Genoulaz 2014).

Inform decision-making: SLCA provides stakeholders, including businesses, governments, consumers, and civil society organizations, with valuable information to inform decision-making processes, product design, and policy development. It helps identify opportunities for social improvement and supports the development of strategies to mitigate negative social impacts (Krajnc and Glavič 2005).

Promote social responsibility: By integrating social considerations into life cycle assessment, SLCA promotes social responsibility and corporate sustainability practices. Encourage companies to address social issues within their operations and supply chains, fostering a more equitable and socially responsible business environment. SLCA is often conducted in conjunction with traditional LCA to offer a detailed evaluation of the effects on society and the environment (Amrutha and Geetha 2020). It helps to ensure that sustainability initiatives address not only environmental concerns, but also social justice and equity considerations, contributing to more holistic and sustainable outcomes. As interest in corporate social responsibility and sustainable supply chain management continues to grow, SLCA is becoming an increasingly important tool to promote social sustainability and advance the broader goals of sustainable development.

4.4 Life Cycle Assessment

As a method for assessing how operations and products affect the environment throughout the duration of their life cycles, LCA arose in response to environmental concerns in the late 1960s and early 1970s. LCA's early growth was influenced by the environmental movement in general, as well as by influential publications such as William Shands' book and reports from the US National Academy of Sciences (Cowan et al. 2010). Formal frameworks for LCA were established in the 1990s with the ISO 14040 series, providing guidelines and principles for conducting studies. Since then, LCA has become widely accepted and applied in various sectors to improve environmental performance. Recent advances include expanding evaluation scopes to include social and economic impacts and integrating LCA into decision-making processes and policy development (Dias-sardinha and Reijnders 2005). Concerns about resource shortages and climate change are growing on a worldwide scale, and LCA remains crucial to promoting sustainable development by reducing environmental footprints. The LCA process typically involves four main stages.

Goal and scope definition: In this first step, the objectives and parameters of the Life Cycle Assessment Study (LCA) are established. This includes identifying the particular system or product that needs to be evaluated, the functional unit (such as the cost per unit of production or service rendered), and the pertinent environmental impact categories that need to be taken into account (Rosen and Kishawy 2012).

Life cycle inventory (LCI): This stage involves collecting data on all inputs such as raw materials, energy, and water, and outputs such as emissions, waste, and emissions related to each stage of the product's life cycle (Shirin and Yadav 2021; Shirin et al. 2023). This means compiling a comprehensive inventory of all relevant inputs and outputs through surveys, data collection from suppliers, and database use (Walker et al. 2002).

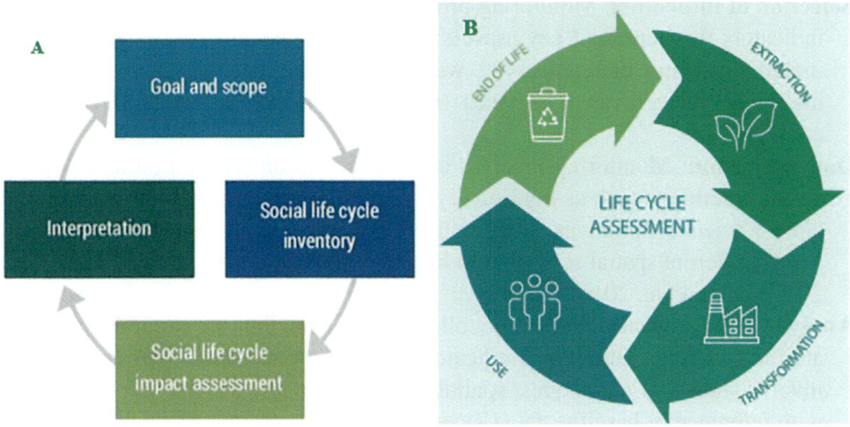

Fig. 4.2 Benefits and limitations of LCA

Life Cycle Impact Assessment (LCIA): Once the LCI data is collected, they are used to assess the potential environmental impacts associated with each stage of the product's life cycle. This involves characterizing emissions and resource consumption data into impact categories, such as global warming potential, acidification, eutrophication, and resource depletion, using established impact assessment methods and models (Lauesen 2019).

Interpretation: The last phase of the (LCA) process is to analyze the impact assessment data to derive conclusions and offer suggestions for enhancing the system or product's environmental performance. This could involve performing sensitivity studies, evaluating various scenarios or design choices, and finding hotspots and life cycle periods with the greatest environmental impacts to find opportunities to improve the environment (Frischknecht et al. 2016).

Businesses, governments, and researchers frequently use lifecycle assessments to help with decision-making, product design, and policy creation. It offers a thorough and rigorous scientific method for understanding how systems and goods affect the environment, assisting in the discovery of opportunities to reduce resource use, lessen pollution, and improve sustainability in general. Furthermore, by providing lucid and uniform data on the environmental impact of products and services, LCA can help stakeholders in making well-informed decisions; benefits and limitations of LCA are shown in Fig. 4.2.

4.5 Monitoring of Ecosystem Health

Monitoring ecosystem health involves systematically observing and assessing the condition and functioning of ecosystems over time. This process is essential for understanding the status, trends, and resilience of natural systems, as well as detecting any changes or disturbances that may occur.

Selection of indicators: Monitoring programs typically begin with the selection of indicators that represent key aspects of ecosystem health. These indicators can include measures of biodiversity, water quality, soil health, air quality, habitat integrity, and ecosystem services (Lecours et al. 2012; Yadav 2015; Yadav et al. 2023).

Data collection: Monitoring involves collecting data on selected indicators using various methods, such as field surveys, remote sensing, laboratory analysis, and citizen science initiatives. Data collection can occur at regular intervals and across different spatial scales, from local to regional or even global (Wang et al. 2020; Sahoo et al. 2016).

Analysis and interpretation: After collecting the data, they are examined to evaluate the state and patterns of indicators of ecosystem health. This analysis can involve statistical techniques, spatial mapping, trend analysis, and comparison with reference or baseline data (Rice 2003).

Identification of trends and patterns: Monitoring data helps identify any trends, patterns, or anomalies in ecosystem health indicators over time. This information provides insight into the drivers of ecosystem change, such as human activities, climate variability, or natural disturbances (Wells et al. 2004).

Assessment of resilience: Ecosystem resilience, or an ecosystem's capacity to tolerate shocks and bounce back, may also be evaluated through monitoring. Measures of species diversity, ecosystem function, and ecosystem structure are a few examples of resilience indicators (Gregory and Van Strien 2010).

Communication and reporting: The findings of ecosystem health monitoring are communicated to stakeholders, policymakers, and the public through reports, presentations, and other outreach efforts. Effective communication helps raise awareness of ecosystem issues and informs decision-making (Chapman et al. 2020).

Adaptive management: Monitoring data can inform adaptive management strategies, in which management actions are adjusted based on new information on ecosystem health. This iterative process helps improve the effectiveness of conservation and management efforts over time (O'Faircheallaigh 2007). Monitoring of ecosystem health is a vital part of environmental management because it provides data that are needed for conservation, biodiversity, and the preservation of ecosystem services, as well as sustainable development. A pictorial representation of this is shown in Fig. 4.3.

4.6 Biodiversity Conservation Strategies

Biodiversity conservation strategies are multifaceted approaches that aim to protect and preserving the variety of life on Earth. These strategies encompass a variety of initiatives designed to address the various threats to biodiversity and promote its sustainable management. Some key biodiversity conservation strategies include (Das et al. 2020).

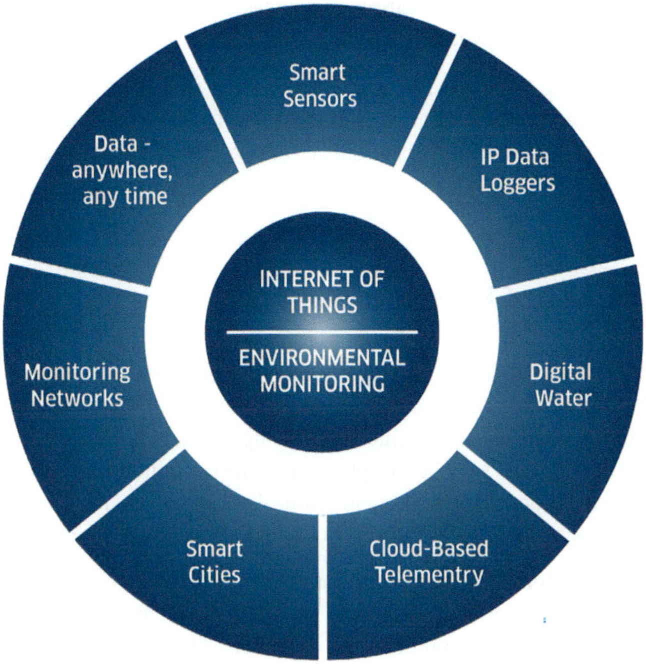

Fig. 4.3 Traditional methods included AI techniques in monitoring

Protected areas: The establishment and effective management of protected areas, including animal reserves national parks and marine protected areas, is necessary to preserve animals and their habitats from human activities such as pollution, poaching, and habitat loss (Kleijn et al. 2020).

Habitat restoration: Putting into practice habitat restoration initiatives to repair damaged ecosystems, reunite fragmented habitats, and improve biodiversity by reestablishing native plants and ecosystems (Pinho 2011; Lindenmayer et al. 2000).

Sustainable land use practices: Encourage the use of sustainable land for activities such as agroforestry and sustainable agriculture, and land zoning to minimize habitat loss, fragmentation, and degradation, while also supporting local livelihoods (Hugé et al. 2013).

Invasive species management: Creating and implementing management and control plans for invasive species that endanger native biodiversity and interfere with ecosystem function (Sandifer et al. 2015).

Sustainable resource management: Sustainable resource management practices should be used by forest, fisheries, and other natural resource sectors to ensure the long-term survival of ecosystems and the species they support (International Finance Corporation 2012).

Community-based conservation: Engaging local communities, indigenous peoples, and other stakeholders in conservation initiatives to promote sustainable resource use, improve conservation awareness, and foster stewardship of natural resources (Johns and Eyzaguirre 2006).

By implementing these biodiversity conservation strategies in an integrated and coordinated manner, for the benefit of current and future generations, stakeholders can cooperate to preserve and sustainably manage the great biological variety of the planet.

4.7 Evaluation of Ecosystem Services

It provides ecosystems for human life, and the evaluation of ecosystem services includes an assessment of human well-being. Ecosystem services are typically divided into four main groups, including policing, fostering, and assisting services. The evaluation of these services requires a multidisciplinary approach that integrates ecological, economic, social, and cultural perspectives. Here is how the evaluation of ecosystem services is typically conducted (Janku et al. 2022).

Identifying ecosystem services: The first step in evaluating ecosystem services is to identify and categorize the services provided by a particular ecosystem. This involves compiling a comprehensive list of services, including regulating (e.g., water purification, climate control), providing (e.g., food, water, and wood), cultural (e.g., recreation, spiritual values), and sustaining (e.g., nutrient cycling, soil formation) (Tolvanen and Aronson 2016).

Assessing service quantification: The availability, distribution, and value of ecosystem services are quantified and measured after they have been recognized. Data collection on ecological processes (such as carbon sequestration and pollination), ecosystem characteristics (such as biodiversity and habitat quality), and the temporal and geographical dynamics of service provision may be necessary to achieve this (Haase et al. 2014).

Valuing ecosystem services: Assigning economic values to ecosystem services helps quantify their importance and integrate them into decision-making processes. Economic valuation methods include market-based approaches (e.g., market prices, cost-based methods), methods for non-market valuation (such as revealed preference and expressed preference approaches), and hybrid approaches that combine both market and non-market valuation methods (Wu 2013).

Mapping ecosystem services: Spatially mapping ecosystem services allow one to visualize the distribution and intensity of service provision across landscapes. Geographic information systems (GIS) and remote sensing technologies are commonly used to map ecosystem services, enabling decision-makers to identify areas of high value or vulnerability and prioritize conservation or restoration efforts accordingly (Söderman et al. 2012).

Stakeholder engagement and preferences: Assessing stakeholder preferences and perceptions of ecosystem services is essential for understanding how services contribute to human well-being and addressing trade-offs and conflicts in resource management decisions. Stakeholder participation methods, such as participatory mapping, surveys, and focus groups, can help to elicit diverse perspectives and values related to ecosystem services (Agboola et al. 2023).

Integrating findings into decision-making: Finally, the evaluation of ecosystem services aims to inform policy and management decisions that sustainably manage and conserve ecosystems while improving human well-being. Integrating ecosystem service assessments into decision-making processes ensures that the full range of benefits provided by ecosystems is considered, leading to more effective and equitable resource management outcomes (Newig et al. 2023).

By systematically evaluating ecosystem services, decision-makers can better understand the relationships between ecosystems and human societies, prioritize conservation and restoration efforts, and promote sustainable development that enhances both ecological and socio-economic resilience.

4.8 Restoration and Rehabilitation of Degraded Ecosystems

Sustainable resource management and the preservation of biodiversity are highly dependent on the rehabilitation and restoration of damaged ecosystems. These mechanisms restore the ecological integrity and functionality of ecosystems that have suffered from pollution, deforestation, habitat destruction, and land degradation, among other human-caused problems and have been revived and improved (Gann et al. 2019). Rehabilitation and restoration of damaged ecosystems involve the following important elements: (a) *Assessment and planning*: The first step in ecosystem restoration is to perform extensive evaluations of the degraded site, including an analysis of soil quality, hydrology, vegetation composition, and biodiversity. Based on these assessments, restoration plans are developed, highlighting goals, objectives, and strategies for ecosystem recovery (McDonald et al. 2016); (b) *Ecological restoration*: Ecological restoration involves implementing actions to repair and rebuild degraded ecosystems and to restore their structure, composition, and function. This could involve employing native plant species and suitable restoration methods to carry out tasks such as revegetation, reforestation, wetland restoration, and habitat building (Aronson et al. 1993); (c) *Soil and water management*: Addressing soil erosion, loss of soil fertility, and water degradation is essential for successful restoration of the ecosystem. While water management strategies, such as reestablishing natural hydrological processes and implementing water conservation measures, increase water quality and availability, soil conservation techniques, such as terracing, contour plowing, and cover cropping, help stabilize soils and prevent erosion (Merino et al. 2019); (d) *Invasive species control*: Invasive species can significantly hinder ecosystem restoration efforts by outcompeting native vegetation and disrupting ecosystem processes. Therefore, controlling invasive species through eradication, containment, or management measures is often a priority in degraded ecosystem rehabilitation projects (Aradottir and Hagen 2013); and (e)

Monitoring and adaptive management: Monitoring the progress of ecosystem restoration activities is crucial to assess effectiveness, identify challenges, and make necessary adjustments. By using adaptive management techniques, restoration professionals can improve their restoration procedures based on input from monitoring data and learn from past mistakes (Singh et al. 2016; Carpenter et al. 2009). By prioritizing restoration and rehabilitation efforts, stakeholders can help reverse the ecological damage caused by human activities and promote recovery of degradation (Alexander et al. 2016).

4.9 Conclusion

In the quest for sustainable development, it emerges as a beacon that illuminates the path toward harmonizing human endeavors with ecological vitality. Through meticulous exploration of biodiversity conservation, ecosystem restoration, and adaptation to climate change, this article underscores the urgency of assessing and nurturing ecological health. Central to its narrative is the recognition of the inseparable bond between human activities and ecological well-being. By delineating the intricate relationships between biodiversity, habitat integrity, and human welfare, the manuscript advocates a paradigm shift toward sustainable practices. It emphasizes the imperative to mitigate environmental degradation and promote resilience in ecosystems to ensure the continuity of vital ecological services. As the world grapples with increasing environmental challenges, the manuscript serves as a call for interdisciplinary collaboration and concerted action. It envisions a future where rigorous evaluation frameworks and innovative approaches empower stakeholders to protect ecological health while advancing sustainable development goals. A group pledge to stewardship that encourages us to accept our responsibility as the planet's guardians. With persistent work and unwavering commitment, we can pave the way for a more resilient and sustainable future in which human well-being and ecological health are closely linked.

4.10 Future Aspect

Future developments in this field of research should see the incorporation of artificial intelligence, big data analytics, and remote sensing, enhancing our ability to monitor and assess ecosystem health with greater precision and efficiency. The importance of understanding and measuring the social, cultural, and economic aspects of ecosystem services will increase, leading to better informed decision-making and policy integration. Climate change adaptation and resilience planning will become increasingly important, with a focus on identifying the impacts of climate stressors on ecosystems and implementing adaptive management strategies. Transdisciplinary collaboration and stakeholder engagement will continue to be essential, fostering partnerships between scientists, policymakers, practitioners, and local communities to co-design research agendas and co-produce actionable insights. Initiatives aimed at improving capacity and educating stakeholders will also provide them with the information and skills necessary to advance sustainable practices, ensuring the long-term success and impact of ecological health evaluations in achieving sustainable development goals. Looking ahead, this field holds significant promise in shaping future environmental management and policy decisions.

Acknowledgments The authors thank NIT Uttarakhand for supporting this work.

References

Agboola OP, Bashir FM, Dodo YA, Mohamed MAS, Alsadun ISR (2023) The influence of information and communication technology (ICT) on stakeholders' involvement and smart urban sustainability. Environ Adv 13:100431. https://doi.org/10.1016/j.envadv.2023.100431

Alexander S, Aronson J, Whaley O, Lamb D (2016) The relationship between ecological restoration and the ecosystem services concept. Ecol Soc 21. https://doi.org/10.5751/ES-08288-210134

Amrutha VN, Geetha SN (2020) A systematic review on green human resource management: implications for social sustainability. J Clean Prod 247:119131. https://doi.org/10.1016/j.jclepro.2019.119131

Aradottir AL, Hagen D (2013) Ecological restoration: approaches and impacts on vegetation, soils and society. In: Advances in agronomy. Elsevier. https://doi.org/10.1016/B978-0-12-407686-0.00003-8

Arcese G, Lucchetti MC, Merli R (2013) Social life cycle assessment as a management tool: methodology for application in tourism. Sustainability 5:3275–3287. https://doi.org/10.3390/su5083275

Aronson J, Floret C, Le Floc'h E, Ovalle C, Pontanier R (1993) Restoration and rehabilitation of degraded ecosystems in arid and semi-arid lands. II. Case studies in Southern Tunisia, Central Chile and Northern Cameroon. Restor Ecol. https://doi.org/10.1111/j.1526-100X.1993.tb00023.x

Arunachalam N, Tyagi BK, Samuel M, Krishnamoorthi R, Manavalan R, Tewari SC, Ashokkumar V, Kroeger A, Sommerfeld J, Petzold M (2013) Community-based control of Aedes aegypti by adoption of eco-health methods in Chennai City, India. Pathog Glob Health 106:488–496. https://doi.org/10.1179/2047773212Y.0000000056

Batterman S, Elsenberg J, Hardin R, Kruk ME, Lemos MC, Michalak AM, Mukherjee B, Renne E, Stein H, Watkins C, Wilson ML (2009) Sustainable control of water-related infectious diseases: a review and proposal for interdisciplinary health-based systems research. Environ Health Perspect 117:1023–1032. https://doi.org/10.1289/ehp.0800423

Borja A, Elliott M, Andersen JH, Berg T, Carstensen J, Halpern BS, Heiskanen AS, Korpinen S, Stewart Lowndes JS, Martin G, Rodriguez-Ezpeleta N (2016) Overview of integrative assessment of marine systems: the ecosystem approach in practice. Front Mar Sci 3:1–20. https://doi.org/10.3389/fmars.2016.00020

Campion BB, Essel G (2013) Environmental impact assessment and sustainable development: a critical review. Environ Nat Resour Res 3:37–51. https://doi.org/10.5539/enrr.v3n2p37

Carpenter SR, Mooney HA, Agard J, Capistrano D, Defries RS, Diaz S, Dietz T, Duraiappah AK, Oteng-Yeboah A, Pereira HM, Perrings C, Reid WV, Sarukhan J, Scholes RJ, Whyte A (2009) Science for managing ecosystem services: beyond the millennium ecosystem assessment. Proc Natl Acad Sci USA 106:1305–1312. https://doi.org/10.1073/pnas.0808772106

Chapman J, Truong VK, Elbourne A, Gangadoo S, Cheeseman S, Rajapaksha P, Latham K, Crawford RJ, Cozzolino D (2020) Combining chemometrics and sensors: toward new applications in monitoring and environmental analysis. Chem Rev 120:6048–6069. https://doi.org/10.1021/acs.chemrev.9b00616

Chardine-Baumann E, Botta-Genoulaz V (2014) A framework for sustainable performance assessment of supply chain management practices. Comput Ind Eng 76:138–147. https://doi.org/10.1016/j.cie.2014.07.029

Chen M, Qian X, Zhang L (2015) Public participation in environmental management in China: status quo and mode innovation. Environ Manag 55:523–535. https://doi.org/10.1007/s00267-014-0428-2

Cowan DM, Dopart P, Ferracini T, Sahmel J, Merryman K, Gaffney S, Paustenbach DJ (2010) A cross-sectional analysis of reported corporate environmental sustainability practices. Regul Toxicol Pharmacol 58:524–538. https://doi.org/10.1016/j.yrtph.2010.09.004

Das S, Pradhan B, Shit PK, Alamri AM (2020) Assessment of wetland ecosystem health using the pressure-state-response (PSR) model: a case study of Mursidabad District of West Bengal (India). Sustainability 12. https://doi.org/10.3390/SU12155932

De Vreese R, Leys M, Fontaine CM, Dendoncker N (2016) Social mapping of perceived ecosystem services supply—the role of social landscape metrics and social hotspots for integrated ecosystem services assessment, landscape planning and management. Ecol Indic 66:517–533. https://doi.org/10.1016/j.ecolind.2016.01.048

Dias-sardinha I, Reijnders L (2005) Performance of large Portuguese companies: a balanced. Bus Strateg Environ 91:73–91

Frischknecht R, Fantke P, Tschümperlin L, Niero M, Antón A, Bare J, Boulay AM, Cherubini F, Hauschild MZ, Henderson A, Levasseur A, McKone TE, Michelsen O, Milà i Canals L, Pfister S, Ridoutt B, Rosenbaum RK, Verones F, Vigon B, Jolliet O (2016) Global guidance on environmental life cycle impact assessment indicators: progress and case study. Int J Life Cycle Assess 21:429–442. https://doi.org/10.1007/s11367-015-1025-1

Gann GD, McDonald T, Walder B, Aronson J, Nelson CR, Jonson J, Hallett JG, Eisenberg C, Guariguata MR, Liu J, Hua F, Echeverría C, Gonzales E, Shaw N, Decleer K, Dixon KW (2019) International principles and standards for the practice of ecological restoration. Second edition. Restor Ecol 27:S1–S46. https://doi.org/10.1111/rec.13035

Gregory RD, Van Strien A (2010) Wild bird indicators: using composite population trends of birds as measures of environmental health. Ornithol Sci 9:3–22. https://doi.org/10.2326/osj.9.3

Haase D, Larondelle N, Andersson E, Artmann M, Borgström S, Breuste J, Gomez-Baggethun E, Gren Å, Hamstead Z, Hansen R, Kabisch N, Kremer P, Langemeyer J, Rall EL, McPhearson T, Pauleit S, Qureshi S, Schwarz N, Voigt A, Wurster D, Elmqvist T (2014) A quantitative review of urban ecosystem service assessments: concepts, models, and implementation. Ambio 43:413–433. https://doi.org/10.1007/s13280-014-0504-0

Hinze J, Godfrey R, Sullivan J (2013) Integration of construction worker safety and health in assessment of sustainable construction. J Constr Eng Manag 139:594–600. https://doi.org/10.1061/(asce)co.1943-7862.0000651

Hugé J, Waas T, Dahdouh-Guebas F, Koedam N, Block T (2013) A discourse-analytical perspective on sustainability assessment: interpreting sustainable development in practice. Sustain Sci 8:187–198. https://doi.org/10.1007/s11625-012-0184-2

Hunsberger CA, Gibson RB, Wismer SK (2005) Citizen involvement in sustainability-centred environmental assessment follow-up. Environ Impact Assess Rev 25:609–627. https://doi.org/10.1016/j.eiar.2004.12.003

International Finance Corporation (2012) Performance Standard 6: biodiversity conservation and sustainable management of living natural resources, p 9

Janku J, Jehlicka J, Hermanová K, Toth D, Maitah M, Kozák J, Vopravil J, Vácha R, Jacko K, Herza T (2022) An overview of a land evaluation in the context of ecosystem services. Soil Water Res 17:1–14. https://doi.org/10.17221/136/2021-SWR

Johns T, Eyzaguirre PB (2006) Linking biodiversity, diet and health in policy and practice. Proc Nutr Soc 65:182–189. https://doi.org/10.1079/pns2006494

Kleijn D, Biesmeijer KJC, Klaassen RHG, Oerlemans N, Raemakers I, Scheper J, Vet LEM (2020) Integrating biodiversity conservation in wider landscape management: necessity, implementation and evaluation. In: Advances in ecological research, 1st edn. Elsevier Ltd. https://doi.org/10.1016/bs.aecr.2020.08.004

Krajnc D, Glavič P (2005) A model for integrated assessment of sustainable development. Resour Conserv Recycl 43:189–208. https://doi.org/10.1016/j.resconrec.2004.06.002

Lattimore B, Smith CT, Titus BD, Stupak I, Egnell G (2009) Environmental factors in woodfuel production: opportunities, risks, and criteria and indicators for sustainable practices. Biomass Bioenergy 33:1321–1342. https://doi.org/10.1016/j.biombioe.2009.06.005

Lauesen LM (2019) Sustainable investment evaluation by means of life cycle assessment. Soc Responsib J 15:347–364. https://doi.org/10.1108/SRJ-03-2018-0054

Lindenmayer, D.B., Margules, C.R., Botkin, D.B., 2000. Indicators of biodiversity for ecologically sustainable forest management. Conserv. Biol. 14, 941–950. https://doi.org/10.1046/j.1523-1739.2000.98533.x

Lecours N, Almeida GEG, Abdallah JM, Novotny TE (2012) Environmental health impacts of tobacco farming: a review of the literature. Tob Control 21:191–196. https://doi.org/10.1136/tobaccocontrol-2011-050318

MacLaren C, Storkey J, Menegat A, Metcalfe H, Dehnen-Schmutz K (2020) An ecological future for weed science to sustain crop production and the environment. A review. Agron Sustain Dev 40. https://doi.org/10.1007/s13593-020-00631-6

Mariani A, Vastola A (2015) Sustainable winegrowing: current perspectives. Int J Wine Res 7:37–48. https://doi.org/10.2147/IJWR.S68003

Mathe S (2014) Integrating participatory approaches into social life cycle assessment: the SLCA participatory approach. Int J Life Cycle Assess 19:1506–1514. https://doi.org/10.1007/s11367-014-0758-6

McDonald T, Jonson J, Dixon KW (2016) National standards for the practice of ecological restoration in Australia. Restor Ecol 24:S4–S32. https://doi.org/10.1111/rec.12359

Merino A, Doni S, Evelpidou N, Ferreira T, García-Arias AI, Masciandaro G, Rodríguez-González PM (2019) Best practices in evaluation and restoration of degraded Mediterranean environments. Monogr. do Ibader Ser. Territ., p 327

Newig J, Jager NW, Challies E, Kochskämper E (2023) Does stakeholder participation improve environmental governance? Evidence from a meta-analysis of 305 case studies. Glob Environ Chang 82:102705. https://doi.org/10.1016/j.gloenvcha.2023.102705

Nicholls E, Ely A, Birkin L, Basu P, Goulson D (2020) The contribution of small-scale food production in urban areas to the sustainable development goals: a review and case study. Sustain Sci 15:1585–1599. https://doi.org/10.1007/s11625-020-00792-z

O'Faircheallaigh C (2007) Environmental agreements, EIA follow-up and aboriginal participation in environmental management: the Canadian experience. Environ Impact Assess Rev 27:319–342. https://doi.org/10.1016/j.eiar.2006.12.002

N.Of, U. Of, A.Z.A. Of, "General guidelines for conducting EIA and SEA studies under the EIA act 2005 Table of contents,"March, 2005.

Parkes MW, Horwitz P (2009) Water, ecology and health: ecosystems as settings for promoting health and sustainability. Health Promot Int 24:94–102. https://doi.org/10.1093/heapro/dan044

Pinho MR (2011) Pauly (1980b) fez uma análise de regressão de M (por ano) contra K (por ano), L∞ (cm) e T (temperatura média da superfície em graus centígrados), com base nos dados de 175 diferentes manancias de peixes, e estimou a relação linear empírica. 14, 1–11

Rice J (2003) Environmental health indicators. Ocean Coast Manag 46:235–259. https://doi.org/10.1016/S0964-5691(03)00006-1

Rosen MA, Kishawy HA (2012) Sustainable manufacturing and design: concepts, practices and needs. Sustainability 4:154–174. https://doi.org/10.3390/su4020154

Sahoo SK, Kumar AV, Yadav AK, Tripathi RM (2016) Metal characterization of airborne particulate matters in a coastal region. Toxicol Environ Chem 98(7):768–777. https://doi.org/10.1080/02772248.2016.1145222

Sandifer PA, Sutton-Grier AE, Ward BP (2015) Exploring connections among nature, biodiversity, ecosystem services, and human health and well-being: opportunities to enhance health and biodiversity conservation. Ecosyst Serv 12:1–15. https://doi.org/10.1016/j.ecoser.2014.12.007

Santos AC, Mendes P, Ribau Teixeira M (2019) Social life cycle analysis as a tool for sustainable management of illegal waste dumping in municipal services. J Clean Prod 210:1141–1149. https://doi.org/10.1016/j.jclepro.2018.11.042

Shirin S, Yadav AK (2021) Assessment and suitability for the dumping of flyash in an abandoned mine. Environ Challenges 4:100111. https://doi.org/10.1016/j.envc.2021.100111

Shirin S, Jamal A, Emmanouil C, Singh VP, Yadav AK (2023) Assessment and characterization of waste material used as backfilling in an abandoned mine. Int J Coal Prep Util 43(8):1402–1410. https://doi.org/10.1080/19392699.2022.2118259

Singh A, Vaish B, Singh RP (2016) Eco-restoration of degraded lands through microbial biomass: an ecological engineer. Acta Biomed Sci 3:133–135

Söderman T, Kopperoinen L, Shemeikka P, Yli-Pelkonen V (2012) Ecosystem services criteria for sustainable development in urban regions. J Environ Assess Policy Manag 14:1250008. https://doi.org/10.1142/S1464333212500081

Tolvanen A, Aronson J (2016) Ecological restoration, ecosystem services, and land use: a European perspective. Ecol Soc 21. https://doi.org/10.5751/ES-09048-210447

Vandana M, John SE, Sunny S, Maya K, Padmalal D (2024) Environmental impact assessment of laterite quarrying from Netravati–Gurpur river basin, South West Coast of India. Environ Dev Sustain 26:909–930. https://doi.org/10.1007/s10668-022-02741-5

Walker P, Horrigan L, Lawrence RS (2002) How sustainable agriculture can address the environmental and human health harms of industrial agriculture. Environ Health Perspect 110:445–456

Wang Z, Yang Z, Shi H, Han F, Liu Q, Qi J, Lu Y (2020) Ecosystem health assessment of world natural heritage sites based on remote sensing and field sampling verification: Bayanbulak as case study. Sustainability 12. https://doi.org/10.3390/su12072610

Wells R, Rhinehart H, Hansen L, Sweeney J, Townsend F, Stone R, Casper DR, Scott M, Hohn A, Rowles T (2004) Bottlenose dolphins as marine ecosystem sentinels: developing a health monitoring system. EcoHealth 1:246–254. https://doi.org/10.1007/s10393-004-0094-6

Wu J (2013) Landscape sustainability science: ecosystem services and human well-being in changing landscapes. Landsc Ecol 28:999–1023. https://doi.org/10.1007/s10980-013-9894-9

Yadav AK (2015) Elemental composition and source apportionment of suspended particulate matters and health risk assessment in mining and nonmining areas of Odisha, India. J Hazard Toxic Radioact Waste 19(3):04014037. https://doi.org/10.1061/(asce)hz.2153-5515.0000248

Yadav AK, Sahoo SK, Kumar AV et al (2023) Identifying size-segregated particulate matter ($PM_{2.5}$, PM_{10} and SPM) sources in an industrial town of India. Aerosol Sci Eng 7(4):455–473. https://doi.org/10.1007/s41810-023-00191-8

Yang T, Siddique KHM, Liu K (2020) Cropping systems in agriculture and their impact on soil health—a review. Glob Ecol Conserv 23:e01118. https://doi.org/10.1016/j.gecco.2020.e01118

Part II
Biomass and Biogas

Chapter 5
Enhancing Biomass Production and Utilization for Sustainable Energy in India

Vidyanath Jha and Chandra Bhushan Kumar Sinha

Abstract All efforts are being made to achieve the targets of reaching Net Zero at the earliest. India also is doing its optimum in eliminating the consumption of fossil fuels and maximizes the development of new and renewable sources. Biomass energy including those from agricultural residues has vast potential of its exploitation in a regulated form. The country today is facing the devastating problem of farm and forest fires for want of a foolproof strategy to contain the same. The enormous phytomass generated by the invasive water hyacinth and other aquatic and land weeds need to be harnessed through preparation of energy bricks that could be used in time of need. Cow and buffalo dung cakes are now being utilized on a large scale for cremation in rural areas. This is being done by digging pits under a closed system that saves on energy consumption that could be avoided through a proper planning. Scientific studies on containing stubble burning in the developed states like Punjab, Haryana, and western Uttar Pradesh are likely to help tackle this problem that crops up every year during the start of winter. National Capital Region bears the brunt at its worst. Carving of energy bricks from *Pinus* needles in the hilly states needs to be geared up on an enhanced scale. This chapter reviews the status of biomass production in India and takes into account the possibilities of their scientific harnessing on an improved scale to provide clean energy solutions in the country.

Keywords Biomass · Plants · Aquatic · Laticiferous · Invasive · Weeds · Biofuel · Bioenergy

V. Jha (✉)
Department of Botany, L.N. Mithila University, Darbhanga, India

C. B. K. Sinha
Department of Biotechnology, L.N. Mithila University, Darbhanga, India

A. K. Yadav (ed.), *Powering Future: Clean Energy Solutions for a Sustainable Planet*, Clean Energy Production Technologies, https://doi.org/10.1007/978-981-96-5055-2_5

5.1 Introduction

The world today is suffering from a number of health problems caused by the all-pervading air pollution and could ill afford the luxury of using fossil fuels any longer for a sustainable life on the planet. It is high time all forms of clean energy, most of them now christened as new and renewable forms of energy, were adopted at a rapid pace. The example could be cited of the Indian Railways that is on the way to fast electrification. Cars, buses, trains, and aeroplanes run on electricity are now becoming a day-to-day affair. It is expected that share of petrol–diesel cars will come down to less than 50% in the coming 4 years. By 2027 every second car will be an electric one. Auto companies are reducing the cost of electric cars under competitive market strategies. Mid-sized cities in India have emerged as big demand centres for EVs. The annual growth rates of high speed electric two-wheeler sales in tier 2 and tier 3 cities have surpassed those in tier 1 cities.

The summer of 2024 witnessed the worst ever heat wave caused by the adverse effect of greenhouse gases released on account of fossil fuels. As against the previous records of heat wave of about 10 days, the year 2024 witnessed its extension by another 10 days. Coupled with a higher humidity level the overall heat index made the situation unbearable. The difference in the night and day temperatures also minimized. It is under such a situation that production and utilization of clean energy has become extremely essential. The UNFCCC has been striving hard to reach the target of net zero carbon emission by 2050. India, however, keeping in view its ground realities of being a developing country, has initiated all possible steps to achieve this target by 2070. India is making all strides to move ahead in the field of new and renewable energy forms including solar, wind, hydro, hydrogen, biomass, etc. India has taken a lead in the formation of Global Solar Alliance. India declared during 26th Conference of Parties held at Glasgow in 2021 its goal to become a net zero economy by 2070. It has now initiated steps in the desired direction. India intends to become a developed economy by 2047. To achieve this target, by 2030, India is likely to produce 500 GW of non-fossil energy by projecting carbon emission reduction by 1 billion tons. Carbon intensity of GDP is likely to be reduced by 45% from 2005 levels. By raising more forest and tree cover, India shall be in a position to create 2.5–3 billion tons of additional sink.

Time is now ripe to harness the vast potentials of biomass energy ingrained in the sea weeds including those of marine algae (Zhao et al. 2022). Sea weeds are also considered to be a rich source of biohydrogen, one of the cleanest sources of energy today (Wyper et al. 2024). Hydrogen is being produced from biomass through thermochemical processes (Alvardo-Flores 2024). Biomass is an organic matter surrounding us that is derived from the forests and croplands. As against the fossil fuels that take millions of years to replenish, biomass resources regrow in a relatively short span of time. Crops, wood, animal waste, seaweed, etc. are the biomass that can be used as an energy source. Since the dawn of human civilization, there has been a practice of cooking food by burning of wood. Sun is the ultimate source of energy on this earth and the plants convert water and CO_2 into carbohydrate through the process of photosynthesis. Carbohydrate-rich foods are good source of energy for all the animals including human beings. As the sources of biomass are unlimited, it is held as renewable source of energy supply. There is always scope to plant trees and grow crops. As such the wastes are always there to be used as biomass. Biomass is one of the oldest and cheapest forms of energy since the dawn of civilization. Biomass is held as a good source of energy and is harnessed in diverse ways (Perea-Morew et al. 2019; Ghosh et al. 2021; Negi et al. 2023; Heeraman et al. 2023). Biomass has a significant potential as green energy source and shares about 14% of the total energy demand on a global scale. India has a vast potential of harnessing biomass energy (Ghosh et al. 2021).

Next to China it is India that annually produces about 900 million metric tons of agricultural biomass. A large portion of this biomass is either burnt or decomposed in uncontrolled manner which reduces the amount of surplus biomass available. The need of the hour is to channel a significant portion of this phytomass for power generation. Biomass energy could be considered as a clean fuel because the plants that are its source capture almost as much CO_2 through photosynthesis as is released when the biomass is burned. This makes biomass a carbon neutral energy source. Biomass is the organic material that makes up plants and it can be converted into energy in a variety of ways. As on 31 March 2023, the estimated potential for renewable biomass energy in India was 28,447 MW (1.3%) against the total renewable energy power capacity of 2,109,654 MW, 13,818 MW (0.66%) from bagasse-based cogeneration in sugar mills of renewable energy was from bagasse-based cogeneration in sugar mills.

5.1.1 Background

Biomass cogeneration is a process that uses a single primary fuel to simultaneously produce electricity and heat. It is a well-advanced technology that has been around for over a century and is more efficient than using separate production process. India has a rich resource base for biomass-based energy from sources like stubble that creates big health problems in the Delhi NCR during the onset of winter (on account of burning of this agricultural residue by the small farmers in Haryana, Punjab, and western Uttar Pradesh). Vast amount of phytomass of grasses and aquatic weeds could be harnessed for the same. Economy of north-eastern states is much dependent on bamboos. Bamboo has been recently removed from the category of forest trees and as such is now open for unhindered use by the common people. Bamboos in major portions of India could also be used to produce cleaner fuels. Bamboo biomass is a renewable resource that can be converted into ethanol and bioenergy (Liang et al. 2023). It can be processed thermally or biochemically into energy products such as charcoal, pellets, and briquettes, as well as utilized as a firewood replacement. Bamboo biomass has various desirable fuel properties, including low ash content and alkali index, and can generate 1 kWh of energy from 1 kg of bamboo. This is comparable to the biomass requirements for wood products, but superior to other biomass sources such as hemp, bagasse, and rice husk. Bamboo has great fuel characteristics and can be turned into solid, liquid, or gaseous biofuels. Humankind today is in the age of artificial intelligence (AI). A recent report has revealed that Google is now pumping out more greenhouse gases than before as it powers data centres needed to support AI. Google's climate changing emissions have increased 48% in the past 5 years, and this is at odds with a touted goal of becoming carbon neutral for the sake of the planet. Total greenhouse gas emissions in 2023 were 13% higher than they were in the previous year. The same was primarily driven by increased data centre energy consumption and its supply chain (ET Energyworld.com, accessed on 6 July 2024).

5.2 Recent Studies on Biomass Utilization

India today has to lose heavily on account of forest fires that are increasing every year on account of climate change caused by global warming. Pine needles, which grow abundantly in India's hilly areas, are a major global problem due to their extreme sensitivity to fire in dry and hot climes. Pinus needles provide a good source of bioenergy (Rana et al. 2023). A case study of Uttarakhand state in India revealed that the state produces a significant 1.67×10^9 kg of pine needles each year. The annual potential emission from forest fires was estimated to be 58.37×10^9 kg of CO2. Halba and Arora (2024) were able to design and evaluate the efficacy of a gasification-based supply chain for pine needles in Uttarakhand's Almora district. Emission calculations revealed that gasification plant capacity scenarios may potentially cut 44.63×10^6 to 46.16×10^6 kg of CO_2 equivalent emissions per year when compared to grid electricity, while satisfying almost 5.5% of Almora district's electricity consumption. Water-logged sites in India are major sites of infestation by the fast growing water hyacinth (*Eichhornia crassipes*) that is a good source of phytomass and could be harnessed for bioenergy on a big scale (Ighodalo et al. 2011; Moki et al. 2020; Nandi Yanto et al. 2023; Wembe et al. 2023). Emergent aquaphyte like *Ipomoea carnea,* growing lavishly in these sites also holds a similar potential (Saikia et al. 2015).

Mango leaves and wood are utilized for cooking and other household purposes. In rural areas, the leaves are collected by sweeping the orchard floor and stored for pretty long time. Mango holds a good potential for use as a biofuel (Das et al. 2013). A weed like *Cannabis sativa* of the family Cannabaceae grows abundantly alongside rural roads and wheat fields. Its leaves are source of intoxicants and are widely used as a mild narcotic known by the local name of *bhang*. The plant could be used as a potential biofuel (Tahir et al. 2020). *Calotropis procera* (Apocynaceae) also grows equally lavishly as a laticiferous plant and is held as a petro crop. Its bio crude is a rich source of triterpenoid hydrocarbon have successfully upgraded its bio crude to useful liquid fuels by developing conversion processes like fluid catalytic tracking. The fuel thus obtained was found to contain significant proportions of polyaromatic hydrocarbon (Korlipara et al. 2009; Rathore and Meena 2010;

Mahmoud et al. 2021). Pérez-Arévalo and Velázquez-Martí (2018) made a complete evaluation of the process of getting biomass from pruning residues of an ornamental moraceous plant called *Ficus benjamina* that generates a large amount of ligneous residues including the mixture of wood and leaves. Nitrogen content was found to be acceptable for its use as fuel. They obtained 19 MJ/kg mean calorific value that was due to the presence of volatile content and lignin. In India, there is an old practice of large scale plantation of moraceous species of *Ficus* on religious grounds. A large part of our country is bearing the onslaught of an invasive weed *Parthenium hysterophorus* that came to India with Mexican wheat during the pre-green revolution days (Bisaria 2010; Tahir et al. 2020). On account of its high fecundity, the weed has spread almost everywhere. The weed grows with minimum input of nutrients in various soil and climate regimes. Plants belonging to Moraceae, Euphorbiaceae, Apocynaceae, and Asclepiadaceae families are generally laticiferous and hence designated as petro plants. Latex in these plants contain liquid hydrocarbon that could be converted into high-grade transportation fuel. Latex bearing plants serve as renewable source of conventional energy. National Highway Authority of India plants majority of laticiferous plants belonging generally to the species of *Thevetia*, *Nerium*, *Alstonia*, etc. along its course. Animals refrain from feeding upon these plants which need to be exploited as biofuels on an organized scale. Investigations have been made on the biofuel potentials of halophytes growing in estuaries alongside the confluence points of rivers with oceans (Abideen et al. 2023).

Punjab, now bifurcated into two states of Punjab and Haryana, is an agriculturally improved zone of India. The two states grow wheat and paddy on a big scale. This zone produces a big amount of biomass. Biomass-based power generation plants use mostly paddy straw. Small farmers in this zone burn the straw as they do not have sufficient time and means to prepare their field for sowing of wheat at a short interval. This makes the situation unsustainable on account of huge number of farm fires that engulf even the national capital Delhi and the entire National Capital Region (Singh et al. 2021). Biofuels are now being used as aviation fuel as well (Baxter 2020). The first biofuel blended flight was put into operation by the US Air Force on 25 March 2010. Virgin Atlantic used algae as feedstock in its first commercial flight in the year 2008. India's first flight using biofuel flew from Dehradun to Delhi conducted during August 2018.

Search for efficient methods of utilization of different categories of biomass for human welfare has now become a global priority. Investigations on this aspect are being made on varying source materials including microorganisms (Zabermawi et al. 2022), algae (Voloshin et al. 2016; Ru et al. 2020), aquatic and other weeds (Alam et al. 2021; Koley et al. 2023), agricultural residues (Rautray et al. 2019), etc. Grasses find flood plains as suitable habitat for their growth. Quite a good number of poaceous and other grasses are rich in biomass that could be exploited as biofuel. An example could be cited of a versatile grass *Rhizopogon zizanioides*. Cellulose in this grass is abundantly found in lignocellulosic form. Subsamran et al. (2018) were able to obtain bioethanol from this grass through hydrolysis with action of crude cellulase enzyme. Regenerative food forests raised at a few places in India are

Fig. 5.1 Vetiver grass in Belwara village of Darbhanga district

proving much helpful in raising above ground and below ground biomasses. A 100-acre Vanya Organic Farm in the Khalghat area of Madhya Pradesh on the bank of river Narmada and a 40-acre Samar Shail Organic farm at Ramnagar, Purnea in Mithila area of Bihar have vetiver as a significant component of the food forests raised by the two activists. Besides growing the food products, the ingrained phytomass helps in restoration of the barren rocky and sandy patches in these regions (Bewtra 2023) (Fig. 5.1).

Like forest fires in hilly states, farm fires in Punjab, Haryana, and western UP are another vexed issue that requires devising proper scientific methods to contain them. Porichha et al. (2021) have analysed alternative methods for valorization of crop residues that are now being converted as bio bricks (Rautray et al. 2019). Freshwater pools in India get infested with macrophytes including the nymphaeids like *Nymphaea*, *Nelumbo*, *Nuphar*, and other aquatics belonging to the species of *Nymphoides*, *Ottelia*, *Trapa*, and a host of other angiosperms. With voluminous leaves containing fairly large amount of biomass, often their flowers/fruits are consumed and other parts are simply left apart. Their old leaves wither out as per natural cycle. There is scope for harnessing their phytomass for biofuel potentials. The case of a rooted-floating macrophyte like *Euryale ferox* also known as (makhana/gorgon nut/fox nut) could be cited as an example for its under exploitation as a

Fig. 5.2 Shows the *E. ferox* plants that have been outpaced by the menacingly overgrowing *M. vaginalis* plants. The latter established a foothold in makhana plots. Dearth of rain during the initial months of the crop helped *M. vaginalis* plant overgrow and thereby block the growth of *E. ferox* plants

biomass crop as well (Jha and Sinha 2024). But in order to harvest its seeds lying at the pond bottom through virtually sweeping the floor, the leaves are cut down and left to disintegrate and decompose in the pond itself. Nothing except seed is taken out of the system. It is imperative therefore that the voluminous and leathery textured leaves are exploited for biofuel purpose as well. The same could be done along with the associated weeds like *Monochoria vaginalis*, *Jussiaea* sp., *Marsilea* sp., etc. (Fig. 5.2).

5.2.1 Biomass-Based Hydrogen Production

World today is looking forward to the production of biomass-based clean energy of which hydrogen has emerged as the forerunner. Waste biomass is now being utilized as potential renewable source for hydrogen production. Nguyen et al. (2024) have scrutinized the recent advancements in the field of production of hydrogen through adoption of pyrolysis and gasification methods.

5.2.2 Varied Forms of Biofuels

Biofuels are fuels derived from biomass and generally include animal waste, algae, or plant matter. Biofuel is a renewable source of energy is derived from replenishable matter and is easy to extract. Ethanol is essentially ethyl alcohol or pure alcohol, and it is the most common alternative biofuel used in cars. Sugarcane and corn are the most common sources. Naturally, ethanol is clean releasing roughly 15% less greenhouse gas emissions. It is usually incorporated with other fuels to reduce carbon emission. Methanol is an alcohol used in motor vehicles. It is produced through the gasification of biomass at extremely high temperature with a catalyst. Biodiesel is a liquid biofuel that is derived from a combination of vegetables and fats and oils from both animals and plants. As compared with diesel as fossil fuel, it has a reduced amount of carbon and higher hydrogen and oxygen contents. It is non-toxic, biodegradable, and safe to handle and transport. Biobutanol is generally derived from algae and bacteria. It could be used directly in standard gasoline engines with no modification as it has a higher energy per unit. Biogas is a gaseous form of biofuel produced through anaerobic breakdown of biomass. It mainly consists of methane gas.

5.2.3 Generations of Biofuel

Food crops constitute the first-generation biofuel and have a negative net energy gain that is accomplished by emitting more CO_2 during production. This is as compared with what the feedstocks did capture during growth. Biofuels are also blamed for inflation in food prices. Biofuels made from non-food crops constitute the second generation and they are cost-competitive as well. Their net positive energy gains are higher. Algae represent the third-generation biofuels that produce more energy per hectare than the conventional plant materials. They generally grow in water bodies or land that is unsuitable for the production of food. Almost the same amount of carbon is absorbed and also emitted. The fourth-generation biofuels generate sustainable energy and provide a way to capture and store CO_2. This process differs from the second- and third-generation production in the sense that carbon dioxide is captured during all stages through oxy-combustion. Old oil and gas fields could be used for storing the CO_2. Such a type of CO_2 capture is an example of carbon negative category. The fifth generation includes biofuels derived from non-food sources through advanced technologies with algae, bacteria, and microorganism as the base material. Their processes are sustainable and environment friendly that provide high energy yields per unit of input. Cellulosic biofuels generally come under this category, processes of fermentation, tissue culture, genetic engineering, etc. are used for obtaining the fifth-generation biofuels (Mushtaq et al. 2023).

5.3 Policies on Biomass Utilization

The Ministry of New and Renewable Energy Sources created the first National Policy in 2009. This National Policy on Biofuels-2018 was developed on the foundation of previous policy successes and establishes a new agenda consistent with the redefined role of emerging advances in the renewable sector. The National Policy on Biofuels-2018 aims to expand biofuel use and minimize petroleum product imports by blending ethanol with petrol to 20% by 2025–26 and biodiesel to 5% by 2030. According to the 2022 Amendments to the National Policy on Biofuels, the aim for ethanol blending in gasoline has been moved from 20% by 2030 to 20% by 2025–26. The strategy has been expanded to incorporate more feedstocks for biofuel production. This comprises bamboo, bagasse, and corn stover. Domestic ethanol consumption has just overtaken domestic production for the ninth year in a row, with 2023 marking the milestone. In India, industrial ethanol is used for a variety of purposes although biofuels and additional neutral alcohol account for the majority of consumption. At present, India's industrial alcohol production capacity stands at 13.80 BL, with 8.75 BL from molasses and 5.05 BL from green livestock. Bioethanol provides a number of economic, social, and environmental benefits, including increased crop demand, increased employment, and lower emissions. According to the Indian government, measures associated with the India Ethanol Blending Programme reduced greenhouse gas emissions by 31.8 million metric tons between the ethanol supply years of 2014 and 2022. Despite problems such as feedstock availability and technology limits, policymakers, and producers are addressing the problem through better legislation, regulations, and technological advancements. The major states in ethanol production are Maharashtra, Uttar Pradesh, and Karnataka. Maharashtra generates 2.68 BL of ethanol annually. Uttar Pradesh and Karnataka produce 2.08 BL and 1.18 BL of ethanol annually, respectively. These states are at the forefront of industrial ethanol production due to their quantity of feedstock. These states are large sugarcane growers, with established distilleries. Corn production in India is predicted to increase by 10 million metric tons over the next 5 years, allowing for greater conversion to ethanol. The government is supporting the domestic manufacture of flexi-fuels and E-20 compliant automobiles in the Indian market to increase ethanol output and use.

According to the aims of the Global Biofuels Alliance (established recently at the 2023 G20 Summit), India plans to treble its biofuel use over the next 5 years. The Global Biofuels Alliance intends to incorporate biofuels into the world's energy mix by increasing the use of sustainable biofuels and allowing technological improvements. This is a tremendous potential for the Indian bioethanol industry to broaden its reach and influence towards a more sustainable future. High-gravity fermentation is an important approach to reduce energy use during ethanol manufacturing. New technologies and new techniques are likely to contribute to stronger bioethanol market dynamics. The second-generation bioethanol is expected to gain traction in the future years. The potential introduction of new generations of yeasts into the Indian market is expected to create numerous chances in this industry. The Govt. of

Table 5.1 Indian standards for biofuel

S. no.	IS number	Year	Specification
1	15464	2022	Blending component in motor gasoline anhydrous ethanol is used
2	15607	2022	Fatty acid methyl esters FAME—biodiesel B-100
3	16087	2016	Biogas–biomethane
4	16531	2022	Blend B8 to B20—biodiesel diesel fuel
5	16629	2017	ED95 automotive fuel—hydrous ethanol
6	16634	2017	Anhydrous ethanol and gasoline blend as E85 fuel
7	17021	2018	E20 fuel is a blend of anhydrous ethanol and gasoline used as fuel for spark-ignited engine-powered vehicles
8	17081	2019	Kerosene type, Jet A-1 containing synthesized hydrocarbons—aviation turbine fuel
9	17821	2022	Positive ignition engine-powered vehicles (ethanol used as fuel)

Adapted from Posted on: 25 Sep 2023, 4:48 PM by PIB Delhi

Bihar has also come up with its Bihar Biofuels Production Promotion Policy 2023 by amending its previous Ethanol Production Promotion Policy 2021. This policy aims to broadbase the coverage by including CBC (compressed biogas)/bio CNG manufacturing units apart from ethanol. The Bihar Biofuel Policy is a significant step forward for the state's future-cleaner, more self-sufficient, and economically prosperous. Bihar is enabling a more sustainable energy landscape while also empowering its people by creating jobs and economic possibilities through biofuel production and use. The Bureau of Indian Standards (BIS) is the primary national authority in charge of developing and executing the standards required for marking and quality certification. It enables the timely development, dissemination, promotion, and implementation of national and regional standards. It ensures technical regulations for goods, services, processes, and practices (Table 5.1).

5.3.1 Significant Highlights of the National Policy on Biofuels-2018

- Classification of basic (1G) bioethanol and biodiesel, and advanced (2G) biofuels such as ethanol, drop-in fuels, and algae-based biofuels (3G).
- To encourage the procurement of intermediate (B-Molasses), sugarcane juice, other sugar-containing materials, and damaged/surplus food grains to broaden the scope of raw materials for ethanol production.
- To focus on research and demonstration in the field of biofuel feedstock generation, enhanced conversion methods from specified feedstock.
- To encourage biodiesel production from non-edible oilseeds, spent cooking oil, short gestation crops, and the establishment of supply chain mechanisms.
- Setting up a national biomass repository by conducting biomass appraisals around the country.

Fig. 5.3 Shows an advertisement issued by the Govt. of Bihar regarding its Biofuel Policy 2023 (Rashtriya Sahara, Dated: 10 March 2024, Patna Edn.)

- To focus on Advanced Biofuels and offer a funding scheme of Rs. 5000 crore over 6 years for 2G ethanol biorefineries, with additional tax advantages and a higher purchasing price than 1G biofuels. Setting up of National Biofuel Coordination Committee (NBCC) under the Ministry of Petroleum and Natural Gas and Working Group on Biofuels.
- By 2030, the proposed objective is to blend 20% ethanol and 5% biodiesel in petrol (Fig. 5.3).

5.4 Efficient Cremation Systems

Brihanmumbai Municipal Corporation (BMC) is likely to implement an ecofriendly wood combustion system at several crematoria in Mumbai and its suburbs. The system was first piloted in 2020 and has proven successful. Fourteen crematoria have been recently identified to use biomass briquettes and pellets as alternative fuels. The combustion system is engineered to optimize energy efficiency and minimize smoke emissions. Advanced features like water scrubbers and cyclonic separators filler out particulates and toxic gases, ensuring cleaner air. The treated air is thus released through a 30 m high chimney. The wood burning system employs a specially designed combustion chamber that minimizes fuel usage. Traditionally, cremations require 350–400 kg of wood. However, this new system reduces wood consumption to 100–125 kg per cremation saving approx. 250 kg of wood each time. This reduction in wood usage also leads to lower carbon emission. Rural people in Bihar are now cremating corpses by either digging pits or in earthen

Figs. 5.4, 5.5, and 5.6 Show cremation under a closed system with dung cake as fuel

containers with cow dung cakes and maize cobwebs as fuel (Tewary 2009; Jha 2022) (Figs. 5.4, 5.5, and 5.6).

A project called GOBAR (DHAN) has been launched to keep villages clean while also enhancing the income of rural households. The initiative aims to focus on controlling and converting cattle dung and solid waste in farms into usable compost and bioCNG (Fig. 5.7).

5.4.1 Utilization of Human Waste

The concept of human waste utilization was enunciated by Padma Vibhushan Bindeshwar Pathak in Bihar who created a silent revolution through his Sulabh Shauchalaya Movement. Bioenergy generated from human excreta was once used for lighting Gandhi Maidan in the state capital "Sulabh" turned out to be a trendsetter in the field of human waste management (Mukherjee and Chakraborty 2016).

5.5 Conclusion

Biomass has since evolved as a promising source of renewable energy for which India has a vast potential. There is a need to extend all possible government and public support to meet the targets of Net Zero emission at the earliest through adopting the best possible ways. Biomass energy could play a significant role in obtaining the goals of energy transition and save the world from the on-slots of the menacing levels of climate change.

Fig. 5.7 Cremation in an earthen container placed over a boat in Kusheshwar Asthan area of Darbhanga district

Acknowledgments The authors thank the Director of the Department of Biotechnology, L.N. Mithila University, Darbhanga, for their consistent moral support.

References

Abideen Z, Ansari R, Hasnain M, Flowers TJ, Koyro H-W, El-Keblawy A, Abouleish M and Khan MA (2023) Potential use of saline resources for biofuel production using halophytes and marine algae: prospects and pitfalls. Front Plant Sci 14:1026063. https://doi.org/10.3389/fpls.2023.1026063

Alam SN, Singh B, Guldhe A (2021) Aquatic weed as a biorefinery resource for biofuels and value-added products: challenges and recent advancements. Clean Eng Technol 4(3):100235. https://doi.org/10.1016/j.clet.2021.100235

Alvardo-Flores JJ (2024) Thermochemical production of hydrogen from biomass: pyrolysis and gasification. Energies 17(2):1537

Baxter G (2020) The use of aviation biofuels as an airport environmental sustainability measure: the case of Oslo Gardermoen Airport. Mag Aviat Dev 8(1):6–17. ISSN: 1805-7578. https://doi.org/10.14311/MAD.2020.01.01

Bewtra P (2023) Regenerative food forest: a case study of Vanya organic farm. Curr Agric Res J 10(3):207–215. https://doi.org/10.12944/CARJ.10.3.05

Bisaria D (2010) Biofuel production using weeds primarily *Parthenium hysterophorus* and flower dump from the Mahakaleshwar temple. In: Asia Pacific conference of chemical engineers, Taipei

Das PS, Ravindran R, Deka D, Jawed M, Dasand D, Goyal A (2013) Bioethanol production from leafy biomass of mango (*Mangifera indica*) involving naturally isolated and recombinant enzymes. Prep Biochem Biotechnol 43(7):717–734. https://doi.org/10.1080/10826068.2013.773342

Ghosh U, Das D, Banerjee D, Karmakar S, Das J (2021) Biomass energy potential in India: a review. Int J Eng Res Technol 9(11):42–45

Halba A, Arora P (2024) Pinus needle gasification based electricity production: understanding the effect of supply chain. Environ Sci Pollut Res. https://doi.org/10.1007/s11356-024-33592-4

Heeraman J, Kumar S, Seeniappan K, Pravin (2023) Paddy residue potential as energy resources—a critical review. In: Proceedings of the international conference on materials for emerging technologies-2021 (ICMET-21). Lovely Professional University, Jalandhar-Delhi. https://doi.org/10.1063/5.0163092

Ighodalo O, Zoukumor K, Egbon C, Okoh S (2011) Processing water hyacinth into biomass briquettes for cooking purposes. J Emerg Trends Eng Appl Sci 2(2):305–307

Jha V (2022) An indigenous method of cremation in earthen containers as a sequel to disaster resilience: a case study of Kusheshwarsthan wetland area in Darbhanga district of north Bihar. Int J Plant Environ 8(2):149–152

Jha V, Sinha CBK (2024) Economic and medicinal significance of Makhana (*Euryaleferox* Salisb.). CGES Newsl 9(1):5–11

Koley A, Mukhopadhyay P, Gupta N, Singh A, Ghosh A, Show BK, Chaudhury S, Hazra AK, Balachandran S (2023) Biogas production potential of aquatic weeds as the next-generation feedstock for bioenergy production: a review. Environ Sci Pollut Res Int. https://doi.org/10.21203/rs.3.rs-2676987/v1

Korlipara P, Atheya N, Bhatnagar AK, Singh KK (2009) Conversion of *Calotropis procera* biocrude to liquid fuels using thermal and catalytic cracking. Fuel 88(5):780–785

Kumar A, Yadav AK (2024) Challenges and opportunities for water quality monitoring and management in India. In: Yadav AK, Yadav K, Singh VP (eds) Integrated management of water resources in India: a computational approach. Springer Nature, Cham, pp 493–511. https://doi.org/10.1007/978-3-031-62079-9_28

Liang Z, Nemenyi A, Kovacs GP, Gyuricza C (2023) Potential use of bamboo resources in energy value added conversion technology and energy systems. GCB Bioenergy 15(8):936–953

Mahmoud AH, El-Bery HM, Ali MM, Aldaby E (2021) Latex-bearing plant (*Calotropis procera*) as a biorefinery for bioethanol production. Biomass Convers Biorefinery 13(6). https://doi.org/10.1007/s13399-021-01479-w

Moki EC, Oyibo MC, Yauri AUB, Rapheal IA, Yahaya Y (2020) Enhancing the properties of water hyacinth biomass briquettes by mercerization process. Int Res J Pure Appl Chem 21(18):43–55

Mukherjee S, Chakraborty D (2016) Turning human waste into renewable energy: scope and options for India. Munich Personal RePEc Archive MPRA Paper No. 73669

Mushtaq Z, Maqbool R, Bhat KA (2023) Chapter 13: Genetic engineering and fifth-generation biofuels. In: Hakeem KR, Bandh SA, Malla FA, Mehmood MA (eds) Environmental sustainability of biofuels. Elsevier, pp 237–251. ISBN: 9780323911597. https://doi.org/10.1016/B978-0-323-91159-7.00015-1

Nandi Yanto A, Ragadhita R, Hofifah SN, Husaeni DFA (2023) Process in the utilization of water hyacinth as effective biomass material. Environ Dev Sustain. https://doi.org/10.1007/s10668-023-036556

Negi H, Sugal DC, Soni R, Giri K, Goel R (2023) Indian scenario of biomass availability and its bioenergy-conversion potential. Energies 16(15):5805. https://doi.org/10.3390/en1615805

Nguyen VG, Nguyen-Thi TX, Nguyen PQP, Tran VD, Ağbulut Ü, Nguyen LH, Balasubramanian D, Tarelko W, Bandh SA, Nguyen DKP (2024) Recent advances in hydrogen production from biomass waste with a focus on pyrolysis and gasification. 54:127–160, ISSN 0360-3199. https://doi.org/10.1016/j.ijhydene.2023.05.049

Perea-Morew M, Samern-Manzano E, Perea-Moreno A (2019) Biomass as renewable energy: worldwide research trends. Sustainability 11(3):863. https://doi.org/10.3390/sull030863

Pérez-Arévalo JJ, Velázquez-Martí B (2018) Evaluation of pruning residues of *Ficus benjamina* as a primary biofuel material. Biomass Bioenergy 108:217–223. https://doi.org/10.1016/j.biombioe.2017.11.017

Porichha GK, Hu Y, Rao KTV, Xu CC (2021) Crop residue management in india: stubble burning vs. other utilizations including bioenergy. Energies 14:4281. https://doi.org/10.3390/en14144281

Rana AK, Guleria S, Gupta VK, Thakur VK (2023) Cellulosic pine needles-based biorefinery for a circular bioeconomy. Bioresour Technol 367. https://doi.org/10.1016/j.biortech.2022.128255

Rathore M, Meena RK (2010) Potential of utilizing Calotropis procera flower biomass as a renewable source of energy. J Phytol 2(1):78–83

Ru ITK, Sung YY, Jusoh M, Effendy M, Wahid A, Nagappan T (2020) Chlorella vulgaris: a perspective on its potential for combining high biomass with high value bioproducts. Appl Phycol. https://doi.org/10.1080/26388081.2020.1715256

Rautray P, Roy A, Mathew DJ, Boris E (2019) Bio-brick—development of sustainable and cost effective building material. In: International conference on engineering design, ICED 19

Saikia P, Gupta U, Barman RS, Kataki R (2015) Production and characterization of bio-oil produced from Ipomoea carnea bio-weed. Bioenergy Res 8(3). https://doi.org/10.1007/s12155-014-9561-2

Shirin S, Yadav AK (2014) Physico-chemical analysis of municipal wastewater discharge in Ganga river, Haridwar district of Uttarakhand, India. Curr World Environ 9(2):536–543. https://doi.org/10.12944/CWE.9.2.39

Singh M, Brar YS, Singh H (2021) Critical assessment of biomass material for power generation in Punjab. India Mater Today Proc 48(3):927. https://doi.org/10.1016/j.matpr.2021.04.414

Subsamran K, Mahakhan P, Vichitphan K, Vichitphan S, Sawaengkaew J (2018) Potential use of vetiver grass for cellulolytic enzyme production and bioethanol production. Biocatal Agric Biotechnol 17:261–268

Tahir N, Tahir MN, Alam M, Yi W, Zhang Q (2020) Exploring the prospective of weeds (Cannabis sativa L., Parthenium hysterophorus L.) for biofuel production through nanocatalytic (Co, Ni) gasification. Biotechnol Biofuels 13:148

Tewary A (2009) Cowdung cremations catch on in Bihar. https://news.bbc.co.uk/south_asia

Voloshin RA, Rodionova MV, Zharmukhamedov SK, Veziroglu TN, Allakhverdiev SI (2016) Biofuel production from plant and algal biomass. Int J Hydrog Energy 41(39):17257–17273. https://doi.org/10.1016/j.ijhydene.2016.07.084

Wembe DB, Djomi R, Konai N, Nekadeu G, Ntamack GE (2023) Experimental study of biogas production from water hyacinth. Sci Technol Energy Transit 78:14. https://doi.org/10.2516/stet/2023010

Wyper UM, Zendehboudi S, Kerton FM (2024) The sea's best kept secret: the use of sea weed as a source of biohydrogen for clean and renewable energy. RSC Sustain 2:1289–1299

Yadav AK, Sahoo SK, Mahapatra S et al (2014) Concentrations of uranium in drinking water and cumulative, age-dependent radiation doses in four districts of Uttar Pradesh, India. Toxicol Environ Chem 96(2):192–200. https://doi.org/10.1080/02772248.2014.934247

Zabermawi NM, Alsulaimany FAS, El-Saadony MT, El-Tarabily KA (2022) New eco-friendly trends to produce biofuel and bioenergy from microorganisms: an updated review. Saudi J Biol Sci. https://doi.org/10.1016/j.sjbs.2022.02.024

Zhao Y, Bourgougnon N, Lanoiselle Z, Lendormi T (2022) Biofuel production from sea weeds: a comprehensive review. Energies 15(24):9395. https://doi.org/10.3390/en15249395

Chapter 6
Integration of Biomass with Solar and Wind Energy in the Future

Neha Bhadauria and Arjun Suresh

Abstract The increased interest in sustainable energy sources has catalysed the exploration and combination of various renewable sources, specifically focusing on reducing fossil fuel dependency and mitigating environmental degradation. This chapter explores the integration of biomass with solar and wind energy, assessing the efficacy of hybrid renewable energy systems (HRESs) as a strategic solution for sustainable energy production. Biomass serves as a stable complement to the intermittent nature of solar and wind, facilitating a consistent energy supply that promotes grid stability and dependability. The discussion begins with an overview of each type of renewable energy, emphasizing the benefits derived from their integration. It further investigates technologies that enable such integration, including cogeneration and the conversion of surplus electricity into bioenergy during periods of high production. The synergy between biomass and other renewables through advanced grid technologies and energy management systems is critically examined. This chapter assesses the economic and environmental advantages of HRES, demonstrating its ability to reduce greenhouse gas emissions and offer cost-effective energy alternatives. The need for supportive policies and incentives to advance integrated renewable systems is addressed, highlighting the importance of government and institutional support. Global case studies are reviewed to highlight the technologies for real-world application and the potential challenges of implementing integrated biomass, solar, and wind systems. These examples shed light on the adaptability of HRES in diverse geographic and economic conditions. In conclusion, with a look at upcoming trends and technological innovations in renewable energy, the chapter outlines future research directions and potential developments. This comprehensive approach not only demonstrates the integration possibilities of biomass with solar and wind, but also envisions a future where renewable energy is integral to global energy frameworks.

N. Bhadauria (✉) · A. Suresh
Save Vibrant Earth Foundation, Noida, India

A. K. Yadav (ed.), *Powering Future: Clean Energy Solutions for a Sustainable Planet*, Clean Energy Production Technologies, https://doi.org/10.1007/978-981-96-5055-2_6

Keywords Hybrid renewable energy systems · Biomass energy · Solar energy · Wind energy · Grid stability

6.1 Introduction

Renewable energy sources (RES) play a crucial role in meeting global energy needs and reducing environmental impact. These sources include biomass, hydropower, geothermal, solar, wind, and marine energies, all noted for their sustainability, local availability, and low environmental impact (Bergmann et al. 2008). Large-scale hydropower, for instance, currently supplies about 20% of global electricity needs, underscoring its significant contribution to renewable energy portfolios (Haines et al. 2006). Wind power, particularly in coastal and windy regions, holds promise as a burgeoning energy source due to its scalability and reliability. Looking forward, the share of RES is projected to expand markedly, potentially constituting 30–80% of global energy by 2100 (Haines et al. 2006). Sustainable development requires robust methodologies and methods to assess and compare the environmental effects of human activities, ensuring that advances in living standards and industrialization do not come at the expense of environmental degradation (Lafforgue et al. 2008). The excessive consumption of fossil fuels not only depletes finite resources but also exacerbates environmental challenges, including climate change and health risks (Thirugnanasambandam et al. 2010). Global efforts are increasingly focused on transitioning towards more sustainable practices, such as minimizing waste, reducing air pollution, promoting distributed energy generation, conserving natural habitats, and curbing greenhouse gas emissions. This shift reflects a broader societal commitment to achieving environmental sustainability while maintaining economic growth and improving quality of life.

6.1.1 Renewable Energy Sources

In today's global landscape, there is an increase in the necessity for sustainable energy technologies to replace traditional electricity generation sources such as fossil fuels driven by increasing demands worldwide, particularly in both developed and developing nations (Rizzi et al. 2014). Fossil fuel-based energy sources have been implicated in significant environmental issues, with the main among them global warming and climate change. The emission of greenhouse gases into the atmosphere from power generation has shown an alarming exponential rise in recent decades. In response to these pressing environmental challenges, renewable energy (RE) technologies have been introduced as viable alternatives for electricity generation (Santika et al. 2019). These technologies encompass a variety of sources, including solar, wind, hydro, biomass, geothermal, and hydrogen energies, known for their environmentally beneficial characteristics and their ability to generate electricity with minimal or negligible emissions of air pollutants. The growing societal consciousness towards maintaining a cleaner environment has further elevated the profile of RE technologies (Banos et al. 2011; Qazi et al. 2019). Beyond their environmental benefits, RE plays a pivotal role in bolstering economic sustainability. By leveraging natural and renewable resources, RE technologies contribute to reducing the overall cost of electricity generation (Kardooni et al. 2016). Moreover, they offer economic opportunities through mechanisms such as decentralized energy production and the potential for consumers to sell surplus electricity back to the grid. Despite these advantages, the integration of RE into mainstream energy systems faces challenges such as intermittency and variability in energy generation. For example, solar photovoltaic systems operate exclusively during daylight hours, wind turbines require adequate wind speeds, and hydro turbines rely on water flow conditions. Addressing these challenges requires on-going research and development efforts to improve the efficiency, reliability, and cost effectiveness of RE technologies (Santika et al. 2019). RE technologies not only offer a path to environmental sustainability but also present economic opportunities crucial to achieving a cleaner and more resilient energy future.

6.2 Biomass Energy

6.2.1 Understanding Biomass and Sources

Biomass energy, sourced from organic materials, is a versatile and renewable energy option. It includes a wide range of biological materials such as plant residues, animal waste, and other organic matter, which can be transformed into heat, electricity, and biofuels (Demirbas 2009). Sustainability stems from its ability to recycle carbon dioxide, making it a greener alternative to fossil fuels (Sims et al. 2010). Agricultural residues are a major source of biomass energy. Agricultural processes produce large amounts of by-products like crop stalks, husks, and straws, which can be converted into energy through combustion, gasification, and anaerobic digestion (Sokhansanj et al. 2006). For instance, rice husks and wheat straw can be used for bioenergy production, helping reduce waste and providing farmers with an additional income stream (McKendry 2002). The forestry industry also generates substantial biomass in the form of wood chips, sawdust, and logging residues. These materials, often considered waste, can be efficiently converted into energy, contributing to both economic and environmental sustainability. The use of forest by-products for energy production highlights the multiple benefits of biomass as a renewable energy source. These materials are ideal for bioenergy applications, particularly in regions with significant forest activities. Wood pellets, derived from compressed sawdust, are commonly used in biomass power plants and residential heating (Demirbas 2009). Livestock farming generates considerable amounts of manure, which can be processed into biogas through anaerobic digestion. This process not only provides a renewable source of energy but also helps to manage waste and reduce methane emissions, a potent greenhouse gas (Holm-Nielsen et al. 2009). Certain industries, such as food processing and paper manufacturing, produce organic waste that can be used as energy. For example, sugar mills can utilize bagasse (the fibrous residue that remains after sugarcane processing) as a bioenergy feedstock (Pandey et al. 2000). Dedicated energy crops, including switch grass, miscanthus, and willow, are specifically cultivated for biomass production. These crops are selected for their high yields and low input requirements, making them efficient and sustainable sources of bioenergy (Lewandowski et al. 2003). In addition, energy crops can be grown on marginal land, preventing them from competing with food production (Heaton et al. 2008).

6.2.2 Technologies for Biomass Conversion

Biomass can be transformed into usable energy through several methods, each suited to different types of biomass source. The most common method is direct combustion, which produces heat and electricity and is used in both small-scale residential heating and large-scale power plants (McKendry 2002). Biomass can be utilized to generate energy through several methods: high temperature partial oxidation produces syngas, a mixture of hydrogen and carbon monoxide, which can be used to generate electricity or as a raw material for the manufacture of chemicals and fuels (Bridgwater 2003). Anaerobic digestion is a biological process that converts organic matter into biogas, primarily methane and carbon dioxide, suitable for heating, power generation, or vehicle fuel (Holm-Nielsen et al. 2009). Pyrolysis involves heating biomass at high temperatures without oxygen, resulting in bio-oil, syngas, and biochar. Bio-oil can be refined into transportation fuels, and biochar can enhance soil quality (Bridgwater 2003). Biomass energy presents a multifaceted solution to energy generation, offering environmental, economic, and social benefits. It aids in waste control and minimizes greenhouse gas emissions. It provides a renewable alternative to fossil fuels, making it a key component of sustainable energy strategies (Sims et al. 2010).

6.3 Solar Energy

6.3.1 Advances in Solar Technology

Concentrated solar power (CSP) systems represent an alternative approach to harnessing solar energy. These systems employ mirrors or lenses to focus sunlight on a small area, generating high temperatures that produce steam to drive turbines and generate electricity (Mehos et al. 2016). A notable advantage of CSP systems is their ability to integrate thermal energy storage, allowing them to continue producing electricity even when sunlight is unavailable. This addresses the intermittency issue commonly associated with solar energy (Neises and Turchi 2019). The adoption of solar energy systems, including CSP, plays a crucial role in reducing greenhouse gas emissions. By replacing electricity generated from fossil fuels, it helps mitigate climate change and reduces air pollution (Hernandez et al. 2014). Furthermore, solar energy systems generally have a lower environmental impact compared to many other energy sources, as they do not emit harmful substances during operation (Tsoutsos et al. 2005).

6.3.2 Challenges and Benefits of Solar Energy

Solar energy offers substantial economic benefits. The cost of solar photovoltaic modules has declined significantly in the past decade, driven by technological progress and economies of scale. This reduction has placed solar energy as one of the most competitive forms of electricity generation in many regions, particularly where solar irradiance is high. In addition, the global solar industry has become a significant source of employment, contributing to economic growth and improving energy security (IRENA 2020). Despite its advantages, solar energy faces several challenges. One major hurdle is its intermittency, since solar power generation is dependent on weather conditions and daylight availability. Integrating solar energy into the power grid requires sophisticated grid management and efficient energy storage solutions to guarantee a reliable and stable electricity supply (Luo et al. 2015). Furthermore, the initial upfront costs for solar installations can be considerable, although these are often mitigated by long-term savings on energy expenditures and available financial incentives (Bazilian et al. 2013). Solar energy stands as a pivotal element in the transition towards sustainable energy systems. Its environmental benefits, economic viability, and technological advances make it an attractive alternative to traditional fossil fuels. Continued progress in solar technology and supportive policy frameworks will be crucial for maximizing its potential and addressing the challenges associated with its widespread adoption.

6.4 Wind Energy

6.4.1 Fundamentals of Wind Power

Wind power, derived from the kinetic energy of wind, is a significant renewable energy source that contributes to sustainable energy systems throughout the world. The core principle of wind power is to transform the kinetic energy of wind into mechanical energy using wind turbines, which then generate electricity. Wind turbines are made up of blades, a rotor, a nacelle that houses the generator and gearbox, and a tower. As the wind flows over the blades, it generates lift, causing the rotor to rotate. This rotational movement drives the generator, producing electricity (Manwell et al. 2010). The factor of wind energy capacity, which measures the actual output relative to its maximum potential output over time, varies depending on the wind speed and consistency at the installation site. Ideal locations for wind farms include coastal areas, open plains, and hilltops, where wind speeds are higher and more reliable. Offshore wind farms, located in bodies of water, leverage stronger and more consistent winds compared to offshore sites, leading to increased capacity factors (Dai et al. 2015).

6.4.2 Environmental and Economic Aspects of Wind Energy

From an environmental perspective, wind power offers substantial advantages. It operates without emitting greenhouse gases during its lifecycle, playing a crucial role in combating climate change and reducing air pollution (Saidur et al. 2011; Yadav et al. 2022). Furthermore, wind turbines have a relatively modest land footprint, which allows co-existence with agricultural activities. However, there are environmental considerations associated with wind power. Turbines can affect local wildlife, especially birds and bats, which might collide with spinning blades. To mitigate these concerns, mitigation strategies, such as careful site selection and advances like ultrasonic deterrents, are being developed (Erickson et al. 2014). Technological advancements and economies of scale have made wind energy increasingly competitive with traditional fossil fuels. According to the International Renewable Energy Agency (IRENA), the levelized electricity cost (LCOE) for onshore wind dropped by 39% from 2010 to 2019, making it one of the most cost-effective options for new electricity generation (IRENA 2020). Offshore wind power, while currently more costly, is benefiting from cost reductions due to advances in technology and the deployment of larger turbine sizes. Wind energy also delivers economic advantages through job creation and local economic growth. The wind industry supports a diverse array of jobs, covering manufacturing, installation, maintenance, and operations. In regions where wind farms are established, they can generate additional income for landowners and boost local tax revenues, which can subsequently fund community services and infrastructure improvements (Slattery et al. 2011). Despite its merits, wind power generation faces several challenges. One primary concern is its intermittency; wind patterns are variable, and availability can fluctuate significantly at different times and locations. Managing this variability requires robust grid management and the integration of complementary energy sources or storage solutions to ensure a reliable power supply (Denholm et al. 2010). Additionally, upfront capital investments for wind turbines and associated infrastructure can be substantial, although these expenses are balanced by low operational costs and the absence of on-going fuel expenses.

6.5 Integrating Biomass, Solar, and Wind Power

6.5.1 The Rationale for Hybrid Renewable Energy Systems

Integrated Renewable Energy Systems (IRES) offer a strategic approach to improve the reliability, economic viability, and environmental sustainability of energy production. The concept involves integrating multiple renewable energy sources to manage the intermittent nature of individual sources, such as solar and wind power (Jacobson and Delucchi 2011). By combining various renewables like biomass, wind, and solar energy, IRES aim to ensure a more consistent and reliable energy supply, thereby reducing dependence on fossil fuels and improving energy security (Mathiesen et al. 2015). A key goal of IRES is to combat climate change by significantly reducing greenhouse gas emissions associated with conventional fossil fuel-based electricity generation (IEA 2019). This integration enables the optimization of energy production based on the availability of renewable resources, minimizing environmental impact while maximizing energy efficiency (IRENA 2020). Furthermore, IRESs offer economic benefits by reducing overall electricity generation costs through efficient use of renewable resources and improved energy storage capabilities (Denholm and Hand 2011). These systems also foster local economic growth by stimulating employment in the renewable energy sector and reducing dependence on imported energy (Fig. 6.1, Table 6.1) (IRENA 2020).

6.5.2 Case Studies of Hybrid Systems

Several successful case studies illustrate the effective implementation of integrated renewable energy systems (HRESs) worldwide. Denmark, for instance, has achieved significant integration of wind power into its energy portfolio, with wind energy

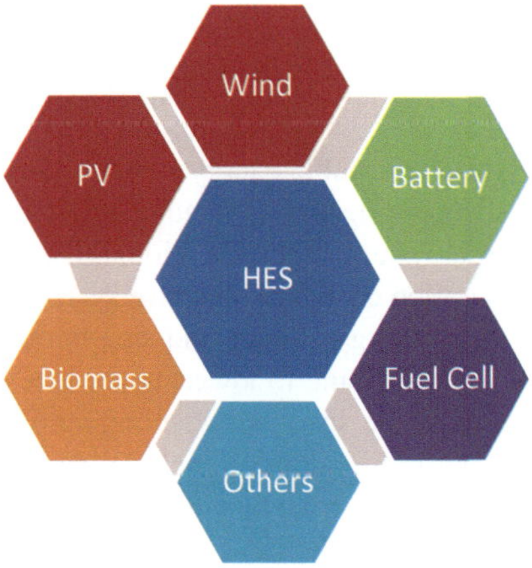

Fig. 6.1 Schematic diagram of the hybrid renewable energy system

Table 6.1 Key components of HRES and their importance in achieving sustainable energy solutions

Component	Description
Renewable sources	Integration of biomass, solar, wind, and potentially other renewable
Energy storage	Batteries, pumped hydro, or other storage technologies
Smart grid solutions	Advanced grid management systems to optimize renewable energy integration
Policy and regulations	Supportive frameworks for renewable energy deployment and investment
Cost reduction	Ongoing efforts to decrease the costs of renewable technologies
Resilience to climate change	Efforts to educate and engage local communities in renewable energy initiatives
Benefits	Enhanced energy security, reduced carbon emissions, economic viability

contributing substantially to electricity generation (Lund et al. 2015). The achievements in this area can be credited to Denmark's proactive renewable energy policies and effective grid management strategies. Similarly, Germany's Energiewende initiative serves as a model for a comprehensive approach to moving towards renewable energy sources such as solar photovoltaics (PV), wind power, and biomass. Germany's dedication to renewable energy has been reinforced by policy frameworks and incentives aimed at expediting the adoption of sustainable energy technologies (IRENA 2020). In the United States, California has implemented ambitious renewable energy policies through the California Renewable Portfolio Standard (RPS), which requires increasing percentages of electricity sourced from renewable sources. California has successfully integrated solar photovoltaic energy, wind energy, and other renewable energy into its energy grid, showcasing the practicality of scaling up renewable energy deployment (California Energy Commission n.d.).

6.6 Technical and Economic Analysis

6.6.1 Economic Viability and Market Trends

Hybrid renewable energy systems (HRESs) offer economic viability by optimizing resource utilization and reducing reliance on fossil fuels (Denholm and Hand 2011). These systems combine biomass, wind, and solar energy to minimize electricity generation costs and improve system efficiency through technological innovations such as energy storage and smart grids (Lund et al. 2015). Technological advancements contribute to lowering the levelized cost of electricity (LCOE) and creating new revenue streams by selling surplus electricity back to the grid. Market trends driving IRES adoption include increasing global investments in renewable energy projects, supported by policies and regulations that promote clean energy transitions (IEA 2019; IRENA 2020). The competitive landscape is shifting towards renewables, with increasing market competition and a growing market share for sustainable energy solutions. Economic viability and evolving market dynamics are accelerating the deployment of integrated renewable energy systems, positioning them as pivotal solutions to achieve sustainable energy goals globally.

6.6.2 Policy and Regulatory Considerations

Policy support is crucial for the widespread adoption of HRES. Governments and regulatory bodies play a pivotal role in providing incentives, subsidies, and supportive frameworks that encourage investments in integrated renewable energy projects (IEA 2019). Examples from countries such as Denmark and Germany demonstrate successful implementations of integrated renewable energy systems (IRES), underscoring the practicality and advantages of these integrated approaches (Mathiesen et al. 2015). In summary, the justification for hybrid renewable energy systems (HRESs) lies in their ability to deliver reliable, economically viable, and environmentally sustainable energy provision. By capitalizing on the synergies among biomass, wind, and solar energy, HRESs present a path to realizing global energy transition objectives, tackling climate change, and fortifying energy security.

6.7 Future Directions

Moving forward, the next steps for hybrid renewable energy systems (HRESs) involve the advancement of energy storage technologies and smart grid solutions to improve system flexibility and efficiency. The policy and regulatory frameworks should be strengthened to support the integration of renewable energy, focusing on long-term goals and financial incentives. Continued cost reductions in renewable technologies are essential to improve the economic viability of HRES. Enhancing resilience to climate change and engaging communities through education and awareness initiatives will also be crucial in promoting widespread adoption of HRES. Collectively, these efforts aim to accelerate the shift towards sustainable energy systems throughout the world.

6.8 Conclusion

Hybrid renewable energy systems (HRESs) represent a critical path towards achieving sustainable energy solutions globally. By integrating various renewable sources such as biomass, wind, and solar power, HRES offers enhanced reliability, economic viability, and environmental sustainability. Furthermore, fostering resilience to the impacts of climate change and engaging communities through education and awareness initiatives are crucial to maximize the benefits of HRES. By tackling these challenges and securing opportunities, integrated renewable energy systems (IRES) can significantly accelerate the transition to a low-carbon future and mitigate global climate change. IRES offers a promising approach to meet increasing energy needs while simultaneously lowering carbon emissions and bolstering energy security. On-going innovation and cross-sector collaboration will be essential to harness all the capabilities of IRES and to secure a sustainable energy future for future generations.

Acknowledgement The authors thank Save Vibrant Earth Foundation experts for proofreading and editing the chapter.

References

Banos R, Manzano-Agugliaro F, Montoya FG, Gil C, Alcayde A, Gómez J (2011) Optimization methods applied to renewable and sustainable energy: a review. Renew Sust Energ Rev 15(4):1753–1766

Bazilian M, Onyeji I, Liebreich M, MacGill I, Chase J, Shah J, Gielen D, Arent D, Landfear D, Zhengrong S (2013) Re-considering the economics of photovoltaic power. Renew Energy 53:329–338. https://doi.org/10.1016/j.renene.2012.11.029

Bergmann A, Colombo S, Hanley N (2008) Rural versus urban preferences for renewable energy developments. Ecol Econ 65(3):616–625

Bridgwater AV (2003) Renewable fuels and chemicals by thermal processing of biomass. Chemical engineering journal, 91(2-3), 87-102

California Energy Commission (n.d.) Renewable energy. https://www.energy.ca.gov/data-reports/renewable-energy

Dai K, Bergot A, Liang C, Xiang WN, Huang Z (2015) Environmental issues associated with wind energy–a review. Renew Energy 75:911–921. https://doi.org/10.1016/j.renene.2014.10.074

Demirbas A (2009) Biorefineries: current activities and future developments. Energy Convers Manag 50(11):2782–2801. https://doi.org/10.1016/j.enconman.2009.06.035

Denholm P, Ela E, Kirby B, & Milligan M (2010) Role of energy storage with renewable electricity generation (No. NREL/TP-6A2-47187). National Renewable Energy Lab.(NREL), Golden, CO (United States)

Denholm P, Hand M (2011) Grid flexibility and storage required to achieve very high penetration of variable renewable electricity. Energy Policy 39(3):1817–1830. https://doi.org/10.1016/j.enpol.2011.01.019

Erickson WP, Wolfe MM, Bay KJ, Johnson DH, Gehring JL (2014) A comprehensive analysis of small-passerine fatalities from collision with turbines at wind energy facilities. PLoS One 9(9):107491. https://doi.org/10.1371/journal.pone.0107491

Haines A, Kovats RS, Campbell-Lendrum D, Corvalán C (2006) Climate change and human health: impacts, vulnerability and public health. Public Health 120(7):585–596

Heaton EA, Dohleman FG, Long SP (2008) Meeting US biofuel goals with less land: the potential of Miscanthus. Glob Chang Biol 14(9):2000–2014. https://doi.org/10.1111/j.1365-2486.2008.01662.x

Hernandez RR, Easter SB, Murphy-Mariscal ML, Maestre FT, Tavassoli M, Allen EB, Barrows CW, Belnap J, Ochoa-Hueso R, Ravi S, Allen MF (2014) Environmental impacts of utility-scale solar energy. Renew Sust Energ Rev 29:766–779. https://doi.org/10.1016/j.rser.2013.08.041

Holm-Nielsen JB, Al Seadi T, Oleskowicz-Popiel P (2009) The future of anaerobic digestion and biogas utilization. Bioresour Technol 100(22):5478–5484. https://doi.org/10.1016/j.biortech.2008.12.046

IEA (2019) World energy outlook 2019. International Energy Agency

IRENA (2020) Renewable energy and jobs – annual review 2020. International Renewable Energy Agency

Jacobson MZ, Delucchi MA (2011) Providing all global energy with wind, water, and solar power, part I: technologies, energy resources, quantities and areas of infrastructure, and materials. Energy Policy 39(3):1154–1169. https://doi.org/10.1016/j.enpol.2010.11.040

Kardooni R, Yusoff SB, Kari FB (2016) Renewable energy technology acceptance in Peninsular Malaysia. Energy Policy 88:1–10

Lafforgue G, Magné B, Moreaux M (2008) Energy substitutions, climate change and carbon sinks. Ecol Econ 67(4):589–597

Lewandowski I, Scurlock JM, Lindvall E, Christou M (2003) The development and current status of perennial rhizomatous grasses as energy crops in the US and Europe. Biomass Bioenergy 25(4):335–361. https://doi.org/10.1016/S0961-9534(03)00030-8

Lund PD, Lindgren J, Mikkola J, & Salpakari J (2015) Review of energy system flexibility measures to enable high levels of variable renewable electricity. Renewable and sustainable energy reviews, 45, 785-807

Luo X, Wang J, Dooner M, Clarke J (2015) Overview of current development in electrical energy storage technologies and the application potential in power system operation. Appl Energy 137:511–536. https://doi.org/10.1016/j.apenergy.2014.09.081

Manwell JF, McGowan JG, Rogers AL (2010) Wind energy explained: theory, design and application. John Wiley & Sons

Mathiesen BV, Lund H, Connolly D, Wenzel H, Østergaard PA, Möller B, Nielsen S, Ridjan I, Karnøe P, Sperling K, Hvelplund FK (2015) Smart energy systems for coherent 100% renewable energy and transport solutions. Appl Energy 145:139–154. https://doi.org/10.1016/j.apenergy.2015.01.075

McKendry P (2002) Energy production from biomass (part 1): overview of biomass. Bioresour Technol 83(1):37–46. https://doi.org/10.1016/S0960-8524(01)00118-3

Mehos M, Turchi C, Jorgenson J, Denholm P, Ho C, Armijo K (2016) On the path to SunShot-advancing concentrating solar power technology, performance, and dispatchability (No. NREL/TP-5500-65688 SAND-2016-2237 R). EERE Publication and Product Library, Washington, DC

Neises T & Turchi C (2019) Supercritical carbon dioxide power cycle design and configuration optimization to minimize levelized cost of energy of molten salt power towers operating at 650 C. Solar Energy, 181, 27-36

Pandey A, Soccol CR, Nigam P, Soccol VT, Vandenberghe LP, Mohan R (2000) Biotechnological potential of agro-industrial residues. II: cassava bagasse. Bioresour Technol 74(1):81–87. https://doi.org/10.1016/S0960-8524(99)00142-X

Qazi A, Hussain F, Rahim NA, Hardaker G, Alghazzawi D, Shaban K, Haruna K (2019) Towards sustainable energy: a systematic review of renewable energy sources, technologies, and public opinions. IEEE Access 7:63837–63851

Rizzi F, van Eck NJ, Frey M (2014) The production of scientific knowledge on renewable energies: worldwide trends, dynamics and challenges and implications for management. Renew Energy 62:657–671

Saidur R, Rahim NA, Islam MR, Solangi KH (2011) Environmental impact of wind energy. Renew Sust Energ Rev 15(5):2423–2430. https://doi.org/10.1016/j.rser.2011.02.024

Santika WG, Anisuzzaman M, Bahri PA, Shafiullah GM, Rupf GV, Urmee T (2019) From goals to joules: a quantitative approach of interlinkages between energy and the sustainable development goals. Energy Res Soc Sci 50:201–214

Sims RE, Mabee W, Saddler JN, Taylor M (2010) An overview of second generation biofuel technologies. Bioresour Technol 101(6):1570–1580. https://doi.org/10.1016/j.biortech.2009.11.046

Slattery MC, Lantz E, Johnson BL (2011) State and local economic impacts from wind energy projects: Texas case study. Energy Policy 39(12):7930–7940. https://doi.org/10.1016/j.enpol.2011.09.047

Sokhansanj S, Kumar A, Turhollow AF (2006) Development and implementation of integrated biomass supply analysis and logistics model (IBSAL). Biomass Bioenergy 30(10):838–847. https://doi.org/10.1016/j.biombioe.2006.04.001

Thirugnanasambandam M, Iniyan S, Goic R (2010) A review of solar thermal technologies. Renew Sust Energ Rev 14(1):312–322

Tsoutsos T, Frantzeskaki N, Gekas V (2005) Environmental impacts from the solar energy technologies. Energy Policy 33(3):289–296. https://doi.org/10.1016/S0301-4215(03)00241-6

Yadav AK, Shirin S, Emmanouil C, Jamal A (2022) Effect of seasonal and meteorological variability of air pollution in Singrauli coalfield. Aerosol Sci Eng 6(1):61–70. https://doi.org/10.1007/s41810-021-00124-3

Chapter 7
Unlocking the Potential of Biomass and Biogas for a Sustainable Planet

Pranali Bharatkumar Dahihande and Pritam Das

Abstract The rise of the worldwide economy has resulted in a considerable increase in the demand for energy globally in recent years. This heightened demand has led to unforeseen price spikes and a surge in greenhouse gas emissions, exacerbating concerns about climate change and environmental degradation. Recognizing the finite nature of fossil fuels and their adverse environmental impact, there has been a growing impetus to explore alternative energy sources. A promising alternative is biomass, which encompasses organic matter derived from animals, plants, and various waste products. Biomass waste can be transformed into bioenergy, a sustainable energy source with several benefits, using cutting-edge technologies. Since the carbon absorbed during plant growth balances out the carbon released during combustion, the generation of biomass bioenergy has no discernible impact on CO_2 emissions. Biofuels, biomass, and biogas represent examples of renewable energy sources that are currently being transformed. Examples of biomass that can be converted into biogas include animal excrement and agricultural waste that has undergone controlled anaerobic decomposition. The anaerobic decomposition of agricultural waste and animal manure are two examples of biomass that can be transformed into biogas. After that, this biogas can be processed further to produce biomethane for use in industry and used to generate power as well as heat. Harnessing the potential of biomass and biogas as renewable resources presents a significant opportunity to mitigate greenhouse gas emissions and foster sustainable energy practices. We may reduce the negative environmental impacts of energy production, guarantee energy security, and promote economic growth by shifting away from fossil fuels and towards alternative energy sources. This transition towards a more

P. B. Dahihande
Department of Processing and Food Engineering, University of Agricultural Sciences, Raichur, Karnataka, India

P. Das (✉)
Department of Soil and Water Engineering, College of Technology and Engineering, MPUAT, Udaipur, Rajasthan, India

A. K. Yadav (ed.), *Powering Future: Clean Energy Solutions for a Sustainable Planet*, Clean Energy Production Technologies, https://doi.org/10.1007/978-981-96-5055-2_7

sustainable energy future underscores the importance of innovation, collaboration, and proactive environmental stewardship.

Keywords Fossil fuel · Renewable energy sources · Biogas · Biomass · Sustainability · Climate change

7.1 Introduction

The expansion of the global economy has led to a higher demand for energy in recent years. The unsustainable use of fossil fuels has triggered unpredictable price spikes and exacerbated greenhouse gas emissions. These problems, along with the rapid depletion of fossil fuels, have driven attention towards alternative, cost-effective, and sustainable energy sources (Moshood et al. 2021). The energy derived from fossil fuels is a significant contributor to environmental issues such as global warming and climate change. The growing consumption of fuel, global warming, diminishing fossil fuel reserves, and rising fuel prices are propelling the search for alternative energy solutions. Biomass is defined as "organic matter of plant or animal origin, produced either directly or indirectly through the process of photosynthesis involving chlorophyll, either spontaneously occurring or cultivated by humans, terrestrial and marine". Over the past several decades, numerous methods have been developed to convert biomass into liquid, solid, and gaseous fuels. The amount of biofuel produced is influenced by both the composition of the biomass feedstock and the conditions of the conversion process, including the size of the biomass particles, the temperature it reaches, how long the reaction takes, and how quickly it is heated. Sustainability is a crucial consideration in the commercialization of biofuels. Earth boasts a vast supply of naturally replenishing carbon sources, with biomass being a prime example. Primary biomass sources for bioenergy production include wood processing residues, forest residues, and agricultural crop residues. Lignocellulose has been shown to be an excellent option for downstream processing of biomass products. Lignocellulose forms an intricate network of hemicellulose and lignin, encasing the cellulose microfibrils (Van Fan et al. 2021; Ma et al. 2019).

The growing need for electricity and heat from clean energy sources presents a significant opportunity for the production of agricultural biogas to thrive (Table 7.1). Furthermore, readily available feedstocks, or substrates, present attractive opportunities for investors, particularly farmers (Ikora and Tomal 2016). Waste from the agri-food sector, as well as substrates derived from plants or animals, can be used as input for agricultural power plants (Piwowar et al. 2016).

Table 7.1 Types of biomass

Types of biomass	Examples
Agricultural residues	Straws, stalks, leaves, husks, shells, etc.
Livestock waste	Solid manure, liquid manure
Energy crops	Switch grass, miscanthus, etc.
Forestry residues	Sawdust, wood chips, and tree branches, etc.
Industrial waste	Industrial waste wood from sawmills, wastes from food processing industry, etc.

7.2 Composition of Lignocellulosic Biomass

Plants are predominantly composed of a trio of biopolymers: hemicellulose, cellulose, and lignin. Agricultural residues, rich in these lignocellulosic components, typically range from 10% to 25% lignin, 20% to 30% hemicellulose, and 40% to 50% cellulose (Kumar et al. 2009). Notably, cellulose serves as the structural backbone of plant cell walls, offering rigidity. Hemicellulose, on the other hand, consists of repeating sugar units like pentoses and hexoses. Lignin, a complex polymer distinct from the others, is formed through the biosynthesis of three specific aromatic alcohols such as coniferyl alcohol, sinapyl alcohol, and p-coumaryl alcohol (Menon and Rao 2012). These alcohols act as a barrier, separating hemicellulose and cellulose within the plant cell wall.

Cellulose In plant biomass, intermolecular hydrogen bonds hold about 40–50% of the cellulose molecules together in their natural condition. Conversely, the inherent tendency of these molecules to form extensive hydrogen bonds within and between themselves (intra- and intermolecular) renders cellulose remarkably rigid and resistant to most organic solvents. Interestingly, cellulose naturally assembles into bundles that further combine to form microfibrils. These microfibrils exhibit a unique architecture, containing both crystalline and amorphous regions (Taherzadeh and Karimi 2008).

Hemicellulose The primary constituents of hemicellulose, which is second most common heterogeneous polymer, are glucuronoxylan, glucomannan, and amounts of other polysaccharides. Mannan is a part of the hemicellulose found in both hardwood and softwood, while arabinan, galactan, and xylan are found in straws and grasses. Though xylans and mannans are the most common, they are categorized based on the three main components of sugar such as glucans, mannans, and xylans (Wyman et al. 2005).

Lignin The third major component of lignocellulosic biomass, lignin, boasts a complex polymeric structure. Lignin strengthens and supports certain tissues in vascular plants and even some algae (Fig. 7.1). It is especially important in the cell walls of woody plants, making them strong and rigid structures like wood and bark. Chemically, lignin is characterized by its cross-linked phenolic polymers, which

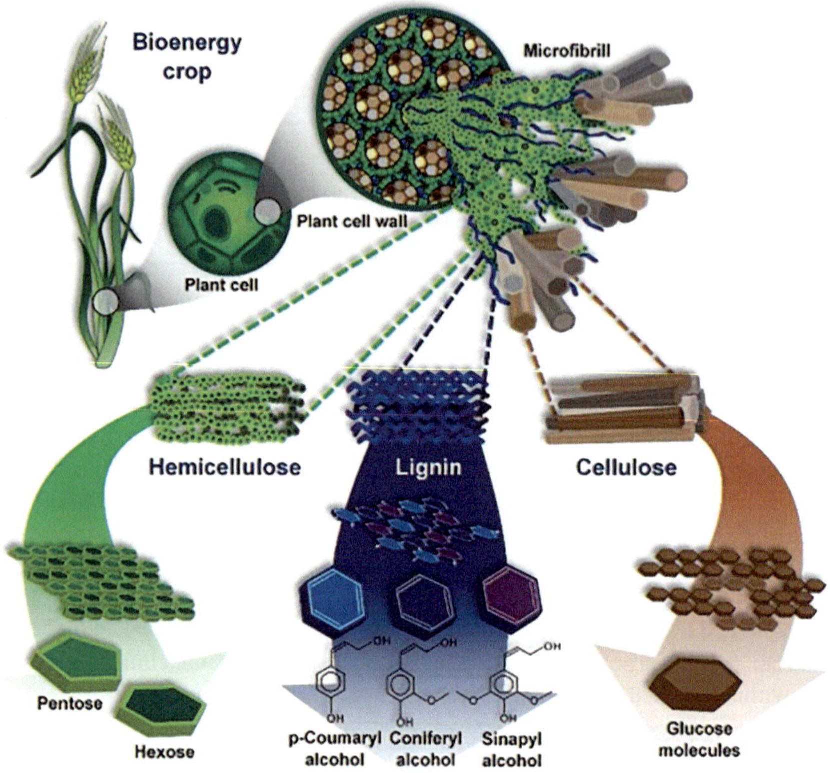

Fig. 7.1 Lignocellulosic biomass structure [reprinted from Hernández-Beltrán et al. (2019) with permission from Applied Sciences. Copyright 2019, MDPI {CC BY}]

impart a rigid and inflexible nature. Unlike cellulose, lignin exhibits a predominantly non-crystalline, amorphous structure. Lignin is a branching long chain polymer composed of three different types of monomers, namely 4-propenylphenol three-dimensional polymers, 4-propenyl-2-methoxy phenol, and 4-propenyl-2.5-dimethoxy phenol (Kikas et al. 2016).

7.3 Biomass Conversion Technologies

7.3.1 Anaerobic Conversion

Anaerobic digestion directly converts organic waste into biogas, primarily composed of CO_2 and CH_4, and small amounts of other gases like hydrogen sulphide (Deublein and Steinhauser 2011). Various biodegradable feedstocks, including food waste, animal manure, agricultural waste, industrial wastewater, and the organic portion of household trash, have been utilized in commercial biogas production. These kinds of systems demonstrate the special potential for simultaneous production of bioenergy and trash stabilization and bioremediation. Lignocellulosic biomass, more especially energy crops and agri-residues, has attracted a lot of attention lately as potential feedstocks for the production of bio-based products and bioenergy. In anaerobic digestion, microorganisms transform the feedstock, sometimes referred to as substrate, without the presence of oxygen. Similar to composting, this conversion method uses oxygen to break down organic matter aerobically while also producing stable, highly valuable compounds for commercial use. It is a widely used, commercially proven method for the treatment of organic wastes with high moisture content, meaning they have between 80% and 90% moisture content. Energy crops (like maize silage) that are grown expressly for feeding the anaerobic digestion plant or organic wastes (like animal manures) are examples of feedstocks for anaerobic digestion. When there are no air present, anaerobic digesters create the ideal environment for bacteria to naturally break down organic materials.

7.3.2 *Process of Anaerobic Digestion*

Hydrolysis Phase The hydrolase enzymes of facultative and obligatory anaerobic bacteria hydrolyse cellulose as well as other carbohydrates, proteins, and lipids into monomers. It only takes a few hours to hydrolyse simple carbs. Lipids and proteins may require several days to hydrolyse.

Acidogenic Phase Numerous anaerobic bacteria, both facultative and obligatory, utilize the monomers generated by hydrolysis processes. Further breakdown of these monomers results in the production of short-chain organic acids, alcohols, CO_2, hydrogen, and C_1–C_5 molecules (such as butanoic acid ($CH_3CH_2CH_2CO_2H$), propionic acid ($CH_3CH_2CO_2H$), acetate (CH_3COO), and acetic acid (CH_3COOH)). Fermentation products are impacted by intermediate hydrogen. If there is enough hydrogen partial pressure, intermediate fermentation products will form.

Acetogenic Phase Bacteria in the acetogenic phase utilize the products generated by the preceding acidogenic phase as their primary substrates. Exergonic H_2 and CO_2 are used by homoacetogenic bacteria in the acetogenic phase to produce acetic acid. Acetogenic and methanogenic bacteria develop simultaneously. Acetate is produced from alcohols and short-chain organic acids. CO_2 is utilized in the process of converting ethanol to acetate, which also yields hydrogen.

Methanogenic Phase Anaerobic conditions are strictly enforced during the methane generation process. Certain substrates are not degraded by all methanogenic bacteria.

7.3.3 Fermentation

Fermentation, an anaerobic biological process, which is the conversion of the simple sugars in biomass feedstock into carbon dioxide and alcohol by a particular class of microorganisms, usually yeasts. After bringing the mixture to a boil, the ethanol (C_2H_5OH) can be removed by cooling it down and allowing it to condense back into a liquid condition. In the industrial production of ethanol, fermentation plays a key role for many countries. This process utilizes starch and sugar-rich crops, such as corn, wheat, sugarcane, and sugar beets. During the breakdown of biomass, enzymes convert starch into simpler sugars. Yeast then utilizes these sugars to generate cellular energy through a process called fermentation. This process also produces ethanol as a by-product.

Sugarcane processing yields bagasse, a versatile resource suitable for direct combustion in boilers or conversion into fuel for gasification processes. The solid fermentation by-products may also be fed to cattle (McKendry 2002). Fuel with a small quantity of ethanol added is known as "bioethanol". The fermentation process used to make ethanol from sugar cane is arguably the most well-known example of fuel fermentation. Fermentation is another way in which a biorefinery might produce a variety of products that can take the place of the hydrocarbons produced in conventional refineries from fossil fuels. Extensive research is being conducted to develop advanced technologies that facilitate the fermentation of lignocellulosic feedstocks through pre-treatment and/or enzyme identification.

Composting and ensiling are two examples of natural microbial processes that can be replicated by solid-state fermentation (SSF). Industrial ethanol production from lignocellulosic biomass necessitates yeast strains with superior ethanol yields and fermentation efficiency to fully utilize the available sugars. Despite *Saccharomyces cerevisiae*'s inability to ferment pentoses by nature, years of intensive research have successfully increased this organism's capacity to use xylose. Over the past 15 years, the majority of studies have gone towards identifying these xylose-fermenting microbes and comprehending the metabolism of xylose while the metabolism of arabinose has received less attention. The efficiency and cost of producing ethanol can be significantly increased by using genetically modified microbes which helps to convert xylose/pentose to ethanol.

The fast growth rates of certain ethanol-producing bacteria have recently sparked renewed interest in their potential for fermentation processes. Furthermore, given the current advancements in biotechnology, these bacteria could be crucial in increasing the efficiency of ethanol production. Compared to other microbes, *S. cerevisiae* has a more robust anaerobic sugar metabolism, can withstand inhibiting industrial substrates more effectively, and can ferment hexoses like glucose, mannose, and galactose which are frequently found in lignocellulosic hydrolysates with high yield and productivity.

7.3.4 *Pyrolysis*

Pyrolysis is thermochemical conversion process that produces pyro-gas, bio-oil, and biochar from biomass when oxygen is not present. Two Greek words "pyro" (fire) and "lysis" (disintegration) create the word "pyrolysis". Pyrolysis is a process that uses heat and mass transfer to break down biomass thermally. The final products are condensable vapours known as bio-oil, solid charcoal known as biochar, and gaseous products known as pyro-gas.

Biomass pyrolysis generally takes place in three stages, firstly, free moisture evaporation; secondly, primary decomposition; and lastly, secondary reactions.

Dehydration Dehydration is the process in which water vapour, or free moisture, is extracted from biomass. It begins at 100 °C and leaves the char's amorphous carbon behind.

Primary Decomposition In primary decomposition, different chemical bonds in biomass are broken during pyrolysis, due to which volatiles are released and rearrangement reactions occur. The three main processes are depolymerization, fragmentation, and formation of char.

Char Formation Thermochemical conversion of biomass, often involving the formation of stable polycyclic aromatic hydrocarbons (PAHs) through the rearrangement of aromatic structures, can generate solid residues as a by-product. This conversion process also releases non-condensable gases as emissions.

Depolymerization In depolymerization process, volatiles are produced when polymers break down into their individual monomers caused lowering the number of monomer units per polymer chain. It is common practice to recover these volatiles as liquid fraction. Reactions for depolymerization take place in the 250–500 °C temperature range.

Fragmentation Fragmentation is the process in which breaking down of bonds between the monomer units of polymers is caused, and it results in the conversion of linear compounds and non-condensable gases. Such ring/bond breakage typically takes place at temperatures higher than 600 °C (Fig. 7.2).

Liquid bio-oil is produced by quickly freezing the intermediate products resulting from the flash decomposition of cellulose, hemicellulose, and lignin. As a result, the liquid has a large number of reactive species, which add to its peculiar characteristics. Usually free-flowing and dark brown in color, pyrolysis oil resembles biomass in terms of its constituent makeup. It is made up of an extremely complicated mixture of oxygenated hydrocarbons and a sizable amount of water from the reaction product and initial moisture. There could be some substantial char as well. Bio-oil is entirely immiscible with petroleum-derived fuels but can mix with polar solvents such as methanol and acetone. In general, by-products like gas and char contain an energy content of 5% and 25%, respectively, of the raw material. The pyrolysis process consumes approximately 15% of the energy from the feedstock; only the char produced contains sufficient energy to sustain this heat requirement.

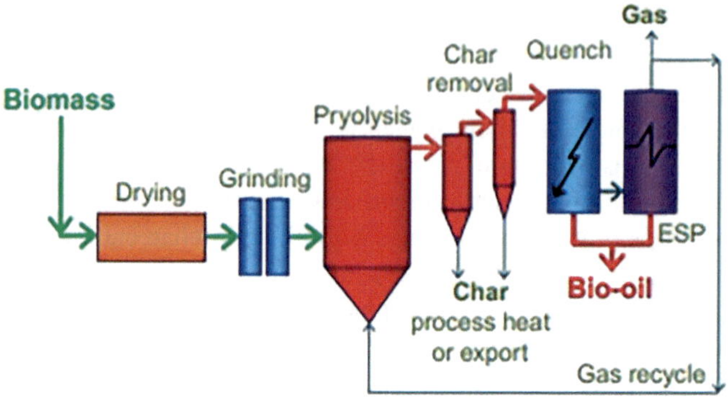

Fig. 7.2 Pyrolysis process [reprinted from Adams et al. (2018) with permission from Greenhouse Gas Balances of Bioenergy Systems. Copyright © 2018 Elsevier Inc.]

Ash from biomass is concentrated mainly in the char, necessitating careful management during combustion. Larger systems can potentially utilize waste heat from char combustion, as well as surplus or by-product gases, for power generation or export purposes.

7.3.5 *Gasification*

During biomass gasification, carbonaceous material is converted into a synthesis gas or gaseous product that is largely consists of carbon monoxide (CO) and hydrogen (H_2), along with minor amounts of nitrogen (N_2), water vapour (H_2O), carbon dioxide (CO_2), methane (CH_4), and other heavier hydrocarbons. The initial stage of the gasification process involves devolatilizing biomass particles, resulting in the formation of char and vapours. Subsequently, the volatiles separate and reorganize, enabling the char to gasify. The char gasification and reforming reactions can create more light gases, like hydrogen, carbon dioxide, and carbon monoxide, when they are supported by a steam environment (Luo et al. 2009; Shirin and Yadav 2014). The resulting gasoline blends serve as intermediary energy sources that can be burned to generate heat and power or further processed into car fuel. Moreover, high-value chemicals like formaldehyde and olefins can be produced using the generated gas as a feedstock (Srirangan et al. 2012). Several factors influence the production and composition of syngas generated in a gasification process. Several aspects affect the outcome of biomass gasification: the type of biomass used (feedstock), the gasification technology employed (reactor), the settings used during the process (operating conditions), and the treatments applied to the resulting gas (downstream conditioning). The starting materials fed into the system include the biomass itself (feedstock) and the gas used to facilitate the conversion (gasifying agent or medium). Gas output conditioning comprises two primary elements: the removal of tar and particle matter from the gas product and cooling (Wongsiriamnuay et al. 2013; Erakhrumen 2012). The gasification reactions are regulated by the working parameters, which include pressure, temperature, and residence time (Wongsiriamnuay et al. 2013).

The Gasification Process Three separate transitions occur in biomass throughout the gasification process. First, biomass is dried out before it reaches a temperature higher than 200 °C, which starts the pyrolysis process and produces char and vapors. The char and vapors are partially oxidized by the presence of oxygen, producing a gas that includes water, CO, H_2, CO_2, tars, ammonia, sulphur, and other contaminants.

Moving beds, fluidized beds (which bubble and circulate), fixed beds (which have an updraft and a downdraft), and entrained flow gasifiers are among the several types of gasifiers. The classification of gasifier technologies hinges on three key aspects: biomass containment within the reactor vessel, the relative flow directions of oxidant and biomass, and the chosen method for reactor heating.

Fixed-bed gasifiers operate with a stationary biomass bed. The reaction zone, termed the reaction front, can either propagate through this bed or be physically moved within the reactor throughout the conversion process. Fixed-bed downdraft, known as a concurrent gasifier, and fixed-bed updraft, known as a countercurrent gasifier, are the two common configurations for fixed-bed gasifiers. Fixed-bed updraft gasifier configurations are the oldest and most fundamental types of gasifiers. Because of the simple reactor principle, these configurations are noted for being quite inexpensive. Biomass gasifiers using fluidized beds can be categorized according to their unique mechanisms of heat transfer and how the gas and solid materials behave within the reactor (fluid dynamics). Two main types are bubbling fluidized beds and circulating fluidized beds. Bubbling beds contain a bed of inert material like sand or alumina. The gasification agent flows through this bed, encountering a balance between the upward force of the gas and the weight of the solid particles. This creates a bubbling motion within the bed. Circulating fluidized beds operate differently. These gasifiers operate at gas speeds greater than what's needed to just barely suspend the particles (minimum fluidization velocity). This higher velocity forces the biomass and inert materials to be continuously mixed and carried around within the gas stream. These entrained particles can be separated from the product gas using cyclones located at the top of the reactor and then reintroduced into the bed at the bottom, creating a continuous circulation loop. Circulating fluidized bed gasifiers generally offer higher conversion rates and efficiency compared to bubbling fluidized beds (Erakhrumen 2012). Biomass gasification shows significant promise in the waste processing industry compared to other current methods. Biomass gasification involves a sequential thermochemical process. The initial step focuses on drying the feedstock to remove moisture. Subsequently, pyrolysis decomposes the biomass in an oxygen-limited environment, generating intermediate products. Partial combustion of these intermediates then occurs, followed by the final gasification stage. During gasification, a gasifying medium (such as carbon dioxide, oxygen, steam, or air) further converts the products into a combustible gas mixture (Yadav et al. 2022a, 2022b). This process effectively upgrades the energy content of the biomass by concentrating the hydrogen content relative to the carbon content, resulting in a product with a higher calorific value. This process yields various valuable products including syngas, heat, electricity, biofuels, fertilizer, and charcoal.

Low-moisture waste is used in biomass gasification to reduce the energy required for drying. Because they have a controlled moisture content, woody biomass and low-moisture herbaceous plants are the best options. Additionally, agricultural wastes like straws, groundnut shells, bagasse, rice husk, and sugarcane debris can also be included.

7.3.6 Combustion

The process of burning fuel with oxygen produces heat, tar, water vapour, alkaline ash particles, and smoke as by-products. This process is known as combustion. Heat is produced when biomass is generally burned. Burning like flames and flammable vapors are volatilized when biomass fuel is ignited in the boiler. The combustion process led to an unanticipated degradation that may be divided into three categories: the vaporous part that contains CO, CO_2, and H_2, the condensable portion that contains water, aldehydes, and ketones, and the tar and alcohol's part that contains furan, phenolic mixtures, and sugar deposits. The residue, which is still carbon char, will ignite as a result of too much air. Depending on specific conditions, the heat generated during combustion can be harnessed for secondary conversion processes to generate electricity. Internal factors like moisture content, ash content, heat loss, and incomplete combustion influence the efficiency of biomass combustion.

7.4 Advantages of Biofuels

Biofuels have a high power output, are highly efficient, and are reasonably priced. Burning biofuel emits extremely little CO_2, making it environmentally beneficial. The feedstock used to produce biofuel absorbs almost the same amount of CO_2 from the atmosphere as it releases during biofuel consumption. Certain biofuels are naturally non-toxic and easily stored. Compared to fossil fuels, biofuels are comparatively less flammable (Abbaszaadeh et al. 2012).

7.4.1 Benefits of Biomass and Biogas

Environmental Benefits Biomass and biogas are considered carbon-neutral because the carbon dioxide released during combustion is recaptured by the plants used as feedstock as they grow. By doing this, climate change is lessened. The methane produced by these technologies is a powerful greenhouse gas and is made from organic waste products that would otherwise add to landfill volumes. The soil can be enhanced with organic matter and nutrients by applying the digestate, or by-product, of anaerobic digestion as fertilizer.

Economic Benefits Reducing dependence on fossil fuels can increase energy security through local biomass and biogas generation. Jobs are created throughout the whole biomass and biogas production value chain, especially in rural areas, from feedstock procurement and processing to plant operation. Farmers can sell biogas or biomass feedstock to generate additional revenue. Additionally, they can lower their operating expenses by using biogas to meet their own energy demands.

Renewability and Sustainability As long as it comes from sustainable sources, biomass is a renewable resource. Regular replenishment is possible with fast-growing energy crops and agricultural leftovers. Efficient handling of organic and agricultural waste reduces contamination to the environment and encourages nutrient recycling. It is possible to produce electricity from biomass and biogas that can be added to the grid. Biogas can be utilized in boilers, and biomass can be burned directly for heating. Vehicle fuel made from upgraded biogas, or biomethane, offers a more environmentally friendly option to petrol and diesel.

7.5 Challenges in Sustainable Bioenergy Production

The generation of bioenergy sustainably faces numerous obstacles. In an attempt to promote sustainable practices in biomass production, numerous governments, international agencies, and nongovernmental organizations at various levels have released a large number of white papers and guidelines. Producing feedstock for bioenergy entails risk. Due to erratic yields, sharp price swings, and seasonal weather uncertainties including floods, droughts, and storms, biomass generation in agriculture carries a higher risk profile than forest-based production. When cultivated to economically feasible levels, several feedstocks for bioenergy have quite high water requirements. If carried out responsibly, increasing the production of biomass and turning it into feedstocks, managing, transporting, and storing it could result in supply chains that boost rural livelihoods and help achieve renewable energy targets.

It may be possible to satisfy renewable energy targets with biomass supplies. However, the potential supply of biomass is influenced by several significant restrictions, including producer profitability, yield gaps, competing land uses, and land availability. Ultimately, to make well-informed decisions about the desired outcomes that should be pursued both now and, in the future, decision-makers at all levels must consider the costs, benefits, drawbacks, and advantages of different types of biomass and production systems.

7.6 Conclusion

The key to achieving a sustainable and affordable supply of bioenergy is the integration of biorefineries. Thus, advancement can be sustained by focusing on a few goals like creating new, affordable technologies for converting biomass, optimizing the engineering of current technologies to reduce energy consumption and boost productivity, lowering the carbon footprint by minimizing waste and by-product generation, and figuring out how complicated the processes are by analysing the chemistry of biomass. The declining cost of crude oil has made biofuels less competitive in recent years. But while fossil fuel supplies are depleting daily and demand for renewable energy sources is rising, biofuels have a bright future.

Developing affordable technology is the most urgent requirement for the production of sustainable energy. By-products and waste materials could be used to create new technologies or combine multiple existing technologies to produce thermal energy, which is another useful energy source. This lessens the environmental impact while also lowering the carbon footprint. Nonetheless, a detailed examination of the functions and traits of various lignocellulosic biomasses is required.

Acknowledgement The authors are grateful to the many who directly or indirectly contributed to this work.

References

Abbaszaadeh A, Ghobadian B, Omidkhah MR, Najafi G (2012) Current biodiesel production technologies: a comparative review. Energy Convers Manag 63:138–148

Adams P, Bridgwater T, Lea-Langton A, Ross A, Watson I (2018) Biomass conversion technologies. In: Greenhouse gas balances of bioenergy systems. Academic Press, pp 107–139

Deublein D, Steinhauser A (2011) Biogas from waste and renewable resources: an introduction. John Wiley & Sons

Erakhrumen AA (2012) Biomass gasification: documented information for adoption/adaptation and further improvements toward sustainable utilisation of renewable natural resources. Int Sch Res Not 2012(1):536417

Hernández-Beltrán JU, Hernández-De Lira IO, Cruz-Santos MM, Saucedo-Luevanos A, Hernández-Terán F, Balagurusamy N (2019) Insight into pretreatment methods of lignocellulosic biomass to increase biogas yield: current state, challenges, and opportunities. Appl Sci 9(18):3721

Ikora J, Tomal A (2016) Determination of the energy potential of biogas in selected farm household. Infrastruct Ecol Rural Areas 3:971–982

Kikas T, Tutt M, Raud M, Alaru M, Lauk R, Olt J (2016) Basis of energy crop selection for biofuel production: cellulose vs. lignin. Int J Green Energy 13(1):49–54

Kumar P, Barrett DM, Delwiche MJ, Stroeve P (2009) Methods for pretreatment of lignocellulosic biomass for efficient hydrolysis and biofuel production. Ind Eng Chem Res 48(8):3713–3729

Luo S, Xiao B, Guo X, Hu Z, Liu S, He M (2009) Hydrogen-rich gas from catalytic steam gasification of biomass in a fixed bed reactor: influence of particle size on gasification performance. Int J Hydrog Energy 34(3):1260–1264

Ma J, Shi S, Jia X, Xia F, Ma H, Gao J, Xu J (2019) Advances in catalytic conversion of lignocellulose to chemicals and liquid fuels. J Energy Chem 36:74–86
McKendry P (2002) Energy production from biomass (part 2): conversion technologies. Bioresour Technol 83(1):47–54
Menon V, Rao M (2012) Trends in bioconversion of lignocellulose: biofuels, platform chemicals & biorefinery concept. Prog Energy Combust Sci 38(4):522–550
Moshood TD, Nawanir G, Mahmud F (2021) Microalgae biofuels production: a systematic review on socioeconomic prospects of microalgae biofuels and policy implications. Environ Chall 5:100207
Piwowar A, Dzikuć M, Adamczyk J (2016) Agricultural biogas plants in Poland–selected technological, market and environmental aspects. Renew Sust Energ Rev 58:69–74
Shirin S, Yadav AK (2014) Physico-chemical analysis of municipal wastewater discharge in Ganga river, Haridwar district of Uttarakhand, India. Curr World Environ 9(2):536–543. https://doi.org/10.12944/CWE.9.2.39
Srirangan K, Akawi L, Moo-Young M, Chou CP (2012) Towards sustainable production of clean energy carriers from biomass resources. Appl Energy 100:172–186
Taherzadeh MJ, Karimi K (2008) Pretreatment of lignocellulosic wastes to improve ethanol and biogas production: a review. Int J Mol Sci 9(9):1621–1651
Van Fan Y, Romanenko S, Gai L, Kupressova E, Varbanov PS, Klemeš JJ (2021) Biomass integration for energy recovery and efficient use of resources: Tomsk Region. Energy 235:121378
Wongsiriamnuay T, Kannang N, Tippayawong N (2013) Effect of operating conditions on catalytic gasification of bamboo in a fluidized bed. Int J Chem Eng 2013(1):297941
Wyman CE, Dale BE, Elander RT, Holtzapple M, Ladisch MR, Lee YY (2005) Coordinated development of leading biomass pretreatment technologies. Bioresour Technol 96(18):1959–1966
Yadav AK, Shirin S, Emmanouil C, Jamal A (2022a) Effect of seasonal and meteorological variability of air pollution in Singrauli coalfield. Aerosol Sci Eng 6(1):61–70. https://doi.org/10.1007/s41810-021-00124-3
Yadav AK, Shirin S, Hopke PK, Pal D, Jamal A (2022b) Process to reduce particulate matter in ambient air using bubbles of sodium palmitate. Chem Eng Technol 45(8):1497–1500. https://doi.org/10.1002/ceat.202100454

Chapter 8
Catalytic Co-hydrothermal Carbonization of Sewage Sludge and Lignocellulosic Biomass for Enhanced Energy Recovery

Shashank Srivastava, Samar Mahata, and Chandra Sekhar Matli

Abstract The lower H/C and O/C ratios in hydrochar, compared to raw biomass, reflect its transformation into a more carbon-rich material through the hydrothermal carbonization (HTC) process, which is a desirable characteristic for solid fuels due to its correlation with increased calorific value and stability. In this chapter, the premise that sewage sludge (SS), despite its high moisture content, is an exemplary feedstock for Co-HTC, facilitating a sustainable method for enhanced energy recovery and superior hydrochar quality has been explored. Our analysis assesses the impact of feedstock composition alongside HTC process parameters to identify which agricultural solid waste among Corn Cob (CC), Yard Trash (YT), and Rice Husk (RH) can act as a more effective co-substrate of lignocellulosic biomass to generate a hydrochar with Higher Heating Value (HHV). The chapter reveals that strategic adjustments to moisture content, temperature, and residence time can significantly increase energy recovery rates, achieving up to 91%. The synergistic role of the introduction of citric acid (CA) and h-chloric acid (HCl) is discussed to counterbalance the additional energy requirements and accelerate the reaction process. Moreover, the study showcases the promising use of orange peels (OP) as a natural catalyst that lowers the pH and increases the fuel energy density of the hydrochar, providing a cost-effective alternative to conventional additives. The chapter underscores the viability of SS in HTC and its dual advantages: reducing watering issues and enhancing energy value, thus contributing to the advancement of sustainable and efficient renewable energy generation. The optimization of natural catalysts and process conditions for Co-HTC can produce superior biofuel attributes, reduce dependence on conventional additives or fossil fuels, and mark a significant advancement in sustainability through biomass.

Keywords Biomass · Catalyst · Co-hydrothermal carbonization · Energy densification · Fuel

---

S. Srivastava (✉) · S. Mahata · C. S. Matli
Department of Civil Engineering, National Institute of Technology, Warangal, India
e-mail: pdf_2022_ced01@nitw.ac.in

A. K. Yadav (ed.), *Powering Future: Clean Energy Solutions for a Sustainable Planet*, Clean Energy Production Technologies, https://doi.org/10.1007/978-981-96-5055-2_8

8.1 Introduction

The burgeoning population and accelerating industrialization have necessitated the establishment of numerous water treatment plants (WTPs) for provision of potable drinking water (Srivastava et al. 2022) and wastewater treatment plants (WWTPs) to manage the vast quantities of wastewater generated. A significant by-product of these plants is sludge, which accumulates at an alarming rate of 13% per year, along with increased water use (Syed-Hassan et al. 2017). In India alone, the production of dry sludge amounts to approximately 395 metric tons annually (Sude et al. 2024), contributing to the global generation of 45 million tons (Zhang et al. 2017). The disposal of sludge presents a considerable challenge due to its substantial volume and semisolid nature, which comprises 80–90% liquid by mass (Sude et al. 2024). Despite its complex composition, sludge is rich in nutrients, making it a potential candidate for land application as fertilizer. However, untreated sludge poses significant risks to biodiversity, ecosystems, and human health due to its contamination with hazardous substances such as pathogens, heavy metals, and organic pollutants.

Recent research has highlighted the potential of using sludge as a raw material for sustainable energy production, particularly in the conversion to biofuel (Mahata et al. 2023). Traditional thermal treatment processes such as incineration, pyrolysis, and gasification are employed to recover energy from organic waste (Srivastava et al. 2018). These methods require dry feedstock, and the drying of sludge can constitute 50–60% of the overall wastewater treatment cost, rendering the process economically unfeasible and environmentally less favorable when considering the life cycle assessment of emissions and sinks (Gaur et al. 2020; Shirin and Yadav 2014). The interplay of environmental and economic factors requires the development of alternative sludge treatment methods to improve sustainability and efficiency in sludge management.

Hydrothermal carbonization (HTC) is emerging as a viable method to convert organic biomass into valuable solid biofuels. This process involves hydrolysis, dehydration, decarbonization, polymerization, condensation, and aromatization, all occurring in an aqueous medium, thus eliminating the need for predrying the sludge (Mahata et al. 2023). Moreover, HTC significantly reduces toxicity by removing unusual organic and inorganic pollutants and harmful bacteria. The resultant by-product, hydrochar, possesses more functional groups and is more amenable to structural modification than the biochar produced via pyrolysis. However, hydrochar derived from sewage sludge presents drawbacks, such as higher ash and lower carbon content, leading to a lower heating value (HHV) than dry sludge. Furthermore, protein hydrolysis during the process diminishes the HHV, further complicating its use as an efficient biofuel (Nicolae et al. 2020).

To address the limitations associated with HTC, co-hydrothermal carbonization (Co-HTC) has been proposed as a potential method for enhancing energy recovery. Research indicates that the addition of other types of biomass can significantly increase the calorific value of sludge (Song et al. 2019). In particular, lignocellulosic biomass (LC) is favored for Co-HTC due to its abundant availability, continuous supply, and renewable nature. The lower ash content and the higher organic component in the LC biomass, including cellulose, hemicellulose, and lignin, enhance recycling efficiency when co-processed with sewage sludge. Furthermore, the higher carbon content and lower nitrogen and sulfur levels in LC biomass contribute to a higher HHV and reduced toxic gaseous emissions (Song et al. 2019; Mahata et al. 2023). During Co-HTC, a synergistic effect on carbon and energy recovery is observed, which is further improved by the use of acids to enhance the characteristics of hydrochar and accelerate the carbonization process (Zhang et al. 2023). The use of acids can facilitate the breakdown of cellulose in the liquid phase while retaining lignin, resulting in improved HHV. For instance, the combination of citric acid with co-biomass, such as food waste and digested yard waste, has been shown to increase energy value to up to 25 MJ/kg through the hydrolysis of biomass into smaller fragments (He et al. 2022). During the reaction, the presence of strong acids promotes the insertion of functional groups and the development of carbon structure, enhancing the surface characteristics of hydrochar.

Our previous research has shown that the composition of co-feedstock is a crucial factor influencing the quality of hydrochar (Mahata et al. 2023). The reaction pathway, primarily governed by Maillard and Mannich processes, can be significantly altered by incorporating different types of biomass, particularly non-LC and LC materials (Kambo and Dutta 2015). Despite these findings, there has been limited exploration of Co-HTC involving diverse types such as agricultural waste, yard waste, and acidic biomass. Consequently, the impact of biomass addition, catalyst use, and sewage sludge on hydrochar properties remains inadequately understood. Moreover, the cost of catalysts such as hydrochloric acid (HCl) and citric acid presents a barrier to large-scale HTC applications. Therefore, this chapter investigates the effects of catalyst and biomass composition on hydrochar characteristics, the interplay between catalyst and biomass mixing ratio, and the potential of substituting natural catalysts, such as orange peel (a natural source of citric acid), for commercial acids like citric and hydrochloric acid.

8.2 Materials and Methods

8.2.1 Sample Preparation

Sewage sludge (SS) was collected from the wastewater treatment plant at the National Institute of Technology Warangal, Telangana, India. Corn cobs (CC), orange peels (OP), rice husks (RH), and yard waste (YW) were obtained from various locations in Warangal City. All samples were sun-dried for 24 h, ground, and sieved through a 75-μm mesh. Subsequently, samples were oven-dried at 105 °C for 24 h before the HTC experiment and stored in zip-lock bags. Commercially available citric acid monohydrate (CA) and hydrochloric acid (HCl) were used as catalysts.

8.2.2 Batch Experiments

A laboratory-scale HTC reactor was employed to investigate the effects of different reaction parameters. Carbonization studies utilized a 0.5 L stainless steel vessel lined with Teflon, following a four-stage experimental procedure:

Stage 1: HTC experiments were carried out on individual samples (SS, RH, YW, and CC) at 220 °C, with a residence time of 2.5 h and a liquid-to-solid ratio (L/S) of 10:1.

Stage 2: SS was mixed with various LC biomasses (RH, YW, CC) in a 1:1 ratio under the same conditions to identify the biomass that yielded the highest energy value.

Stage 3: Experiments were repeated with variations in one process parameter from the base condition (T-220 °C, t-2.5 h, L/S-10:1).

Stage 4: Different acid catalysts were tested with the base conditions. For instance, M-CA (0.3 M) and M-OP (0.33 g/g of mixed feedstock) were added to the base mixture.

8.2.3 Analytical Methods

The volatile matter (VM) and the ash content were evaluated according to ASTM D1762 standards (ASTM 2024), and fixed carbon (FC) was calculated from the difference. An elemental analyzer measured carbon (C), hydrogen (H), nitrogen (N), and sulfur (S) contents, with oxygen (O) content deduced by subtracting C, H, N, S, and ash contents from 100%. The higher heating value (HHV) of raw samples and hydrochar was determined using an oxygen bomb calibrator (He et al. 2022). Fourier transform infrared (FTIR) spectroscopy (Nicolet IS 10, SCINCO Co.) analyzed functional groups within the 4000 cm^{-1} to 400 cm^{-1} range. The mass yield of hydrochar (MY), energy densification (ED), energy recovery efficiency (ERE), and carbon retention (CR) were calculated with the help of the following equations (Eqs. 8.1, 8.2, 8.3, and 8.4):

$$\mathrm{MY} = \frac{M_{\mathrm{DryHydrochar}}}{M_{\mathrm{DryWaste}}} \times 100 \tag{8.1}$$

$$\mathrm{ED} = \frac{\mathrm{HHV}_{\mathrm{Hydrochar}}}{\mathrm{HHV}_{\mathrm{DryWaste}}} \tag{8.2}$$

$$\mathrm{ERE}(\%) = \mathrm{ED} \times \mathrm{MY} \tag{8.3}$$

$$\mathrm{CR} = \frac{C_{\mathrm{DryHydrochar}}}{C_{\mathrm{DryWaste}}} \times \mathrm{MY} \tag{8.4}$$

8.3 Results and Discussion

8.3.1 *Transformation of Biomass in HTC and the Advantages of Co-HTC Process*

The physicochemical properties of hydrochar (HC) and co-hydrochar (Co-HC) produced from SS and other biomass are detailed in Table 8.1. Due to the significant metallic content in sewage sludge, its ash content is notably higher than LC biomass. After HTC, the ash content increased from 37% to 44%, attributable to the reduced yield of hydrochar and the conversion of volatile matter (VM) into fixed carbon (FC) and ash. This increase in ash content is a critical parameter for hydrochar as it poses challenges during combustion and decreases the higher heating value (HHV) (Kambo and Dutta 2015).

The ash content decreased for Co-HC produced with LC biomass, with the lowest value of 24.5% observed when SS was mixed with CC due to the dilution effect. The fixed carbon (FC) content increased from 7% to 11%, which was less than the reduction of VM from 56% to 45%, indicating that some VM was converted into liquid and gaseous phases. The high FC content in LC biomass led to a significant increase in FC in hydrochar after Co-HTC. Additionally, carbon content (C) also increased after HTC and Co-HTC. The maximum carbon content of 37.8% was achieved when SS was mixed with CC, surpassing the carbon content in raw SS and SS-HC.

Table 8.1 Elemental composition of raw feedstock, HC and Co-HC

	Mix	Elemental composition (%)[b]					Atomic ratio		HHV
Sample[a]	ratio	C	H	N	S	O	H/C	O/C	MJ/Kg
R-SS	–	25.18	4.14	3.55	1.02	28.60	1.98	0.85	10.48
R-CC	–	44.68	6.40	2.07	0.64	41.25	1.72	0.69	17.48
R-RH	–	34.37	4.79	2.66	0.85	39.13	1.67	0.85	14.32
R-YW	–	43.38	5.72	1.54	0.26	41.10	1.58	0.71	17.12
HC-SS	–	29.01	4.31	3.47	0.91	17.90	1.78	0.46	13.32
HC-CC	–	47.58	6.28	2.50	0.34	36.00	1.58	0.57	21.52
HC RH	–	34.2	4.51	1.49	0.18	34.82	1.58	0.76	13.22
SS:CC	1:1	37.86	4.01	3.10	0.53	30.00	1.27	0.59	15.20
SS: RH	1:1	33.9	4.71	2.66	0.44	30.61	1.67	0.68	13.14
SS: YW	1:1	37.00	4.66	3.81	0.64	25.69	1.51	0.52	15.61

[a] *T* temperature, *t* time, *L:S* Liquid-Solid ratio, *SS* sewage sludge, *CC* corn cobs, *RH* rice husk, *YW* yard waste, *HC* hydrochar, *R* raw

[b] *C* carbon, *H* hydrogen, *N* nitrogen, *S* sulfur, *O* oxygen content

8.3.2 Selection of an Appropriate LC Biomass for Co-HTC Process

The use of LC biomass in co-hydrothermal carbonization (Co-HTC) significantly reduces the ash content, with the lowest value of 24.5% observed when sewage sludge (SS) is mixed with CC, due to the dilution effect. Moreover, the overall FC in hydrochar increases substantially, with a maximum carbon content of 37.8% achieved in the SS and CC blend, enhancing the suitability as a high-grade fuel.

The higher heating value (HHV) is another crucial factor in using hydrochar as biofuel, and it improved after HTC and when mixed with various types of biomass (Fig. 8.1). When SS was mixed with CC and YW, the HHV increased by 19% and 18%, respectively. The increased FC, C content, and HHV after HTC and Co-HTC indicate that combining sludge with LC biomass is suitable for producing high-grade fuel. Based on HHV, the preferred sequence of biomass for use as co-feedstock is CC > YW > RH. Consequently, for further investigation, CC was used as a co-feedstock with SS.

8.3.3 Understanding and Optimizing Co-HTC Process Parameters for Refining Fuel Properties

The experimental results indicated that the co-hydrothermal carbonization (Co-HTC) process significantly enhanced the elemental composition of hydrochar under varying temperatures, residence times, liquid-to-solid (L:S) ratios, and

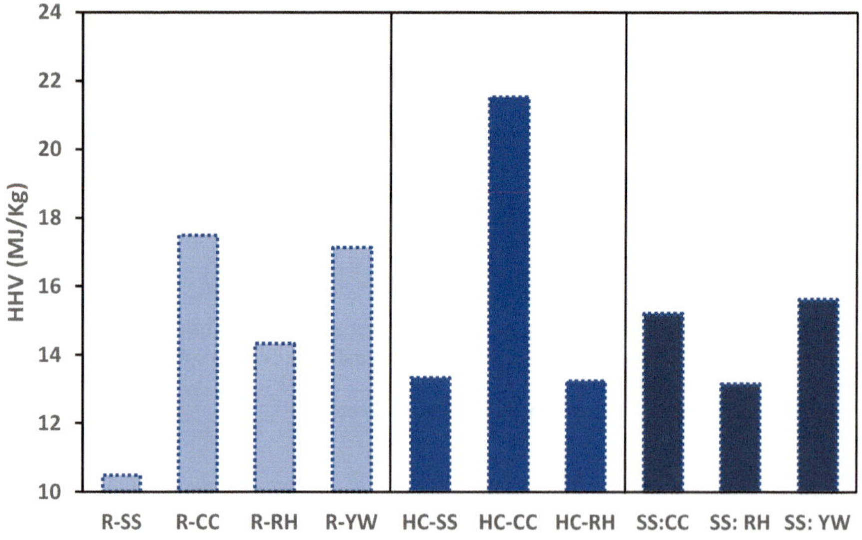

Fig. 8.1 HHV of HC and Co-HC for the feedstocks studied

Table 8.2 Elemental analysis to assess the effect of process parameters

Sample[a]	Parameters	Elemental composition (%) C	H	N	S	O	Atomic ratio H/C	O/C	HHV (MJ/Kg)
SS:CC (1:1)	$t = 2.5$ h	37.86	5.0	3.93	0.59	32.0	1.58	0.63	14.49
SS:CC (1:2)	$t = 2.5$ h	40.8	5.62	1.44	0.41	30.73	1.65	0.56	13.2
SS:CC (2:1)	$t = 2.5$ h	35.13	4.72	3.87	0.68	25.30	1.61	0.54	16.2
M (1 h)	$t = 1$ h	36.32	4.86	2.39	0.63	32.8	1.61	0.68	14.64
M (5 h)	$t = 5$ h	41.84	4.9	2.37	0.74	22.15	1.41	0.40	17.79
M (8 h)	$t = 8$ h	37.59	4.84	2.52	0.73	27.12	1.55	0.54	14.14
M-T (180 °C)	$T = 180$ °C	35.81	5.17	3.71	0.51	31.8	1.73	0.67	14.82
M-T (250 °C)	$T = 250$ °C	41.03	4.52	4.78	0.61	25.86	1.32	0.47	17.91
M-L:S (6:1)	L:S = 6:1	33.09	5.50	2.83	0.46	35.92	1.92	0.81	13.58
M-L:S (14:1)	L:S = 14:1	44.22	5.40	3.21	0.56	22.41	1.47	0.38	19.0

[a]Here, *M* mixture of SS:CC (1:1), *T* temperature, *t* time, *L:S* Liquid-Solid ratio

feedstock ratios (Table 8.2). Notably, when the sewage sludge (SS) to corn cobs (CC) ratio was altered from 1:2 to 2:1, the carbon content increased by 40% and 21% compared to hydrochar produced solely from SS. An increase in the L:S ratio from 6:1 to 14:1 led to a rise in carbon content from 34.10% to 44.64%, representing a 50% increase over hydrochar derived from SS alone. It can be observed from Fig. 8.2 that although temperature had a lesser impact on carbon sequencing than residence time, a maximum carbon content of 41.84% was achieved with a residence time, and 41.03% was achieved at 250 °C. The carbon content generally increased with temperature, as at 200 °C, water acts as a catalyst that facilitates hydrolysis and ionic condensation reactions.

At 250 °C, the carbon content further increased as proteins and fats in SS dissolved into the liquid phase, producing acidic compounds that enhance the polymerization and condensation processes (Baxter 2005). The decrease in process water pH at higher temperatures supports this observation. The results suggest that Co-HTC promotes efficient hydrochar production because the cellulose and lignin in the LC biomass dehydrate and disintegrate faster than the proteins present in the SS. Consequently, Co-HC exhibited a higher carbon content in comparison to HC-SS alone. However, an inverse effect was observed at an 8-h residence time, where the carbon content decreased to 36.32% (Fig. 8.2).

Additionally, the sulfur content in Co-HC was substantially lower compared to that of HC-SS alone. At a 1:3 blending ratio of SS-to-CC, the sulfur content decreased from 3.47% to 1.34%, indicating a reduced risk of SO_2 emissions during combustion. The higher nitrogen content in SS facilitated the Maillard reaction between nitrogen-containing feedstock and LC biomass. At a 1:1 ratio, the synergistic effects remained relatively unchanged, likely due to the combination of positive and negative effects arising from the Maillard reaction and the impact of pH.

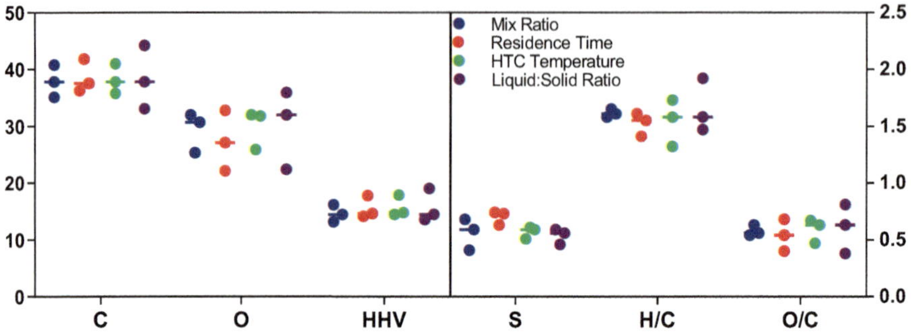

Fig. 8.2 Effect of the change in the process conditions of Co-HC on (**a**) fixed carbon content, oxygen content and HHV, (**b**) sulfur content and H/C & O/C ratios

Table 8.3 Elemental analysis of Co-HC carried out under acidic conditions

	Elemental composition (%)					Atomic ratio		HHV
Sample[a]	C	H	N	S	O	H/C	O/C	(MJ/Kg)
M-HCL (0.1 mol/L)	40.84	5.33	1.78	0.64	28.61	1.57	0.53	16.4
M-HCL (0.3 mol/L)	40.75	4.95	2.38	0.55	29.37	1.46	0.54	16.2
M-HCL (0.5 mol/L)	42.1	5.3	1.75	0.51	28.94	1.51	0.52	17.3
M-CA (0.1 mol/L)	40.0	4.45	2.31	0.74	29.6	1.34	0.56	15.5
M-CA (0.3 mol/L)	41.12	4.95	2.31	0.54	29.28	1.44	0.53	16.8
M-CA (0.5 mol/L)	44.1	5.0	2.01	0.47	27.12	1.36	0.46	18.0
M-OP (0.33 g/1 g)	42.11	5.1	3.18	0.71	30.0	1.45	0.53	17.12
M-OP (0.5 g/1 g)	44.22	5.51	3.21	0.56	28.7	1.5	0.49	18.19

[a]Here, *M* mixture of SS:CC (1:1), *CA* citric acid, *OP* Orange peels

8.3.4 Role of Acids in HTC

The influence of a catalyst on the elemental properties of hydrochar is a subject of significant interest, particularly in the context of utilizing OP as a catalyst source. In this investigation, hydrochloric acid (HCl) and citric acid monohydrate (CA) were employed as additional acids to evaluate the fuel and elemental properties of hydrochar. The results indicate that acidic conditions lead to an increase in carbon content and a decrease in the oxygen (O) and hydrogen (H) content (Table 8.3). This trend is desirable for char production, as higher carbon and lower oxygen contents in hydrochar confer advantageous fuel characteristics.

Specifically, the carbon content increased from 40.8% to 42.1% and 43% after the addition of 0.5 mol/L HCl and CA, respectively. Concomitantly, oxygen and hydrogen content also decreased under acidic conditions. This observation is consistent with the findings of He et al. (2022), who reported a significant increase in carbon content after adding citric acid and acetic acid to their experimental setup. The authors argue that acidic conditions encourage hydrolysis and decomposition, which in turn promote the synthesis of hydrochar.

Furthermore, the results suggest that the increase in carbon content and the reduction in the amount of oxygen and hydrogen can be attributed to an acidic environment that fosters decarboxylation and dehydration reactions. When HCl and CA were used, the maximum nitrogen loss was observed to be 57% and 50%, respectively. This finding indicates that acidic conditions increase nitrogen hydrolysis, which dissolves into the liquid phase (Qiang Zhang et al. 2023). Moreover, the use of acid catalysts in HTC has been shown to reduce carbon loss and provide a high potential for carbon recovery compared to that in uncatalyzed conditions.

Therefore, the effect of catalysts on the elemental properties of hydrochar is a critical factor in the production of char with advantageous fuel characteristics. The use of acidic conditions, such as those created by HCl and CA, promotes hydrolysis and decomposition, leading to an increase in carbon content and a decrease in oxygen and hydrogen content. These conditions also increase the hydrolysis of nitrogen, which dissolves into the liquid phase, and reduce carbon loss, thereby providing a high potential for carbon recovery. Future research in this area should continue to explore the use of acidic catalysts to optimize the fuel properties of hydrochar.

8.3.5 Increasing the Yield of Hydrochar

Hydrochar yield (HY) is a parameter of immense interest for overall energy production and is highly dependent on feedstock selection, composition, process parameters, and various other conditions as elaborated here.

8.3.5.1 HY and Process Conditions

Experiments demonstrated that the yield is highly dependent on the process conditions. Notably, hydrochar yield (HY) increases significantly from 76% to 83% as the SS-to-CC ratio shifts from 1:2 to 2:1. This phenomenon can be attributed to the higher stability of the LC biomass compared to that of the SS during Co-HTC. The compositional and structural differences between SS and LC biomass, such as CC, also play an important role. LC biomass primarily comprises cellulose, hemicellulose, and lignin, which are complex carbohydrates and aromatic polymers that provide structural integrity and resistance to thermal degradation. During the hydrothermal carbonization process, these components exhibit a relatively stable decomposition pathway (Yang et al. 2022), contributing to the formation of solid hydrochar with a high carbon content.

On the contrary, sewage sludge contains a higher proportion of easily degradable organic matter, including proteins, lipids, and carbohydrates. These components are more susceptible to hydrolysis and subsequent breakdown under hydrothermal conditions, leading to the formation of volatile and water-soluble products instead of solid hydrochar. As a result, the overall yield of hydrochar is lower when the SS content is higher.

When the SS-to-CC ratio is adjusted from 1:2 to 2:1, the increased proportion of LC biomass stabilizes the hydrothermal process. The presence of lignocellulose promotes the formation of a more substantial solid phase because of its resistance to complete thermal breakdown, thereby enhancing the yield of hydrochar. Furthermore, interactions between the components of SS and CC during Co-HTC can influence the reaction pathways. The fibrous nature of LC biomass can provide a matrix that traps more of the SS-derived carbon, further increasing the solid yield.

Moreover, the synergistic effects of co-processing SS and CC can lead to the formation of a more homogeneous hydrochar with improved structural properties (Yu et al. 2022). The optimal SS-to-CC ratio thus leverages the advantages of both feedstocks, achieving higher hydrochar yields and potentially improving the quality of the resultant product in terms of energy content and carbon sequestration capabilities.

8.3.5.2 Effect of Temperature and Biomass Composition

Figure 8.3a illustrates a reduction in hydrochar yield from 84% to 65% as the Co-HTC temperature rises from 180 °C to 250 °C. This behavior is influenced by the distinct thermal decomposition characteristics of cellulose, hemicellulose, and lignin, which are the primary components of lignocellulosic biomass.

Cellulose starts to break down at 180 °C, and hemicellulose at 220 °C, leading to smaller molecules and volatile compounds. Lignin remains stable until above 250 °C. With increasing temperatures, solid hydrochar yield decreases as a result of increased decomposition into liquids and gases. Therefore, the reduction in hydrochar yield from 84% to 65% as the Co-HTC temperature increases from 180 ° C to 250 ° C can be explained by the sequential thermal decomposition of cellulose, hemicellulose, and lignin. The stability of lignin at lower temperatures and the early decomposition of cellulose and hemicellulose at moderate temperatures govern the yield (Yang et al. 2022) and composition of the hydrochar, with higher temperatures favoring the production of nonsolid products.

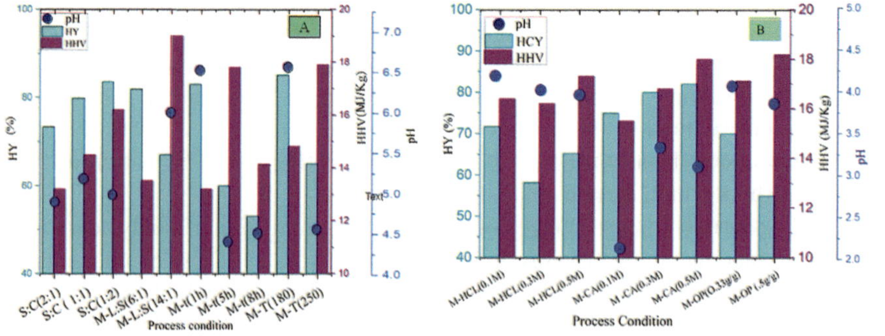

Fig. 8.3 Change in HY and HHV values of Co-HC (**a**) for different process conditions and (**b**) on the introduction of catalysts

8.3.5.3 Liquid-Solid Ratio (L:S) and HY

Increasing the L:S ratio from 6:1 to 14:1 leads to a 36% reduction in hydrochar yield. Water acts as a catalytic reactant during subcritical carbonization, facilitating hydrolysis and cleavage reactions. The higher liquid content promotes faster reaction rates, organic element degradation, and better diffusion in the liquid phase. Furthermore, moisture facilitates decarboxylation reactions, gradually increasing CO_2 generation through degradation of the carbonyl and carboxyl functional groups. These conditions improve the formation of saccharides and organic acids, which readily dissolve into process water, ultimately lowering the hydrochar yield. The pH change induced by organic acids accelerates the dehydration and decarboxylation reactions (Zhang et al. 2023), resulting in reduced solid product and improved solid-liquid product.

8.3.5.4 Influence of Catalyst on HY

The impact of a catalyst on hydrochar yield is depicted in Fig. 8.3b. Under base conditions, the yield is 79.48%. However, when 0.5 M HCl is used as a catalyst, the maximum HY decreases by 26% compared to the base condition. Further increases in HCl concentration lead to gradual reduction in HY. Similar trends were observed in hydrothermal carbonization studies using HCl as a catalyst (Dai et al. 2017; Sarrion et al. 2022). The presence of H^+ and Cl^- ions in the aqueous medium enhances the hydrolysis of cellulose, forming monosaccharides. HCl, a strong metallic acid, facilitates carbon transfer into the liquid phase through mineral salt solubilization.

8.3.6 Utilization of Orange Peels as Natural Catalysts

When citric acid is introduced as a catalyst during HTC, it has a unique impact on the yield of hydrochar. An increase in citric acid concentration from 0.3 M to 0.5 M moles results in an incremental increase in hydrochar yield by 3.5% and 7%, respectively. This enhancement is consistent with the findings of Ameen et al. (2022), who observed a similar trend when citric acid was used and ascorbic acid as catalysts. The acid environment created by the addition of citric acid accelerates the hydrolysis-dehydration-coalification sequence, which is crucial for the transformation of biomass into hydrochar. During the HTC process, citric acid undergoes a transformation, producing various organic acids such as formic and acetic acid or decomposing into citraconic and itaconic acids. These resultant acids play a pivotal role in the reaction mechanism, particularly influencing the hydrolysis, dehydration, condensation, and polymerization processes. Consequently, the process becomes autogenic, which means that it generates enough heat internally to sustain the reaction without external energy input, and this facilitates the formation of secondary char (He et al. 2022).

In an innovative approach, OP was utilized as a substitute for conventional catalysts. By adding 0.33 and 0.5 g of OP per gram of co-feedstock, the study evaluated the acidic effect on hydrochar yield. The hydrochar yield was found to decrease significantly, by up to 70% and 55%, and the pH levels dropped correspondingly to 4.6 and 4.1. The actual HY was notably lower than the calculated one, indicating that the acidic components present in the OP have a substantial influence on the hydrolysis process. These components aid in the dissolution of organic compounds in the liquid phase, which is a critical step in the HTC process. However, it can be seen from Fig. 8.3 that a direct relationship between hydrochar yield and pH conditions could not be ascertained.

The improvement in higher heating value (HHV) by catalyst use is depicted in Fig. 8.3b, which shows HHV values for hydrochar with different concentrations of citric acid (CA) and hydrochloric acid (HCl). In contrast to the trends observed for hydrochar yields, HHV increased in the presence of an acidic environment. The results indicated that the addition of HCl led to a slight increase in the HHV of the hydrochar. Specifically, when HCl was introduced as a catalyst at a concentration of 0.1 M, the HHV increased by 8%. However, as the concentration of HCl increased further, HHV began to decrease, which aligns with reported findings (Sarrion et al. 2022).

Furthermore, the use of CA as a catalyst resulted in a 14% increase in the HHV of the hydrochar. An even more significant improvement was observed with the use of OP as a catalyst, which increased HHV by 21%. Among all of the conditions tested, the use of OP in the co-hydrothermal carbonization (Co-HTC) process resulted in the maximum energy value. This suggests that OP is the most effective catalyst for enhancing the HHV of hydrochar.

The observed increase in HHV when using catalysts can be attributed to the accelerated dehydration and decarboxylation processes facilitated by the acidic environment (Jadhav et al. 2021). These processes lead to the removal of oxygen-containing functional groups from the biomass, resulting in a hydrochar with a higher carbon content and, consequently, a higher HHV.

The sequence of increasing HHV, based on maximum values obtained under different conditions, is as depicted in Eq. 8.5:

$$\begin{aligned}&\text{Liquid to Solid ratio}\left(\text{L : S}\right) > \text{Orange Peels}\left(\text{OP}\right) >\\&\text{Citric Acid}\left(\text{CA}\right) > \text{Hydochloric Acid}\left(\text{HCl}\right) > \text{Temperature} > \text{Time}\end{aligned} \tag{8.5}$$

This ordering highlights the significant role of catalysts, particularly natural catalysts such as OP, in enhancing the energy content of hydrochar. The catalytic enhancement of HHV can be primarily linked to the promotion of key thermochemical reactions, underscoring the importance of catalyst selection in optimizing the Co-HTC process for maximum energy yield.

8.3.7 Energy Densification and Recovery

The variability of HY under various process parameters, catalysts, and pH levels is shown in Fig. 8.4. The maximum energy densification ratio (EDR) obtained were 1.27, 1.28, and 1.42 at 5 h of residence, 250 °C temperature, and moisture content (L:S) of 14:1, respectively, compared to the base condition (1.08). This implies that elevated temperature, time, and moisture content promote improved ED. Severe conditions decreased the content of cellulose and hemicellulose while increasing lignin and aromatics, consequently raising the HC energy value of the HC.

Upon the addition of catalyst, the highest EDR of 1.23 and 1.28 resulted from 0.5 Mole of HCl and citric acid, respectively. Interestingly, an EDR of 1.3 was achieved using orange peel as an acid catalyst substitute, suggesting that catalysts produce ED comparable to higher severity conditions. For large-scale applications, acid usage can be expensive and can be effectively replaced by OP for equivalent energy optimization.

Figure 8.4a illustrates the energy retention efficiency (ERE) under various process conditions. However, it can be observed from Fig. 8.4a, b that most cases with high EDR showed lower hydrochar yields. Therefore, optimizing process conditions for both EDR and energy retention efficiency is essential, accounting for HHV and hydrochar yield. The highest ERE was found at 91%, 89%, and 87% for moisture content, temperature, and residence time, respectively. Citric acid achieved 90% ERE, indicating its potential for energy recovery optimization, thus minimizing additional energy consumption.

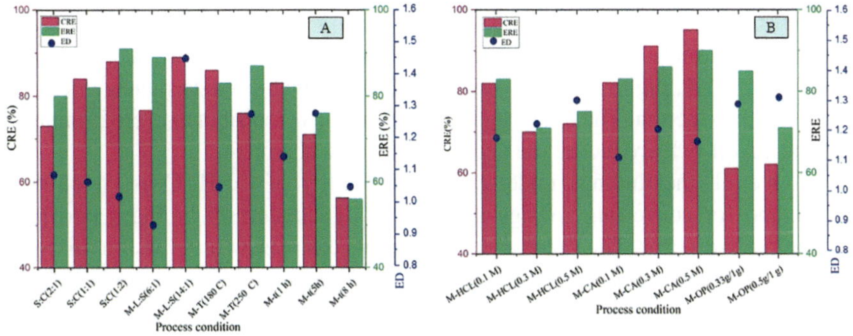

Fig. 8.4 Analysis of CRE, EDR, and ERE (**a**) for different process conditions and (**b**) on the introduction of catalysts

8.4 Conclusions and Perspectives

This chapter explored co-hydrothermal carbonization (Co-HTC) of sewage sludge (SS) and corn cobs (CC) to optimize energy recovery and improve hydrochar properties. Some critical insights revealed from the investigation are as follows:

1. Optimal feedstock composition:
   A balanced ratio of SS and CC significantly enhances the carbon content of the hydrochar. The optimal SS:CC ratio of 1:1 was identified, providing the best compromise between high carbon content and efficient energy recovery.
2. Process parameters:
   The study confirmed that increasing the liquid-solid ratio (L:S) from 6:1 to 14:1 and operating at a temperature of 250 ° C with a residence time of 5 h significantly improve the carbon content and the heating value (HHV) of the hydrochar. These parameters facilitate the efficient decomposition of the feedstock and enhance the quality of hydrochar.
3. Catalyst effectiveness:
   Citric acid (CA) and hydrochloric acid (HCl) were effective in enhancing hydrochar properties by increasing carbon content and reducing the oxygen and hydrogen content. Natural catalysts such as OPs also proved beneficial in lowering pH and improving fuel properties, making them a cost-effective alternative to chemical additives.
4. Elemental and fuel properties:
   The Co-HTC process yielded hydrochar with improved elemental composition, notably a higher carbon content and a lower sulfur content, reducing potential SO_2 emissions. The hydrochar exhibited a higher HHV and a fixed carbon content, indicating its suitability as a superior biofuel.
5. Environmental and energy benefits:
   The optimized Co-HTC process demonstrated an enhanced energy recovery of up to 91%, highlighting its potential for sustainable waste management and energy production. The use of natural catalysts also contributes to environmental sustainability by reducing the need for synthetic chemicals.

   Overall, the findings underscore the viability of Co-HTC as an effective method for converting sewage sludge and lignocellulosic biomass into high-quality hydrochar. The process not only enhances energy recovery but also produces biofuels with improved environmental properties. Future studies should focus on scaling up the process and exploring the long-term stability and performance of hydrochar in various applications.

Acknowledgments The authors acknowledge the support of the Central Research Instrumentation Facility (CRIF), NIT Warangal, in conducting sophisticated analyses.

References

Ameen M, Zamri NM, May ST, Azizan MT, Aqsha A, Sabzoi N, Sher F (2022) Effect of acid catalysts on hydrothermal carbonization of Malaysian oil palm residues (leaves, fronds, and shells) for hydrochar production. Biomass Convers Biorefinery 12:103–114. https://doi.org/10.1007/s13399-020-01201-2

ASTM (2024) ASTM D1762-84 - standard test method for chemical analysis of wood charcoal. GlobalSpec. West Conshohocken, PA, 2024: American Society for Testing and Materials (ASTM) International. https://doi.org/10.1520/BOS-2024

Baxter L (2005) Biomass-coal co-combustion: opportunity for affordable renewable energy. Fuel 84:1295–1302. https://doi.org/10.1016/j.fuel.2004.09.023

Dai L, Yang B, Li H, Tan F, Zhu N, Zhu Q, He M, Ran Y, Guoquan H (2017) A synergistic combination of nutrient reclamation from manure and resultant hydrochar upgradation by acid-supported hydrothermal carbonization. Bioresour Technol 243:860–866. https://doi.org/10.1016/j.biortech.2017.07.016

Gaur RZ, Khoury O, Zohar M, Poverenov E, Darzi R, Laor Y, Posmanik R (2020) Hydrothermal carbonization of sewage sludge coupled with anaerobic digestion: integrated approach for sludge management and energy recycling. Energy Convers Manag 224:113353. https://doi.org/10.1016/j.enconman.2020.113353

He M, Zhu X, Dutta S, Khanal SK, Lee KT, Masek O, Tsang DCW (2022) Catalytic co-hydrothermal carbonization of food waste digestate and yard waste for energy application and nutrient recovery. Bioresour Technol 344:126395. https://doi.org/10.1016/j.biortech.2021.126395

Jadhav A, Ahmed I, Baloch AG, Jadhav H, Nizamuddin S, Siddiqui MTH, Baloch HA, Qureshi SS, Mubarak NM (2021) Utilization of oil palm fronds for bio-oil and bio-char production using hydrothermal liquefaction technology. Biomass Convers Biorefinery 11:1465–1473. https://doi.org/10.1007/s13399-019-00517-y

Kambo HS, Dutta A (2015) Comparative evaluation of torrefaction and hydrothermal carbonization of lignocellulosic biomass for the production of solid biofuel. Energy Convers Manag 105:746–755. https://doi.org/10.1016/j.enconman.2015.08.031

Mahata S, Periyavaram SR, Akkupalli NK, Srivastava S, Matli C (2023) A review on co-hydrothermal carbonization of sludge: effect of process parameters, reaction pathway, and pollutant transport. J Energy Inst 110:101340. https://doi.org/10.1016/j.joei.2023.101340

Nicolae SA, Heather A, Modugno P, Luo H, Szego AE, Qiao M et al (2020) Recent advances in hydrothermal carbonisation: from tailored carbon materials and biochemicals to applications and bioenergy. Green Chem 22:4747–4800. https://doi.org/10.1039/d0gc00998a

Sarrion A, de la Rubia A, Coronella C, Mohedano AF, Diaz E (2022) Acid-mediated hydrothermal treatment of sewage sludge for nutrient recovery. Sci Total Environ 838:156494. https://doi.org/10.1016/j.scitotenv.2022.156494

Shirin S, Yadav AK (2014) Physico-chemical analysis of municipal wastewater discharge in Ganga river, Haridwar district of Uttarakhand, India. Curr World Environ 9(2):536–543. https://doi.org/10.12944/CWE.9.2.39

Song Y, Zhan H, Zhuang X, Yin X, Wu C (2019) Synergistic characteristics and capabilities of co-hydrothermal carbonization of sewage sludge/lignite mixtures. Energy Fuels 33:8735–8745. https://doi.org/10.1021/acs.energyfuels.9b01766

Srivastava S, Brighu U, Rajput H (2018) Waste-to-energy recovery potential: a case study at Jaipur railway station. Environ Qual Manag 27:107–113. https://doi.org/10.1002/tqem.21558

Srivastava S, Saxena K, Brighu U, Gupta AB (2022) Comparative evaluation of pilot-scale reactors based on pulsating floc blanket clarification and conventional clariflocculation technologies in simultaneous treatment of natural organic matter and turbidity. Water Supply 22:6945–6958. https://doi.org/10.2166/ws.2022.267

Sude G, Rajpal A, Tyagi VK, Sharma K, Mutiyar PK, Panday BK, Pandey RP, Kazmi AA (2024) Evaluation of sludge quality in Indian sewage treatment plants to develop quality control indices. Environ Sci Pollut Res 31:17578–17590. https://doi.org/10.1007/s11356-023-25320-1

Syed-Hassan SSA, Wang Y, Hu S, Su S, Xiang J (2017) Thermochemical processing of sewage sludge to energy and fuel: fundamentals, challenges and considerations. Renew Sustain Energy Rev 80:888–913. https://doi.org/10.1016/j.rser.2017.05.262
Yang X, Wang B, Song X, Yang F, Cheng F (2022) Co-hydrothermal carbonization of sewage sludge and coal slime with sulfuric acid for N, S doped hydrochar. J Clean Prod 354:131615. https://doi.org/10.1016/j.jclepro.2022.131615
Yu S, Yang X, Zhao P, Li Q, Zhou H, Zhang Y (2022) From biomass to hydrochar: evolution on elemental composition, morphology, and chemical structure. J Energy Inst 101:194–200. https://doi.org/10.1016/j.joei.2022.01.013
Zhang Q, Kai M, Zhao B, Yi L (2023) The role of the Mannich reaction in nitrogen migration during the co-hydrothermal carbonization of bovine serum albumin and lignin with various forms of acid–alcohol assistance. Molecules 28. https://doi.org/10.3390/molecules28114408
Zhang Q, Hu J, Lee DJ, Chang Y, Lee YJ (2017) Sludge treatment: current research trends. Bioresour Technol 243:1159–1172. https://doi.org/10.1016/j.biortech.2017.07.070

Part III
Solar Energy

Chapter 9
Illuminating the Path to Sustainable Tomorrow with Solar Energy

Pritish S. Doshi, Dhruv N. Salot, and Sachin V. Jadhav

Abstract Solar energy plays a vital role in the modern era, especially due to the importance of renewable energy. Unlike finite fossil fuels like coal and crude oil, solar energy can be collected without depleting natural resources. The utilization of solar energy has multiple pros and a few cons; therefore, it can act as a method via which we can achieve a sustainable tomorrow. Solar energy helps us reduce our dependency on fossil fuels (Pollutants), and it also helps us reduce the overall emission of greenhouse gases which helps us mitigate climate change. Technological advancements continue to drive efficiency improvements and cost reductions in solar technology. The inception of solar energy utilization dates back to approximately the seventh century BC, when ancient civilizations used a simple magnifying glass to converge solar rays, thereby generating sufficient heat to start a fire. A few decades later, solar energy was used to heat bathhouses and homes in extreme conditions. Solar energy was also used to heat water for domestic use in ancient cultures, working on the same principle Clarence Kemp developed the first commercial water heater. Recent advances include concentrating solar power technology which generate electricity by producing steam to drive a turbine.

Russel Ohl was the scientist who played a pivotal role in the transition of solar cell technology by using silicon instead of selenium wafers. The efficiency of the first silicon solar cell was around 6% but with the help of advancing technology, the latest solar cell has an efficiency of about 47.6%. Today, solar energy is being utilized not just for energy generation but also in innovative ways like the production of hydrogen fuel by a solar-powered electrolyzer and the treatment of wastewater by solar detoxification and solar disinfection.

Keywords Solar energy · Photovoltaic · Concentrating solar power · Solar disinfection · Solar farms

P. S. Doshi · D. N. Salot · S. V. Jadhav (✉)
Department of Chemical Engineering, Institute of Chemical Technology, Mumbai, Maharashtra, India
e-mail: sv.jadhav@ictmumbai.edu.in

A. K. Yadav (ed.), *Powering Future: Clean Energy Solutions for a Sustainable Planet*, Clean Energy Production Technologies, https://doi.org/10.1007/978-981-96-5055-2_9

9.1 Introduction

The escalating global population is causing an exponential growth in energy demand. By 2050 the energy demand is going to increase twofold and by the end of the century be more than triple (Lewis et al. 2005). Due to the exhaustive and polluting nature of fossil fuels, they are deemed an inferior option compared to their renewable alternatives. Renewable energy is crucial for future generations as it provides a sustainable way to meet increasing energy needs while protecting the environment. Unlike finite fossil fuels, renewable sources like solar, wind, and hydropower are abundant and produce minimal greenhouse gases, aiding us in the fight against climate change. Investing in renewable energy now ensures a cleaner, healthier planet for those to come, decreases reliance on depleting resources, and encourages technological advancements that boost economic stability and energy security. Transitioning to renewable energy is vital for a resilient and sustainable future. Solar energy is the largest renewable energy source available to us, offering immense potential to sustain global energy needs. The sun emits an enormous amount of energy every day, far more than the total energy consumption of the entire world. We can generate electricity and heat without depleting natural resources or emitting greenhouse gases by harnessing this energy through technologies like photovoltaic panels and solar thermal systems. The period from the 1950s to the present day has witnessed remarkable progress in solar energy utilization technologies (Fig. 9.1).

9.2 Historical Overview

In 1954, the first practical solar cell was developed in Bell laboratories. This solar cell was made of silicon and was capable of generating 60 W/m^2 but this setup was accompanied by many losses and had an overall efficiency of 6% (Chapin et al. 1954). Due to the oil crisis in 1973, in the 1970s European architects and engineers collaborated in developing new technologies to use natural resources for heating and cooling of buildings (Medici 2018), e.g., ventilated walls, solar chimneys, Trombe walls, etc. Trombe wall uses passive heating and natural ventilation to effectively use energy. There have been many modifications to this application, e.g., zigzag Trombe wall, water Trombe wall, composite Trombe wall, and fluidized Trombe wall. Solar chimney uses radiation from the sun to heat the air under a roof which then runs a turbine generating electricity (Schlaich and Schiel 2003). During the 1980s period, a large number of solar water heating applications were used in houses and offices consisted of a solar collector, circulation pump, insulated tank, and a controller. Solar thermal collectors usually consisted of flat plate collectors and were placed on the rooftops. In 1992, the University of South Florida developed a 15.9% efficient thin-film photovoltaic cell made of cadmium telluride, breaking the 15% barrier for the first time. In 1996, the Icare was built which was the first

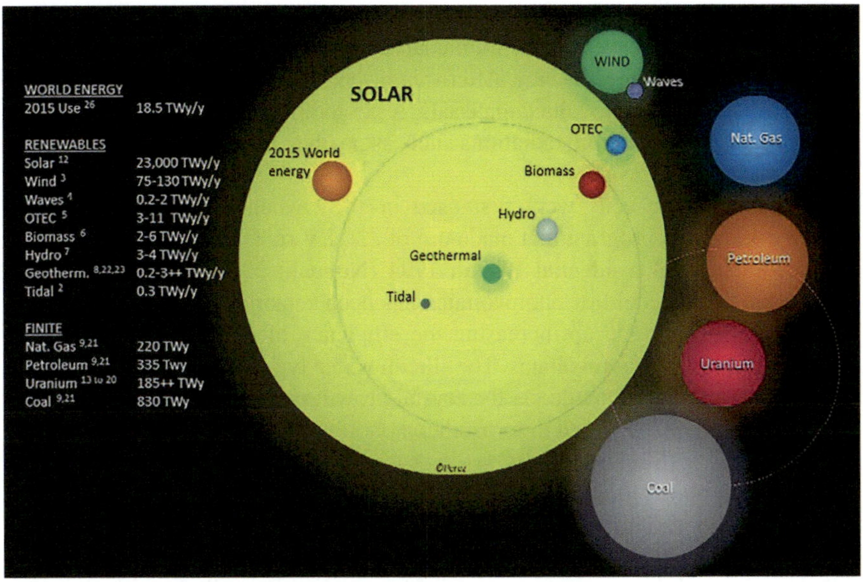

Fig. 9.1 Solar energy compared to other energy sources [reprinted from Perez et al. (2015) {CC BY}]

solar-powered plane and it flew over the country of Germany. In 2002, the largest solar power facility in the Northwest, the 38.7-kilowatt White Bluffs Solar Station, became fully functional. Since then, multiple small-scale and macro-scale advancements have contributed to the renewable energy industry. The small-scale developments are the most important as small advancements often serve as the foundation for larger breakthroughs.

9.3 Recent Advancements and the Future in Solar Innovation Technology

9.3.1 Photovoltaics

Photovoltaic devices operate based on the principle of the photovoltaic effect. They convert solar energy into electrical energy using semiconductor materials. The photovoltaic effect can be defined as the generation of voltage and electric current in a material upon exposure to light. When light is incident on a material that exhibits the photovoltaic effect, the charged carriers (mainly electrons) get excited to a higher energy level producing an electrical potential that gives rise to an electric current. It was first demonstrated by Edmund Becquerel in 1839.

Considering the photovoltaic effect, Charles Fritts invented the first solar cell in 1884. He used a thin layer of selenium coated with a thin film of gold but unfortunately, it had very low efficiency. After this invention, the photodiode became the main component of photovoltaic materials. A photodiode is a semiconductor diode that is sensitive to photon radiation, such as visible light, infrared, ultraviolet, X-rays, and gamma rays.

The photovoltaic cells were first used in 1989 when the German Research Ministry initiated a program that made use of 220 PV roofs (Palz 2013). Similarly, Japan installed 539 residential PVs in 1994 (Noguchi 2005). The technological development in the field of photovoltaics has been remarkable, and multiple new technologies have come up increasing the efficiency of photovoltaics. There are three generations of photovoltaics. The silicon wafer-type photovoltaics are termed as 1G. They are mainly made via the crystal growth method. Metallurgical silicon is crystallized and then doped to form silicon photovoltaics. They have an efficiency of about 22–27% (Shubbak 2019).

Thin-film photovoltaics are termed as 2G photovoltaics. They are formed via a deposition method. Thin layers of photovoltaic material are deposited which form a heterojunction barrier. The thickness of these layers is usually below 2 μm. These cells have a much better temperature coefficient and can easily be pasted on thin flexible materials due to negligible thickness. Their efficiency is about 19–24% (Shubbak 2019).

The 3G photovoltaics are still in the R&D phase and have not been mass-produced. They mainly comprise multilayer cells generally made of amorphous silicon or gallium arsenide. Multiple layers with different bandgaps ensure that a much wider spectrum can be absorbed. These p-n layers include materials like Ge, GaInAs, GaInP, and GaAs. They do have an efficiency of approximately 46%. Organic solar cells are also being developed using conductive organic polymers but the efficiency of these cells is nearly 15% and are therefore not used (Shubbak 2019).

9.3.1.1 Future Goals

The current photovoltaic cells which are produced on large scale are efficient, but there is scope of improvement in their efficiency. The photovoltaics are also extremely bulky and consume a lot of space, so development in the basic design of the photovoltaic is necessary and the search for new photovoltaic materials which are efficient will always be an important area of research (Fig. 9.2).

9.3.2 Concentrating Solar Power

The practice of using solar energy to heat water for domestic use can be traced back to ancient cultures. In 1767, the world's first solar collector was made by a Swiss Scientist Horace-Benedict de Saussure. This device consisted of an insulated box

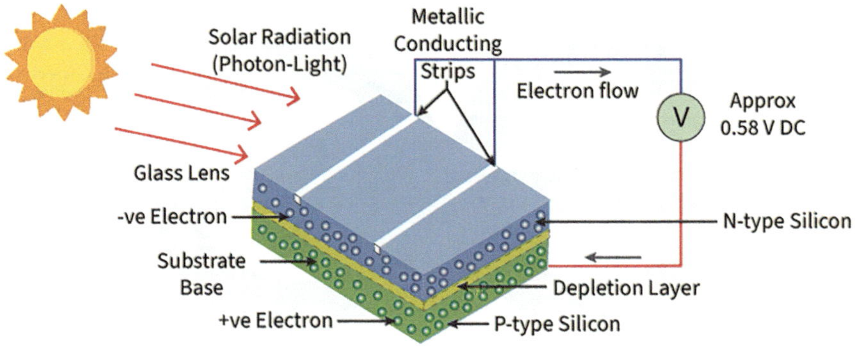

Fig. 9.2 Basic diagram of photovoltaic cell

with three layers of glass to absorb solar radiation. Clarence Kemp patented the first commercial for the first solar water heater named "Climax." The apparatus consisted of metal tanks painted black, placed inside a wooden box, and covered with a glass top. This configuration allowed the system to trap heat effectively, resulting in water temperatures reaching up to 38.8 °C (Norton 2011).

William Bailey modified Kemp's design by dividing it into two components: a solar thermal collector for capturing solar radiation and a storage tank to store the hot water produced. This was the first time the principle of thermosyphon was used to circulate water in the collector and tank without the aid of pumps (Raisul Islam et al. 2013).

Solar radiation is concentrated on water bodies using reflectors, this results in steam generation which can be used to drive a turbine to produce electricity.

There are two types of solar thermal collectors, i.e., concentrating and non-concentrating. The non-concentrating type mainly consists of:

1. Flat plate collectors: A solar radiation absorber plate is covered with a glass layer and inside the housing with the help of tubes cold water comes from one side and hot water from the other. Used for domestic purposes to provide water at 80 °C.
2. Evacuated tube collectors: Concentric tubes are used, and light passes through the transparent outer tube and is absorbed by the inner coated tube which leads to heating. A vacuum is created in the inner tube to not allow heat transfer.

Concentrated solar thermal collectors are divided into three types:

1. Compound parabolic collector: Two parabolic mirrors are used to concentrate light onto the tube placed along the focal line.
2. Parabolic trough collector: Concave mirror used to concentrate light on the tube placed along the focus.
3. Linear Fresnel collector: It consists of flat or slightly parabolic mirrors that focus the solar rays on the receiver tube placed along the focal plane of the reflectors.
4. Central receiving system/solar tower: Solar energy is concentrated by a heliostat field into the central receiver on the top of a tower (Faisal Ahmed et al. 2021).

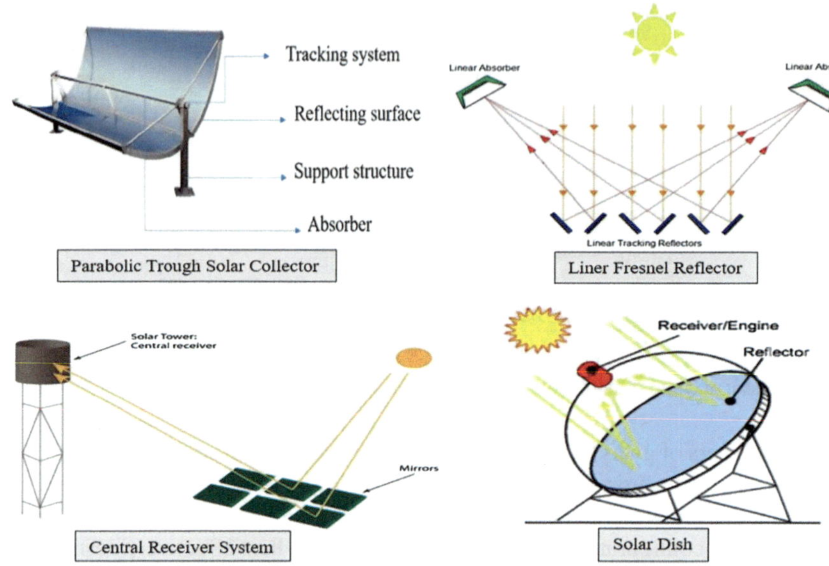

Fig. 9.3 Concentrating solar power technologies [reprinted from Ramde et al. (2020) with permission from Sustainability. Copyright 2020, MDPI {CC BY}]

9.3.2.1 Future Goals

The main goal is to build passive houses which not only fulfils the energy requirement of that particular house but also of that particular society and vicinity. Along with building passive houses, the aim is to make flexible building insulation systems, indoor climate systems, and modular construction elements that utilize solar energy to a maximum extent (Fig. 9.3).

9.3.3 Using Solar Energy for Wastewater Treatment

The global demand for freshwater is continually rising due to population growth, industrial activities, and agricultural needs. Solar energy can play a crucial role in wastewater treatment by providing a sustainable and renewable power source for various treatment processes, reducing reliance on conventional energy sources, and lowering operational costs. There are multiple ways in which solar power can aid different methods of treating wastewater. They are as follows:

1. Solar desalination

   Solar desalination or solar distillation is a method that uses solar power to convert solar water into potable water even fit for irrigation. The most common example of solar desalination is a solar still. It is an enclosed region that allows the evaporation and condensation of water to take place (also called direct solar desalination). In indirect solar desalination, solar energy is initially collected and later used to carry out humidification and dehumidification processes. The first patent for solar distillation was granted in 1870 to Norman Wheeler and Walton Evans. In 1872, Charles Wilson, a Swedish Engineer started a solar distillation plant (Delyannis 2003). The USA started using solar desalination in the early 1950s, and they set up multiple pilot plants at different locations. All these plants aimed to solve water scarcity issues in isolated and remote areas. Currently, there are multiple desalination plants located all over the globe. Almost 22 million m^3 of fresh water is produced using desalination out of which only 1% is obtained using solar power (Kalogirou 2023).

2. Solar disinfection and detoxification

   Solar disinfection is a method that harnesses solar power to make extremely contaminated water safe for human consumption. It uses photovoltaic cells to generate electricity. When this electric current is passed through contaminated water, electrolysis occurs, producing oxidative free radicals.These free radicals help kill germs and pathogens, effectively disinfecting the water. It was first discovered in the early 1980s at the American University of Beirut. Multiple small-scale advances have been seen in the field of solar disinfection, and various alums have also been developed that are used along with solar disinfection to eliminate turbidity.

   Solar detoxification is a process that destroys harmful pollutants by thermal reactions using solar energy as a source. This process is aided by a catalyst which is activated when UV rays are incident on it. The catalyst is responsible for the formation of radicals that break down pollutants into non-toxic by-products like carbon dioxide and water (Shirin and Yadav 2014). The first field test of solar decontamination (detoxification) was held in 1991 in California (Anderson and Clyne 1991).

Solar wastewater treatment is still a small-scale process and therefore the aim is to convert this R&D operation into an industrial operation. Further, it is desirable to make solar wastewater treatment more economically viable, and the formation of new and efficient catalysts for wastewater treatments will also be a constant area of research in the future.

9.3.4 Generation of Green Hydrogen

Using fossil fuels as an energy source is the leading source of emitting greenhouse gases like CO_2 which cause global warming, pollution, respiratory diseases, etc. Hydrogen can be used as an alternative source that produces water as the only by-product released. About 120.7 kJ/g is the energy content of hydrogen (Savla et al. 2020). Hydrogen can be produced using solar energy through two methods: (1) photobiological and (2) photoelectrolysis.

1. Photobiological

   Microorganisms like cyanobacteria, green algae, purple non-sulfur bacteria (PNSB), and dark fermentative bacteria are used to produce H_2. Photobiological production of H_2 requires electrons from electron-donating species, and this process can be categorized into oxygenic and non-oxygenic H_2 production. In the oxygenic H_2 production, water is split into H_2 and oxygen (biophotolysis) by eukaryotic microalgae and cyanobacteria in water using [FeFe] hydrogenase and [NiFe] hydrogenase, with ferredoxin and NAD(P)H as the electron-donating species (Ghirardi et al. 2007). Non-oxygenic/photofermentation process is carried out by PNSB where organic acids and sunlight are used to form H_2 with nitrogenase as a catalyst.

   Due to the formation of oxygen in the oxygenic hydrogen production method, the catalyst (hydrogenase) is deactivated by diffusion of oxygen to the catalytic centers whereas in the case of PNSB bacteria, no oxygen is produced, hence no deactivation of nitrogenase by bacteria (Poudyal et al. 2015). In 1933, Majory Stephenson reported that the reduction of methylene blue with hydrogen was catalyzed by cell suspension Later in 1939, Hans Gaffron noted that green algae (*Chlamydomonas reinhardtii*) produced H_2 under certain conditions. Anastasios Melis in the late 1990s observed that in the absence of sulfur, algae produce H_2 rather than O_2 (normal photosynthesis) and hydrogenase is the enzyme responsible for this reaction (Melis et al. 2000).

2. Photoelectrolysis

   Photoelectrolysis/photoelectrochemical splitting of water to form hydrogen and oxygen involves a photovoltaic system, where the electricity generated from sunlight by a photovoltaic cell is coupled with an electrolyzer which splits water into H_2 and O_2. Another method is directly splitting water into H_2 and O_2 using sunlight by placing a semiconductor in the solution. The reduction of water to H_2 or oxidation to O_2 depends on the band gap between the conduction and the valence band of the semiconductor material in such a way that it can utilize the broad spectrum of sunlight. The semiconductor material should also remain stable in the electrolyte and corrosion-free. TiO_2 and WO_3 are stable in an acidic medium but are only active in the short wavelength spectrum. CdTe/InP have small band gaps but corrode or become inert (Currao 2007). Photoelectrochemical water splitting using TiO_2 was discovered by Fujishima and Honda in 1972.

The aim is to increase the activity of enzymes, which will increase the production of green hydrogen. Developing new strains to increase the efficiency of hydrogen production is the area to undertake for future goals (Fig. 9.4).

9.3.5 *Solar Farms*

Solar farms also known as solar parks or photovoltaic power stations are a humongous network of photovoltaic panels connected in a grid system to harness solar energy. They are designed to supply power at the utility level. ARCO Solar built the

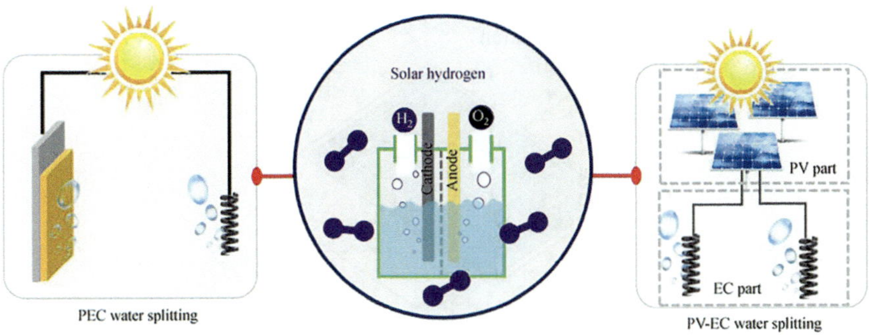

Fig. 9.4 Solar hydrogen production concentrating solar power technologies [reprinted from Wang et al. (2021) with permission from Frontiers in Energy. Copyright © 2021, Higher Education Press]

first 1 MW solar park at Lugo, California in 1982 (Arnett et al. 1984). In 1984, a 5.2 MW solar farm was built in the Carissa Plains (Wenger et al. 1991). Soon multiple solar farms were constructed all around the globe in countries like the USA, China, India, France, and many more. The advancement in technology has helped us make multiple solar farms that can generate almost 100 MW and also a few plants that operate at 1 GW capacity. Solar farms are of different types. An agrivoltaic solar farm is a farm that is used for both agriculture and harnessing solar power (Dinesh and Pearce 2016). Similarly, a solar landfill is a landfill that is converted into a solar farm by installing an array of panels (Barone 2022).

Type of solar array arrangements:

1. Fixed arrays

   Solar panels when installed at a particular elevation and inclination to increase the overall efficiency of the solar power harnessed can be called a fixed array system of solar panels. They highly depend on the local climate and the geological landscape. Even though these kinds of arrays increase efficiency, the complex design and the heavy installation costs outweigh the increase in efficiency.
2. Dual-axis trackers

   These kinds of arrays are solar panels that have a tracker which is capable of monitoring the daily movement of the sun throughout the day and also through different seasons. These trackers help align the solar panels in the normal direction of the incident solar rays. These arrays are quite spaced out to prevent inter-shading. They are mainly stationed in places where we get almost direct sun rays (the regions near the equator). Single-axis trackers are similar to dual-axis trackers but they track the movement of the sun only in one dimension and not through different seasons.

It is important to reduce the installation and maintenance costs of solar farms so that the common man can also use his land to install a solar farm. Also, making the solar panels space efficient by developing a more effective design could be useful (Fig. 9.5).

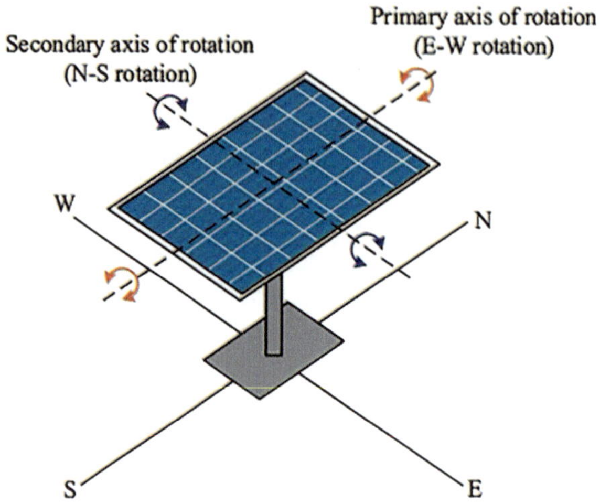

Fig. 9.5 Solar panel with a dual-axis tracker technologies [reprinted from Jamroen et al. (2020) with permission from Sustainable Energy Technologies and Assessments. Copyright 2019 Elsevier Ltd]

9.3.6 Transparent and Flexible Solar Cells Using Graphene

Graphene is a two-dimensional single layer of carbon atoms that is extremely strong, almost entirely optically transparent, flexible, and conductive material but it is not good at collecting electrical current generated by the solar cell, hence graphene oxide can be useful for solar panels. Due to 90–100% optical transparency, graphene can also be used as a replacement for indium tin oxide which is expensive as transparent electrodes in polymer-based solar cells. This transparent electrode using a graphene-silicon heterojunction solar cell was first created in 2010 (Li et al. 2010).

Properties of graphene-based solar cells can be tuned according to the doping mixture used. In 2012, flexible solar cells were made by chemically doping graphene multilayer of organic solar cells by poly(3-hexylthiophene) and [6,6]-phenyl-C61-butyric acid methyl ester, i.e., P3HT and PCBM, respectively, with a power conversion efficiency of approximately 2.5% (Lee et al. 2012). In 2015, graphene was coated with PEDOT: PSS polymer creating a semi-transparent perovskite solar cell with an efficiency of 12% (Hong et al. 2008). In 2018, a super flexible bis(trifluoromethanesulfonyl-amide) doped graphene electrode for perovskite solar cell was reported with a power conversion efficiency of 18.9% (Heo et al. 2018).

The further research can be aimed at making all the photovoltaics graphene-based, as graphene significantly improves the efficiency of a photovoltaic to a high extent.

9.4 Conclusions

Concluding the chapter, we surely know that the rising population coupled with the environmental impacts of fossil fuels necessitates the use of renewable energy sources like solar energy. Solar energy is abundantly available and virtually inexhaustible, offering a pathway to fulfil our energy needs. Solar technologies have progressed from the experimental stage in 1950s to the practical applications like photovoltaic cells, solar water heating, etc. In photovoltaic cells, recent advancements include methods of increasing their efficiency. Concentrating solar power has helped in the industrial application of solar energy to generate electricity by driving a turbine, solar energy can also be applied for desalination or disinfection of wastewater generated. Ongoing research aims to increase the efficiency of solar cells and use of novel materials like graphene for flexible and transparent applications and hydrogen production technologies.

Acknowledgments The authors declare no conflict of interest. This work did not receive any specific grant from public, commercial, or not-for-profit funding agencies. The authors acknowledge the Institute of Chemical Technology, Mumbai, for the resources provided in writing this chapter.

References

Anderson JV, Clyne RJ (1991) Solar detoxification technology: using energy from the sun to destroy hazardous waste. In: Challenge of the environment: new technologies - new markets, Reston, Virginia, 5–6 September 1991. https://www.nrel.gov/docs/legosti/old/4474.pdf

Arnett JC, Schaffer LA, Rumberg JP, Tolbert REL (1984) Design, installation and performance of the ARCO solar one-megawatt power plant. In: Photovoltaic solar energy conference; proceedings of the fifth international conference, Athens, Greece, October 17–21, 1983 (A85-11301 02-44), pp 314–320. https://www.semanticscholar.org/paper/Design%2C-installation-and-performance-of-the-ARCO-Arnett-Schaffer/525a81de5b4dd9f881fe61573c7828b444ca193e

Barone E (2 June 2022) U.S. landfills are getting a second life as solar farms. Time. https://time.com/6183376/landfills-becoming-solar-farms/

Chapin DM, Fuller CS, Pearson GL (1954) A new silicon p-n junction photocell for converting solar radiation into electrical power. J Appl Phys 25(5):676–677. https://doi.org/10.1063/1.1721711

Currao A (2007) Photoelectrochemical water splitting. Chimia 61(12):815. https://doi.org/10.2533/chimia.2007.815

Delyannis E (2003) Historic background of desalination and renewable energies. Sol Energy 75(5):357. https://doi.org/10.1016/j.solener.2003.08.002

Dinesh H, Pearce JM (2016) The potential of agrivoltaic systems. In. Renew Sust Energ Rev 54:299. https://doi.org/10.1016/j.rser.2015.10.024

Faisal Ahmed S, Khalid M, Vaka M, Walvekar R, Numan A, Khaliq Rasheed A, Mujawar Mubarak N (2021) Recent progress in solar water heaters and solar collectors: a comprehensive review. Therm Sci Eng Prog 25:100981. https://doi.org/10.1016/j.tsep.2021.100981

Ghirardi ML, Posewitz MC, Maness PC, Dubini A, Yu J, Seibert M (2007) Hydrogenases and hydrogen photoproduction in oxygenic photosynthetic organisms. In. Annu Rev Plant Biol 58:71. https://doi.org/10.1146/annurev.arplant.58.032806.103848

Heo JH, Shin DH, Song DH, Kim DH, Lee SJ, Im SH (2018) Super-flexible bis(trifluoromethanesulfonyl)-amide doped graphene transparent conductive electrodes for photo-stable perovskite solar cells. J Mater Chem A 6(18):8251. https://doi.org/10.1039/c8ta02672f

Hong W, Xu Y, Lu G, Li C, Shi G (2008) Transparent graphene/PEDOT-PSS composite films as counter electrodes of dye-sensitized solar cells. Electrochem Commun 10(10):1555. https://doi.org/10.1016/j.elecom.2008.08.007

Jamroen C, Komkum P, Kohsri S, Himananto W, Panupintu S, Unkat S (2020) A low-cost dual-axis solar tracking system based on digital logic design: design and implementation. Sustain Energy Technol Assess 37:100618. https://doi.org/10.1016/j.seta.2019.100618

Kalogirou SA (2023) Solar energy engineering: processes and systems. Academic Press

Lee S, Yeo JS, Ji Y, Cho C, Kim DY, Na SI, Lee BH, Lee T (2012) Flexible organic solar cells composed of P3HT:PCBM using chemically doped graphene electrodes. Nanotechnology 23(34):344013. https://doi.org/10.1088/0957-4484/23/34/344013

Lewis NS, Crabtree G, Nozik AJ, Wasielewski MR, Alivisatos P, Kung H, Tsao J, Chandler E, Walukiewicz W, Spitler M, Ellingson R, Overend R, Mazer J, Gress M, Horwitz J, Ashton C, Herndon B, Shapard L, Nault RM (2005) Basic research needs for solar energy utilization. Report of the basic energy sciences workshop on solar energy utilization, April 18–21, 2005

Li X, Zhu H, Wang K, Cao A, Wei J, Li C, Jia Y, Li Z, Li X, Wu D (2010) Graphene-on-silicon Schottky junction solar cells. Adv Mater 22(25):2743. https://doi.org/10.1002/adma.200904383

Medici P (2018) The Trombe Wall during the 1970s: technological device or architectural space? Critical inquiry on the Trombe Wall in Europe and the role of architectural magazines. SPOOL 5(1). https://doi.org/10.7480/spool.2018.1.1938

Melis A, Zhang L, Forestier M, Ghirardi ML, Seibert M (2000) Sustained photobiological hydrogen gas production upon reversible inactivation of oxygen evolution in the green alga *Chlamydomonas reinhardtii*. Plant Physiol 122(1):127. https://doi.org/10.1104/pp.122.1.127

Noguchi M (2005) Communication approaches to marketing solar photovoltaic homes in Japan. https://www.researchgate.net/publication/228931649_Communication_Approaches_to_Marketing_Solar_Photovoltaic_Homes_in_Japan

Norton B (2011) Solar water heaters: a review of systems research and design innovation. Green 1(2). https://doi.org/10.1515/GREEN.2011.016

Palz W (2013) Introduction to solar power for the world. Sol Power World 4:323–334. https://doi.org/10.4032/9789814411882

Perez R, Perez M, Renné D (2015) A fundamental look at supply side energy reserves for the planet. IEA SHC Sol Update 62:4–6. http://www.wise-uranium.org/stk.html?src=stkd03e

Poudyal RS, Tiwari I, Koirala AR, Masukawa H, Inoue K, Tomo T, Najafpour MM, Allakhverdiev SI, Veziroğlu TN (2015) Hydrogen production using photobiological methods. In: Compendium of hydrogen energy: hydrogen production and purification, vol 1, pp 289–317. https://doi.org/10.1016/B978-1-78242-361-4.00010-8

Raisul Islam M, Sumathy K, Ullah Khan S (2013) Solar water heating systems and their market trends. Renew Sust Energ Rev 17:1. https://doi.org/10.1016/j.rser.2012.09.011

Ramde EW, Tchao ET, Fiagbe YAK, Kponyo JJ, Atuah AS (2020) Pilot low-cost concentrating solar power systems deployment in Sub-Saharan Africa: a case study of implementation challenges. Sustainability (Switzerland) 12(15). https://doi.org/10.3390/SU12156223

Savla N, Shinde A, Sonawane K, Mekuto L, Chowdhary P, Pandit S (2020) Microbial hydrogen production: fundamentals to application. In: Microorganisms for sustainable environment and health. Elsevier, pp 343–365. https://doi.org/10.1016/B978-0-12-819001-2.00017-6

Schlaich J, Schiel W (2003) Solar chimneys. In: Encyclopedia of physical science and technology. Academic Press. https://doi.org/10.1016/b0-12-227410-5/00699-2

Shirin S, Yadav AK (2014) Physico-chemical analysis of municipal wastewater discharge in Ganga river, Haridwar district of Uttarakhand, India. Curr World Environ 9(2):536–543. https://doi.org/10.12944/CWE.9.2.39

Shubbak MH (2019) Advances in solar photovoltaics: technology review and patent trends. Renew Sustain Energy Rev 115:109383. https://doi.org/10.1016/j.rser.2019.109383

Wang Z, Gu Y, Wang L (2021) Revisiting solar hydrogen production through photovoltaic-electrocatalytic and photoelectrochemical water splitting. Front Energy 15:596–599. https://doi.org/10.1007/s11708-021-0745-0

Wenger HJ, Schaefer J, Rosenthal A, Hammond B, Schlueter L (Oct 1991) Decline of the Carrisa Plains PV power plant: the impact of concentrating sunlight on flat plates. In: The conference record of the twenty-second IEEE photovoltaic specialists - 1991. https://ieeexplore.ieee.org/document/169280

Chapter 10
Renewable Solar Energy Potentials and Policies of Bangladesh

Zelina Sultana and Md Mofazzal Hossain

Abstract The climate and environment of Bangladesh are very fortunate to use solar energy as a renewable energy source. In one way, it diminises the use of fossil fuels and carbon emissions, and in another way, it increases the economic growth and sustainability of the planet. Although worldwide the use of solar energy is increasing to generate electricity, Bangladesh only uses 2.8% of solar energy out of a total of 3.8% of the renewable energy sources in of the country. This keeps Bangladesh too far from reaching its promise to the Climate Vulnerable Forum that it will increase its renewable energy sources by 30% and 40%, respectively, in 2031 and 2041, and to achieve sustainable development goals (SDGs). As Bangladesh is a sunshine country with an average of 4–5 kWh/m^2 of daily solar radiation, it can potentially keep its promise of using renewable energy by adopting comprehensive plans and policies relating to the use of solar energy. The main objective of this part is to analyze the potential of using solar energy as renewable energy in Bangladesh by highlighting current policies and strategies.

Keywords Bangladesh · Policies · Renewable energy · Solar energy · Sustainability

Z. Sultana (✉)
Faculty of Business and Law Anglia Ruskin University, Cambridge, UK

Department of Law, Jagannath University, Dhaka, Bangladesh
e-mail: zelina@law.jnu.ac.bd; zs17@aru.ac.uk

M. M. Hossain
Department of Law, Jagannath University, Dhaka, Bangladesh

A. K. Yadav (ed.), *Powering Future: Clean Energy Solutions for a Sustainable Planet*, Clean Energy Production Technologies, https://doi.org/10.1007/978-981-96-5055-2_10

10.1 Introduction

The location of Bangladesh is in the subtropical zone (situated between 20°34′ and 26°38′ north latitudes and 88°01′ and 92°41′ east longitude) which is an ideal area for harnessing solar energy, and Bangladesh receives daily solar radiation of between 4.2 and 5.5 kWh/m^2, or 1862.5 kWh/m^2 annually (Islam et al. 2017). The geographic location of this country is favorable to using its sunlight to produce solar energy which might be a potential source of renewable energy (renewable energy, hereinafter RE) and act as a sustainable and environmentally friendly energy source in the country. Undoubtedly, Bangladesh has a resource limitation but has an increased demand for energy, so it can select solar energy as an alternative power source due to its availability and cheap cost (Uzzaman 2023). Additionally, it can be produced in consumer homes, which is more environmentally friendly and distinct from other forms of energy production such as fossil fuel-based power plants, hydropower, and wind turbines. It also helps general energy consumers turn them into producers which help to grow a sense of liability and assists in creating a new consumer group termed "Prosumer" (Farghali et al. 2023).

Solar energy is comparatively cheaper to produce than other renewable energy sources, which cost only $36/MWh, while wind turbines cost $38, hydropower costs $64/MWh, geothermal energy costs $75/MWh, and biomass costs $114/MWh (Uzzaman 2023). Additionally, the location of other energy sources is complex, far from people's residences, and close to environmental entities, which are the constraints for decentralization of energy production and damaged natural entities. Thus, alternative energy production through the use of solar power is no exception, considering its price and environmental benefits. Moreover, the world is now achieving its goal of producing zero-carbon emissions, which has triggered every country to use renewable energy sources (hereinafter RES). RES is a sustainable and ecologically supportive source of energy that can produce near-zero emissions and air toxins. It is crucial for preserving ecological equilibrium and combating current climate change.

However, the first formal journey of the RE began in the 1970s with the United Nations Stockholm Declaration on the Human Environment. The UN and other agencies have since emphasized the use of RE resources and have taken some initiatives, including the World Commission on Environment and Development Report, Nairobi Program of Action, IPCC, WMO, COP3, Kyoto Protocol, and Resolution 65/151 (Karim et al. 2019). Recently, the environmental and legal experts of the European Court of Justice have advocated for a shift toward RES to maintain development activities and to strike a balance between ecology and development (Gautam and Sinhab 2019). To address energy scarcity and environmental issues, a shift toward non-conventional sources known as RES, including hydropower, bioenergy, nuclear, wind, and solar energy, has been made (Gautam and Sinhab 2019).

Since 2008, Bangladesh has been working on promoting green energy to address international commitments by introducing plans and policy frameworks, including the National Solar Energy Action Plan 2021–2041, the Power System Master Plan 2010–2016, the Renewable Energy Policy 2008, etc. In 2012, Bangladesh established the Sustainable and Renewable Energy Development Authority (SREDA) to enhance RE. Besides, the country's central bank, "The Bangladesh Bank," has already introduced the "Solar Energy, Biogas, and Effluent Treatment Plant" in 2009 to promote the RE sector through sustainable financing projects (Ijaz 2024). In 2020, the bank introduced the "Sustainable Finance Policy for Banks and Financial Institutions" program, requiring banks to provide 2% of all loans to green and renewable energy projects (Uzzaman 2023). Additionally, a private company named "The Infrastructure Development Company Limited" (hereinafter IDCOL) is the only renewable energy-focused financial institution in Bangladesh that has funded 322 MWp (megawatt peak) of solar projects (Uzzaman 2023).

Until now, Bangladesh only used 2.8% of solar energy out of a total of 3.8% of renewable energy sources in the country (BPDB 2023). This keeps Bangladesh too far from reaching its promise to the Climate Vulnerable Forum that it will increase its renewable energy sources by 30% and 40%, respectively, in 2031 and 2041, and to attain sustainable development goals (SDGs) and other global goals [Mujib Climate Prosperity Plan MCPP, 2021 (Ministry of Environment, Forest and Climate Change Government of the People's Republic of Bangladesh 2022)]. However, the country's limited fossil fuel reserves are insufficient for large-scale development activities. The government is developing energy sources to increase energy production, including initiating the controversial Rampal power plant, considering the necessity of a constant and affordable supply of renewable energy yet too far from succeeding with the refinancing of sustainable energy (Ijaz 2024; Karim et al. 2019). The government promises to expand its RE production, including solar energy, but the total on-grid installed capacity of renewable energy power origination is <3.5% of the total grid capability, which is <1% and puts the country at the bottom of the list of countries producing RE.

Thus, the core aim of this part is to highlight the potentiality of using solar energy as RE in Bangladesh by highlighting current policies and strategies. The study is based on secondary data and literature and finds that the published scholarship is mainly based on scientific discussion of solar energy in Bangladesh, which limits the focus on the legal issues. For this, adopting a qualitative literature review and policy analysis, we analyzed published literature and existing policies to highlight the potential of this RE in Bangladesh through a legal lens and also try to shed light on the limitations in executing solar energy projects and solar waste management. Thus, it finds that Bangladesh has huge prospects for employing solar energy. It is currently regulated by the Bangladesh Renewable Energy Policy 2008, which has inherent weaknesses like a lack of strategic planning and net metering. However, it also finds that Bangladesh has some limitations, like funds and land, to execute solar widely, which limitations can be removed by implementing the existing legal regime. After analyzing the potentiality of solar energy, the study suggests implementing a robust action plan, limiting the cost and tariff of solar equipment, removing some legal barriers for investors, ensuring coordination, and embracing solar waste management strategies with the implementation of existing policies for using solar energy as a renewable energy source will help to step ahead for a sustainable planet.

10.2 Solar Energy Used

Currently, Bangladesh lacks RE production compared with other energy production sources. According to the Bangladesh Power Development Board (BPDB), only 0.38% of power is produced from renewable sources, and 99% of power is manufactured from domestic fuels (natural gas, coal, and hydro 62.17%), petroleum fuels imported (diesel and furnace oil 28.14%), and energy exchange (with India 0.91%) (BPDB 2023). SREDA has published the National Database of Renewable Energy, where total renewable energy is only 1309.047 MWp and approximately 1074.964 MWp (Table 10.1) comes from solar technology (SREDA 2024). In all previous years, the production of power has been at a minimal level and comparatively lower than at present, though the country has the potential to produce and use solar energy.

In Bangladesh, 12 utility scales, including solar home systems, solar irrigation pumps, rooftop systems under NEM, solar drinking water systems, solar mini-grids, solar-powered telecom towers, charging stations, streetlights, and other applications like vehicles, have been used to produce renewable energy (RE) from solar plants. Among all of them, solar park and solar home systems are the two most popular solar energy systems in Bangladesh. Until recently, the process has not been applied countrywide, and the power contribution from solar RE is not mentionable in comparison with the general energy sources. Renewable energy programs, including solar energy systems, are undertaken under private and public initiatives (SREDA 2024). However, each of these has its own features and limitations, which are stated in Table 10.2.

Table 10.1 Solar energy use in Bangladesh through different sources

Technology	Quantity	Off-grid MWp	On-grid MWp	Total MWp
Solar Park	11	0	536	536
Rooftop Solar Except NEM	236	27.778	50.299	78.077
Net Metering Rooftop Solar	2364	0	108.474	108.474
Solar Irrigation	3381	54.616	2.742	57.359
Solar Home System	6,037,689	263.793	0	263.793
Solar Minigrid	28	5.805	0	5.805
Solar Microgrid	0	0	0	0
Solar Nanogrid	2	0.001	0	0.001
Solar Charging Station	15	0.275	0.016	0.291
Solar Street Light	297,691	17.104	0	17.104
Solar Powered Telecom BTS	1933	8.06	0	8.06
Solar Drinking Water System	82	0.095	0	0.095

Source: SREDA 2024

Table 10.2 Used solar technology in Bangladesh with limitations

Technology	Characteristics	Limitations
Solar Park (536 MWp) Total: 68	• Large scale • Commercial purpose • Utilization of non-agricultural land, targeting unproductive agricultural land and railway wasteland for installation	• Large amount of land acquisition • Need to develop land and secure its protection from erosion • Time consuming • Costly • Harsh weather conditions • Dust accumulation • Longer implementation period • Gap in supply chain • Limited local expert human resources
Rooftop Solar Except NEM (78.077 MWp) Total: 236	• Installed for domestic, commercial, and industrial electricity connections • Availability of plenty open rooftops • To address the space constraint issue for installing solar panels	• Most are unused • These panels are not connected to the main circuit of the building • Only connected to common utilities like security lights and staircase lighting • Lack of awareness leads to the wastage of rooftop solar panels in residential and commercial building
Net Metering Rooftop Solar (108.474 MWp) Total: 2364	• To use rooftop solar for commercial purpose • Sharing and selling additional generated electricity through national grid by net metering • Prospect 400 MW in 42 million sq. ft. roof-top space of textile mills	• Net metering tariff is not executed properly • Net metering technology is not sufficiently implemented here • Owners of different industry are being not aware and interested to invest a huge money initially • Owners are confused about the effectiveness of net metering systems
Solar Irrigation (57.359 MWp) Total: 3381	• Provide environment friendly solution in irrigation facility to rural off-grid areas • Reduce dependency on fossil fuel and electricity • Saves millions in foreign currency	• Not working at night and rainy or cloudy days • Lacking technical knowledge to maintain the solar system • Lacking expert in this field • Initial cost is high • Not suitable for large land field irrigation
Solar Home System (263.793 MWp) Total: 6,037,689	• To provide electrification to households and where electrification is challenging and costly • Saved consumption of 1.14 million tons of kerosene • It is a small-scale renewable energy project • Used for domestic and commercial purposes	• Storage capacity is limited • Lifespan of battery is not long • Technical support is costly and not easy access • No scope to use or share additional generated solar electricity • Primary investment is high and not affordable for poor people • Long time duration of rainy and cloudy weather reduces solar energy generation

(continued)

Table 10.2 (continued)

Technology	Characteristics	Limitations
Solar Minigrid (5.805 MWp) Total: 28	• Installed in remote areas where there is a probability to expand the grid in near future such as remote island located in the estuary of the mighty rivers in Bangladesh • Small commercial and households' users can get the opportunity to avail the grid quality electricity • To promote business venture in the project areas	• To maintain the balance between storage and supply on demand is difficult • Limited in a particular area • No opportunity to use additional produced electricity • Maintenance in remote area is difficult and not cost effective
Solar Charging Station (0.291 MWp) Total: 15	• Bangladesh is planning to increase the number of electric vehicles, requiring electrification of the transport sector • To reduce grid burden	• Not cost-effective • Technological barriers • Lack of technical human resources • Charging station totally depend on import from developed country
Solar Street Light (17.104 MWp) Total: 297,691	• Solar LED street lighting, designed to replace 1000 km of streetlights across the six Bangladesh City Corporations • Land acquisition or resettlement is not required	• Battery storage capacity only 11 h, decreased by lapse of time • Difficult to maintain and keep clean for full efficiency • Most of streetlights in remote areas can be stolen • Not provide enough light in rainy and cloudy days
Solar Powered Telecom BTS (8.06 MWp) Total: 1933	• Reduced diesel cost • Lower energy bills • Lower transportation cost • Reduced fossil fuel consumption so less pollution	• The main disadvantage of this system is that if the AC power source is not operational for over 2 h, the entire system will be shut down • Isolated areas are not suitable for the implication of this system • The electric utility grid must be available
Solar Drinking Water System (0.095 MWp) Total: 82	• To ensure safe drinking water in remote area specially in coastal area • Extracted and supplied from surface pond water and reservoir of groundwater	• Expensive initial cost • Dependent on weather and sunlight exposure • Challenging maintenance in remote areas • Limited power storage capacity

Source: Authors prepared based on secondary scholarships

10.3 Prospects and Limitations

Solar energy is promising due to the vast sunlight resources of Bangladesh. Despite the global trend toward renewable energy, Bangladesh struggles to effectively utilize solar energy, including biogas and biomass (Das et al. 2018). It is a sustainable, renewable, and clean energy source derived directly from the sun's rays (Hussain et al. 2024). Bangladesh's natural climatic conditions present opportunities for solar energy production, potentially addressing energy issues in many regions mentioned by Solargies. Solar photovoltaic (PV) and concentrated solar power (CSP) are promising technologies, and solar-powered household systems can use solar radiation to generate electricity. The designs of these systems are based on the number of sunny hours throughout the year in Bangladesh (Mridul et al. 2021).

Bangladesh's superior sunlight availability is a key factor in attracting investments in solar energy. The selection criteria for solar plants include solar radiation, average temperature, and average sunshine hours. In 2010, it was discovered that 1.7% of the country's land might tentatively be used for grid-connected solar projects (Mondal 2010). According to the study, the nation has a potential for 50,174 megawatts (MW) of solar electricity, reflecting the yearly mean worth of solar radioactivity and the competence of the solar panels. In a different study in 2019, Bangladesh might install 341,000 MW of solar and wind power overall. It has a collective rooftop solar, utility-scale solar, and floating solar potential of 191,000 MW (Teska et al. 2019). Similarly, it has a solar probability of 240,000 MW, according to the National Renewable Energy Laboratory (NREL), which uses only 1.5% of the country's terrestrial resources for solar energy (U.S. Agency for International Development 2011).

Considering the prospect, the government is encouraging the use of solar radiation in a number of ways, such as through solar-powered street lights in eight city corporations, drinking water distribution, cold storage, solar-powered telecom, and heaters. In the commercial and industrial sectors, SREDA has begun installing solar water heaters, and the government intends to amend building rules to include facilities for solar-powered heaters (SREDA 2024). BREB has installed 14 solar charging stations to save grid electricity. In order to show the possibilities for such alternatives, IDCOL supplied 138 telecom BTSs in off-grid locations with a solar power solution (IDCOL 2019). Thus, modern solar home systems and innovations are increasingly attractive in Bangladesh. It has successfully introduced solar power usage through the SHS program, providing electricity access to 15 million rural households and 45 million city dwellers in exchange for government subsidies (IDCOL 2019). Solar power technology has a zero-carbon footprint and is economically viable due to its affordability, efficiency, and ability to mitigate poverty, like other Asian and African countries.

Despite all these probabilities, there are some limitations to expanding this in Bangladesh, which needs the utmost priority when making plans relating to solar energy production. Lack of adequate funding and demand for terrestrial requirements for renewable energy projects, especially solar PV, are the reasons for less expansion (Ijaz 2024). Bangladesh has the potential and possibility to initiate solar energy power plants considering the limitations like the necessity of large land, pressure to meet the climate goal related to food and wildlife preservation, managing the solar energy waste, and maintaining the balance of demand for food, energy, and ecological needs (Ijaz 2024). However, some robust and effective plans and policies are essential to consider those issues, including solar waste management, as well as embracing some effective policies relating to solar energy plan development, tariff control in importing equipment for solar power plants, and investor-friendly rules (Ijaz 2024; Shirin et al. 2023). The next part will focus on existing policies and plans relating to solar energy and meeting the challenges of establishing solar energy plants in Bangladesh.

10.4 Extant Legal Regime

Bangladesh has laws relating to the production, supply, and maintenance of energy but lacks a strategy for implementation. However, this part will focus on the plans and policies of Bangladesh focused on energy regulation. The earlier laws and policies of Bangladesh were mainly focused on general energy regulation (mentioned in Table 10.3) and, after some time, shifted their concentration to RE production and passed some plans and policies on RE (Table 10.4). However, considering the prospect and possibilities of solar energy in Bangladesh, policymakers and policies need to concentrate on both the prospects and challenges of this project, such as financial or funding, solar waste management, and land acquisition.

Table 10.3 Legal instruments of Bangladesh to focus on energy

Year	Name	RE provisions
1995	Bangladesh Environment Conservation Act (BECA)	No
1996	Private Sector Power Generation Policy of Bangladesh	No
1996	National Energy Policy Upgraded in 2004 to include RE	Yes
2003	Bangladesh Energy Regulatory Commission Act	No
2007	Policy Guideline for Power Purchase from Captive Power Plant	No
2008	Policy Guideline for Small Power Plant (SPP) in Private Sector	No
2008	Policy Guideline for Enhancement of Private Participation	No
2009	The Bangladesh Climate Change Strategy and Action Plan	Indirectly yes
2015	Bangladesh Energy and Power Research Council Act	No
2015	Nationally Determined Contribution of Bangladesh	Indirectly yes
2015	Energy Efficiency and Conservation Master Plan up to 2030	Indirectly yes
2016	Second Power System Master Plan	Yes
2018	Electricity Act (The previous electricity Act 1910 repealed)	No
2021	Perspective Plan of Bangladesh 2021–2041	Yes
2021	Eighth Five Year Plan (EFYP): 2021–2025	Yes
2022	The Mujib Climate Prosperity Plan (MCPP): Decade 2030	Yes
2018	Bangladesh Delta Plan 2100	Yes

Source: Authors after studying laws/policies relating to energy/renewable energy

Table 10.4 Legal instruments of Bangladesh to focus on renewable energy

Year	Name
2008	Bangladesh Renewable Energy Policy
2012	Sustainable and Renewable Energy Development Authority Act (SREDA)
2020	Sustainable Finance Policy for Banks and Financial Institutions (SFPBFI)
2022	The Renewable Energy Policy (Draft)

Source: Authors after studying laws and policies relating to renewable energy

10.4.1 Energy-Focused General Legal Regime

After independence in 1971, Bangladesh first adopted the Electricity Act, 1910, to regulate its electricity arena, which was repealed by the Electricity Act, 2018. Since 2008, Bangladesh has had no policies or plans relating to RE, and the country first initiated an RE policy in the stated year. However, the Bangladesh Renewable Energy Policy (2008) aims to uphold the use of RE technologies throughout the nation by enhancing the technical, financial, and legal environment. It sets the groundwork for the Sustainable and Renewable Energy Development Authority (henceforth SREDA) and directs institutional structures. The policy describes the stages for developing programs, technologies, and resources, as well as possible financial, legal, and regulatory incentives.

Moreover, shifting energy resources is connected to sustainable energy consumption and environmental protection (Gautam and Sinhab 2019). Realizing this fact, Bangladesh established the Sustainable and Renewable Energy Development Authority (SREDA) in 2014 to fulfill the objectives of the SREDA Act (minimizing the use of fossil fuels). It organizes government initiatives, harmonizes products, tests novel technologies, promotes investor-friendly conditions, carries out research and development, builds capacity, increases consciousness, and creates connections with local, national, and worldwide organizations. It is willing to provide technical assistance for the Clean Development Mechanism (CDM), which requires substantial funding and technical support. It will verify land suitability for renewable resources and associated technologies, ensuring their longevity and quality assurance before implementation. While RE policies are likely to play a noteworthy role in promoting RE, most expectations have not been met yet.

10.4.1.1 The National Energy Policy (NEP) 1996 Upgraded in 2004

It was updated in 2004 with an emphasis on the national adoption of renewable energy (RE), with a target of 10% renewable power share by 2020. The policy consists of only two RE-supportive policy instruments, which should have been more comprehensive. It first sets a portfolio standard for RE in Bangladesh, aiming for a 5% RE share by 2010 and 10% by 2020. This policy provides an exemption from corporate income tax as well as other financial incentives for renewable energy project sponsors.

10.4.1.2 The Nationally Determined Contribution of Bangladesh, 2015 and Energy Efficiency and Conservation Master Plan up to 2030 (2015)

Both of these policies indirectly encourage the use of renewable energy by reducing general energy consumption. With international assistance, Bangladesh's commitment to addressing climate change is outlined in the previous approaches, which established both unconditional and conditional goals for reducing GHG emissions in the transportation, power, and industrial zones. To accomplish this, a later project in partnership with SREDA seeks to cut primary energy use per GDP by 20% by 2030. Nonetheless, these regulations obliquely encourage the nation to employ more renewable energy sources. These policies are primarily target-based, focusing on energy security and phasing out fossil fuels. However, prioritizing carbon emission reduction is not strong enough.

10.4.1.3 Power System Master Plan 2016

The Bangladeshi government's second Power System Master Plan (PSMP) intends to achieve the target of making the nation a high-income nation by 2041. The plan includes five value-added plans of which the advanced deployment of green energy is mentionable. Additionally, the plan uses the energy-GDP elasticity method to assess past trends in energy use and forecast future energy demand for three different scenarios of economic growth.

10.4.1.4 The Bangladesh Delta Plan 2100 and Perspective Plan of Bangladesh 2021–2041

The Delta Plan 2100, which was accepted in 2018, is a longstanding (100-year) visionary strategy. It proposes a 30% renewable energy target by 2041 and demonstrates the considerable potential for land recovery near estuaries and riverbanks. The plan suggests that the land gained through natural reclamation can be used for urbanization or industrialization, highlighting the importance of RE in the country's energy landscape.

With an emphasis on the energy industry, the General Economic Division of the Planning Commission (GOB) has announced its Perspective Plan for Bangladesh 2021–2041. One of the main goals is to reach 115 GW of generation capacity, with 35% of the energy mix coming from imports and renewables. The plan also emphasizes sustainable energy sources. The plan nevertheless deviates in selecting the low-cost primary fuel supply infrastructure and the least-expensive approach for expanding power generation.

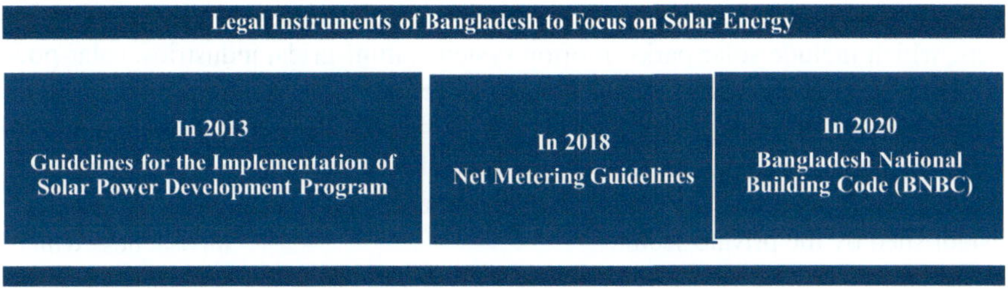

Fig. 10.1 Legal instruments focus on solar energy (Prepared by Authors)

10.4.2 Renewable Energy-Focused Legal Regime Including Solar Energy

Besides the Act SREDA, Bangladesh has already embraced two guidelines, (Fig. 10.1), particularly for solar energy, four legal instruments for RE (Table 10.4) and six plans and policies (Table 10.3) containing provisions for RE expansion. It has also made some amendments to its Bangladesh National Building Code (BNBC), which requires buildings to supply a portion of their electric load from renewable electricity. Residential buildings are required to supply 3% of their total electric load from solar or other renewable sources, while non-residential buildings have a quota of 5%. The level of reflection on renewable energy support mechanisms in the BNBC is high, but the rest of the six support mechanisms are absent. The BNBC now requires buildings to supply a portion of their electric load from renewable sources, with residential buildings supplying 3% and non-residential buildings supplying 5%.

10.4.2.1 Net Metering Guidelines, 2018

The Net Metering Guidelines have been initiated in July 2018 by the Power Division of the Ministry of Power, Energy, and Mineral Resources (MPEMR) to boost the widespread installation of rooftop solar PV systems. With the help of a bi-directional smart meter, customers can connect their RE systems to the distribution grid and take advantage of the net metering mechanism, which turns them into "prosumers." If there is surplus electricity generated by the RE system, it is consumed by the prosumer and fed into the grid through the net meter.

10.4.2.2 Strategies for Implementing Solar Power Development Program 2013

To fulfill the 10% target of power production from RES by 2020, SREDA prepared the "Guidelines for the Implementation of Solar Power Development Program" in October 2013. Two categories of projects were included in the 500 MW solar power generation plan announced by the GOB: commercial and social solar power projects. The private sector is mostly responsible for developing commercial projects, which include solar parks, rooftop systems, mini-grids, industries, solar-powered irrigation pumps, and private solar power projects. Social initiatives include the solar electrification of government offices, off-grid government buildings, educational facilities, religious centers, and rural health centers. While social projects are created and funded by grants from the government, commercial projects are mainly established by the private sector.

10.4.2.3 Sustainable Finance Policy for Banks and Financial Institutions (SFPBFI) 2020

The Bangladesh Bank published justifiable funding rules for banks and economic institutions (SFPBFI) in 2020, which include a green finance taxonomy and an analysis of a green bond standard. The policy was influenced by the government's progress plan, a universal obligation to sustainability, and the bank's planned objectives. The Sustainable Finance Taxonomy and Sustainable Linked Finance are part of the policy, inclusive of sustainable green products, schemes, proposals, and finance focused on sustainability fields. Green finance is one of its components, along with agriculture, CMSME (cottage, micro, small, and medium enterprises), and socially responsible finance linked to sustainability. However, this Act paves the way for financing solar energy through the promotion of renewable energy (Ijaz 2024).

10.4.2.4 The Renewable Energy Policy (Draft) 2022

This draft addresses a wide range of issues linked to renewable energy targets and is more comprehensive and ambitious. The draft outlines a proposal for creating a master plan for the development of renewable energy, together with possible project locations, resource analyses, and a range of grid integrations. A framework for power generation companies like the Infrastructure Development Company (IDCOL), the Power Grid Company of Bangladesh (PGCB), and the Bangladesh Energy Regulatory Commission (BERC) is also outlined. The updated draft promotes ideas like green energy and the market for energy storage while putting a strong emphasis on private sector investments and climate goals. It includes a wider range of energy sources and presents novel technologies such as biofuel, waste-to-electricity, and regional clean energy markets. Additionally, it establishes a 10-year policy period, or until a new policy is established, making goal-setting possible. Overall, the revised draft aims to uphold international norms and practices while promoting clean energy (Moazzem and Hridoy 2023). It also contains exemptions from import tariffs and VAT on all materials and equipment used in the production of renewable energy, as well as concessions for the captive electrical consumption of power producers. The revised draft also introduces regulatory policies like the Renewable Purchase Obligation (RPO) and Renewable Energy Certificate (REC) at the generation, distribution, utility, and consumer levels.

10.4.3 Legal Regime for Solar Waste Management

The solar home system in Bangladesh has added the maximum amount of waste production and materials regained from the consumed panels. It is estimated that solar PV waste in Bangladesh could reach 5.496 million tons between 2025 and 2060, including glass, plastic, aluminum, silicon, and copper (Tasnim et al. 2022). Only in Dhaka does the accumulation of 10–20 thousand tons of solar waste in landfills every year pose risks (Antara 2023). Antara also mentioned that the process of breaking down the batteries, which often involves burning, releases polluted air into the environment, further polluting air, soil, and waste. Thus, the lack of a proper recycling process for solar waste in Bangladesh caused a significant environmental and human health impact. Therefore, proper policies and recycling processes are crucial to preventing this issue and protecting the environment and human health.

10.4.3.1 Solar E-waste Management Policy

Environmental laws, policies, and rules, such as the Environment Conservation Act 1995, the Environmental Court Act 2000, and the Environmental Conservation Rules 1997, fail to include the country's e-waste management. The e-waste management strategy first appeared in the National 3R (Reduce, Reuse, and Recycle) strategy; subsequently, some other policies have been included, such as the Strategy for Waste Management, 2010, the draft E-waste Management Rules, 2017, and the National Environment Policy Bangladesh, 2018. These policies promote the 3Rs method, but implementation remains a challenge (Tasnim et al. 2022). The Department of Environment released its rules relating to hazardous waste management in 2021, but they do not apply to solar waste, leading to concerns about a significant increase in solar e-waste (Table 10.5).

Since none of the rules specifically address the disposal of e-waste produced by solar or renewable energy sources, Bangladesh's environment and public health are not protected against solar waste. The environment department in Bangladesh

Table 10.5 Waste management legal regime in Bangladesh

Policy regime	Strategic Management Features
National 3R Strategy for Waste Management, 2010 (DoE/MoEFCC)	E-waste recycling must be regulated due to hazardous constituents in waste components, and governments should promote public–private partnership for e-waste recycling projects
E-waste management rules 2017 (Draft) (DoE/MoEFCC)	The 3Rs method is used to manage e-waste, while the Extended Producer Responsibility (EPR) policy ensures producers treat and dispose of electronic items, but end-of-use solar panels are not included
National Environment Policy Bangladesh 2018 (DoE/MoEFCC)	Section 3.23.7 of the 3R policy focuses on electronic waste management and e-pollution, excluding e-waste from renewable energy sectors

Source: Authors after studying policies relating to waste management

proposed a plan to address solar waste management. Before executing solar panels countrywide, it is essential to embrace policies for solar waste management by initiating some provisions relating to solar waste management through the e-waste management rules and the National 3R strategy for waste management.

10.4.3.2 Policy of Land Acquisition

A shortage of agricultural land and maintaining food security for large populations make it extremely difficult to build massive solar farms. Thus, researchers have advised a technique called "agrivoltaics," or transparent solar panels, which are designed to float atop reservoirs, lakes, rivers, and other bodies of water (Ijaz 2024; Supti 2022). A land acquisition policy for RE is essential in this respect. The legislation regarding the acquisition and requisition of immovable property allows the state to acquire the property for public purposes. And here, public purpose is not defined, so any declaration regarding the establishment of the establishment of a solar project for public purpose is invalid. The government can acquire unlimited land or property for solar projects. With a focus on researcher recommendations, the government will undertake a land assessment to find potential land uses for utility-scale renewable energy projects and select Khas land, railway unused land, or land with lesser agricultural value for this purpose (Ijaz 2024; USAID Bangladesh Advancing Development and Growth through Energy 2023).

10.5 Suggestions

The Renewable Energy Policies and Plans of Bangladesh aim to harness RE resources and technologies in rural, peri-urban, and urban areas, encourage public and private sector investments, develop sustainable energy supplies, and scale up RE contributions to electricity production. However, the government faces policy challenges in the RE sector, including a lack of private sector incentives, inconsistent policies, a lack of feed-in tariffs, fossil fuel subsidies, weak environmental regulations, the absence of RE or solar energy waste management, and an overall weak implementation culture. Thus, to make a successful attempt at starting RE, the forthcoming policies should address the following:

10.5.1 Integrated Action Plan

The Bangladeshi government is entrusted with developing an open, comprehensive, and publicly available action plan for RE, with a particular emphasis on solar energy and a long-term outlook. The plan should address existing restrictions and problems, ensuring stakeholder participation. It should be integrated with national development plans like power generation, transmission, and economics and should be regularly updated to keep up with other important development plans. The target for renewable energy generation should be achievable through the action plan. Because it depends on the weather, renewable energy has a lower generation factor than conventional power. The goal needs to be stated as a percentage of overall demand, or in GWh. Based on the availability of resources, the techno-economic assessment, the social analysis, and the involvement of stakeholders, the type and level of the future target should be decided (USAID Bangladesh Advancing Development and Growth through Energy 2023).

10.5.2 Enhance Coordination

SREDA needs to enhance its institutional capacity to promote sustainable energy, minimize coordination gaps, and accelerate project implementation. It should launch a "one-stop service" for the private sector to obtain project information and resources. SREDA should collaborate with academic institutions and related stakeholders to improve policies, business models, financing instruments, incentive programs, and waste management (Alam 2023). The government should promote community engagement to mitigate land acquisition impacts on local individuals and marginalized groups and consider creating employment opportunities for displaced individuals. Public and private collaboration and participation are crucial for achieving the goals. Regular coordination is essential to ensuring the plan remains effective and accessible.

10.5.3 Reductions of Cost and Tariff and Some Legal Barriers

The government should reduce the cost of installation of solar projects like in other countries and reduce tariffs on imported parts for RE. Bangladesh's high investment costs necessitate a payback period shorter than the agreement's duration. Tariffs should be adjusted to accommodate the diversity of RES, such as solar. The tariff should promote innovative technologies like rooftop solar photovoltaic panels. Regarding legal barriers, the net energy metering (NEM) guidelines should eliminate the 10-megawatt maximum capacity restriction, allowing consumers to install 70% of the sanctioned load. This would allow factories in industrial parks, economic zones, and export processing zones to install solar systems; here also, net metering should be properly implemented (Khatun et al. 2022).

10.6 Concluding Remarks

Bangladesh has the potential to initiate solar energy projects to expand its RE sources to fulfill its international commitment to reduce carbon emissions and increase RE use toward its journey to achieve the SDGs and other global goals. The paper focused on the current solar energy used by the country, its possibilities for enhancement, and some limitations relating to it. Limitations like the land crisis and fund crisis could be minimized with the help of existing laws and policies, but solar waste management needed special care by including a provision in existing rules and plans or by initiating new policies for solar waste management. The paper found that the laws, policies, and plans of Bangladesh are supportive of expanding solar energy resources, but high costs and tariffs, a lack of integrated action plans, and a lack of synergy among different stakeholders are now burning issues against the successful attempt at solar project expansion. Thus, the paper suggested initiating some robust action plans and implementing them accordingly, including reducing the cost and tariff. Then it will be easy for Bangladesh to contribute to establishing a sustainable planet.

Acknowledgments The authors express their gratitude to the editors for allowing us to contribute to their edited book. We also thank our prestigious institution and all who directly and indirectly assisted us in completing and finalizing this research.

References

Alam S (2023) Towards a rooftop solar transition in Bangladesh. The Institute for Energy Economics and Financial Analysis (IEEFA). https://ieefa.org/resources/towards-rooftop-solar-transition-bangladesh. Accessed 20 May 2024

Antara NF (2023) Is a mountain of solar waste on the horizon?. Dhaka Tribune. https://www.dhakatribune.com/bangladesh/320412/is-a-mountain-of-solar-waste-on-the-horizon. Accessed 24 June 2024

Bangladesh Power Development Board (BPDB) (2023) Annual report 2022–23. https://bpdb.portal.gov.bd/sites/default/files/files/bpdb.portal.gov.bd/page/771c9a89_a06c_4c2f_9b8c_699d17ed769a/2023-12-20-09-13-bc35a2ac1abda167c7c48437e4dc1061.pdf

Das A, Halder A, Mazumder R, Saini V, Parikh J, Parikh K (2018) Bangladesh power supply scenarios on renewables and electricity import. Energy 155:651–667. https://doi.org/10.1016/j.energy.2018.04.169

Farghali M, Osman A, Chen Z et al (2023) Social, environmental, and economic consequences of integrating renewable energies in the electricity sector: a review. Environ Chem Lett 21:1381–1418. https://doi.org/10.1007/s10311-023-01587-1

Gautam R, Sinhab S (2019) Outlook on the Indian scenario of solar energy strategies: policies and challenges. Energ Strat Rev 24:331–341. https://doi.org/10.1016/j.esr.2019.04.005

Hussain MN, Zaman MR, Halim MA, Ali MS, Ali Khan MY (2024) A comprehensive review of renewable and sustainable energy sources with solar photovoltaic electricity advancement in Bangladesh. Control Syst Optim Lett 2(1):1–7. https://doi.org/10.59247/csol.v2i1.59

IDCOL (2019) Annual report 2018. https://idcol.org/annual_report/IDCOL-Annual-Report_2018_new.pdf

Ijaz H (2024) Refinancing sustainable energy projects in Bangladesh. The Daily Star. https://www.thedailystar.net/anniversary-supplement-2024/innovation-key-for-the-future/news/refinancing-sustainable-energy-projects-bangladesh-3547451#. Accessed 12 June 2024

Islam MR, Sarker PC, Ghosh SK (2017) Prospect and advancement of solar irrigation in Bangladesh: a review. Renew Sust Energ Rev 77:406–422. https://doi.org/10.1016/j.rser.2017.04.052

Karim ME, Karim R, Islam MT, Sukki FM, Bani AN, Muhtazaruddin MN (2019) Renewable energy for sustainable growth and development: an evaluation of law and policy of Bangladesh. Sustain For 11(20):1–30. https://doi.org/10.3390/su11205774

Khatun F, Bari E, Al Kabir F, Centre for Policy Dialogue (CPD) and Delegation of the European Union to Bangladesh (2022) Green investment in Bangladesh. https://www.eeas.europa.eu/sites/default/files/documents/2023/Green%20Investment%20in%20Bangladesh_Report_30%20Nov%202022%20-%20Clean%20Version.pdf. Accessed 5 May 2024

Ministry of Environment, Forest and Climate Change Government of the People's Republic of Bangladesh (2022) Mujib climate prosperity plan 2022–2041. https://moef.portal.gov.bd/sites/default/files/files/moef.portal.gov.bd/publications/f6c2ae73_30eb_4174_9adb_022323da1f39/Mujib%20Climate%20Prosperity%20Plan%202022-2041.pdf. Accessed 10 May 2024

Moazzem KG, Hridoy MA (2023) Renewable energy policy (draft) 2022: a comprehensive assessment. Policy brief 2023. CPD. https://cpd.org.bd/resources/2023/07/Renewable-Energy-Policy-Draft-2022-A-Comprehensive-Assessment.pdf. Accessed 10 May 2024

Mondal MA (2010) Implications of renewable energy technologies in the Bangladesh power sector: long-term planning strategies. Inaugural-dissertation, Rhenish. Friedrich Wilhelm University of Bonn. Faculty of Agriculture. https://d-nb.info/1010643290/34. Accessed 5 June 2024

Mridul D, Azad AK, Biswas M, Datta A, Risty K (2021) Current scenario of solar energy production in Bangladesh and future potentiality. Int J Sci Eng Res 12:1234–1241. https://doi.org/10.14299/ijser.2013.01

Shirin S, Jamal A, Emmanouil C, Singh VP, Yadav AK (2023) Assessment and characterization of waste material used as backfilling in an abandoned mine. Int J Coal Prep Util 43(8):1402–1410. https://doi.org/10.1080/19392699.2022.2118259

SREDA (10 May 2024) National database of renewable energy. https://ndre.sreda.gov.bd/index.php?id=8

Supti SA (2022) Renewable energy and policy in Bangladesh. In: IEEE 13th annual ubiquitous computing, electronics & mobile communication conference (UEMCON). New York, USA

Tasnim S, Rahman MM, Hasan MM, Shammi M, Tareq S (2022) Current challenges and future perspectives of solar-PV cell waste in Bangladesh. Heliyon 8(2):1–9. https://doi.org/10.1016/j.heliyon.2022.e08970

Teska S, Morris T, Nagrath K, World Future Council, Dhaka (2019) 100% Renewable energy for Bangladesh: access to renewable energy for all within one generation. https://www.worldfuturecouncil.org/wp-content/uploads/2019/10/100-Renewable-Energy-for-Bangladesh.pdf. Accessed 9 May 2024

U.S. Agency for International Development (March 2011) Renewable energy implementation action plan: recommendations for Bangladesh. https://www.usaid.gov/energy/sure/bangladesh/re-action-plan. Accessed 10 May 2024

USAID Bangladesh Advancing Development and Growth through Energy (BADGE) (Feb 2023) Enewable energy policy in Bangladesh: recommendations for acheiving vision 2041. https://pdf.usaid.gov/pdf_docs/PA00ZVVS.pdf. Accessed 11 May 2024

Uzzaman SA (2023) Transforming solar energy landscape in Bangladesh. The Confluence. https://theconfluence.blog/transforming-solar-energy-landscape-in-bangladesh/. Accessed 1 June 2024

Chapter 11
Understanding Solar Irradiance and Its Utilization for Electricity Generation

Zoya Mahtab, Amit Kumar Dutta, and Akhilesh Kumar Yadav

Abstract Energy is the ability to do work. It exists in various forms, such as kinetic (energy of motion), potential (stored energy), thermal (heat energy), chemical (energy stored in bonds), and so on. Solar energy comes from the Sun and can be harnessed to generate electricity. Every day, the Earth receives a massive amount of energy from the Sun through its rays. Solar power is a renewable energy source, meaning it won't run out.

The **solar constant** refers to the amount of solar energy reaching a square meter of space just outside Earth's atmosphere, which is approximately 1.4 kW/m^2. However, as sunlight passes through the atmosphere, some of its energy is absorbed and scattered by gases, reducing the power that reaches the surface.

Solar energy can be used in different ways, such as heating water or generating electricity. **Semiconductor materials** like silicon (Si) are used in solar panels to convert sunlight into electricity. These panels contain multiple solar cells, also called **ray collectors**, which work together to produce more electricity than a single cell. During rainy weather or at night, the electricity generated during the day can be stored in **nickel-cadmium (Ni-Cd)** batteries or other storage systems. These batteries provide power when sunlight is unavailable, ensuring a steady energy supply.

Keywords Energy · Solar energy · Earth · Solar cell · Solar panel

Z. Mahtab (✉) · A. K. Dutta
Amity Institute of Biotechnology, Amity University Jharkhand, Ranchi, India

A. K. Yadav
Department of Environmental Engineering and Management,
Chaoyang University of Technology, Taichung, Taiwan

Department of Electronics and Communication Engineering, Manipal University Jaipur, Jaipur, India

Department of Civil Engineering, Manipal University Jaipur, Jaipur, India
e-mail: yadavbasti@gmail.com

A. K. Yadav (ed.), *Powering Future: Clean Energy Solutions for a Sustainable Planet*, Clean Energy Production Technologies, https://doi.org/10.1007/978-981-96-5055-2_11

11.1 Introduction

11.1.1 History

The world's energy demand is on the rise and is projected to double by the middle of this century. To address this immense need and the environmental challenges we face, we can look to the increasing use of renewable energy sources, particularly solar power, as a sustainable solution (Robyns et al. 2012). The year-long solar power capacity is 23,000 TWy, while the consumption of Earth's power over 100 years is estimated at 1600 TWy (Robyns et al. 2012). In contrast, the limited power supplies amount to 1655 TWy, and the most optimistic estimates of yearly conversion from non-solar renewable resources only reach 94 TWy (Perez et al. 2011). Almost all the energy utilized for electrical power generation comes from solar sources, such as fuel, carbon, gasoline, and wood, which were turned out through light synthesis. Breezes, including surfs, are influenced by climate balance (Sorensen et al. 2008; Smil 2004; Bithas and Kalimeris 2016). In history, advanced power forms have driven social development (Hefner III 2009). Solar energy, a clean and ancient resource, has a rich and diverse history of energy conversion. Almost 1.74 lacs of terawatts (TW) of energy hit the Exosphere, with about 30% being reflected backwards to space. The sun energy obtained consists of wavelengths in the noticeable UV and nearby rays. The solar power for electricity generation is limited by aspects such as sky and geomorphology. Part of the earth near to the periphery is more sensitive to sunrays with respect to areas near to the poles. The vital role in ancient cultures is evident in Egyptian history. Pharaoh Akhenaten IV established a monotheistic state religion centered on the worship of Aten, the Sun god (Roberts et al. 2006).

The Grand Pyramid is the last remaining structure from the great wonders of ancient times. It is said to have been built as steps to the Sun, according to ancient Egyptians writings. Across the globe, many ancient cultures have their own gods associated with the Sun. As the usefulness of solar energy became apparent, it transitioned from a metaphysical concept to practical applications in architecture (Roberts 2006). Aristotle (384 BC to 322 BC) taught methods for using the Sun for heating and cooling, laying the foundation for present-day techniques. In 80 BC to 15 BC, Vitruvius a Roman designer introduced an idea regarding how to orient a house properly. It has been designed with a wide opening made from mica, and translucent stripes or clear crystals to hold in sunlight (Perlin 2022). Since ancient times, the sun has been used to dry and preserve food. For several years, residents of Egypt used a technique of sun drying for the storage of seeds, cereals, etc. Ancient people quickly learned to use the energy to light fires. As early as the seventh century BC, magnifying glasses concentrated sunlight, and by the third century BC, mirrors were used to intensify sunbeams for torches (Ngô and Natowitz 2016). In 287 BC, to 212 BC, a Greece citizen and a researcher Archimedes of Syracuse reportedly set large boats of their opponent from the city of Rome on fire using a "burning mirror", which focused sunrays onto the approaching ships (Ngô et al. 2016).

Between 1452 and 1519, the brilliant Italian polymath Leonardo da Vinci was recognized as a man of the renaissance and a renowned architect. Beyond his famous portraits, such as the *Mona Lisa*, he made significant contributions to various scientific fields, including engineering, chemistry, mathematics, and physics. He was also environmentally conscious, conducting studies on the use of solar power, particularly aimed at solar heating.

Notably, he conducted research on the effects of sunlight, demonstrating awareness of light patterns similar to crepuscular rays on bent metal sheets. This work contributed to the development of the first practical applications of hollow glass solar collectors intended for heating water. Da Vinci proposed innovative ideas for heat-illuminating systems and an operating steam engine, as well as developing techniques for welding brass through concentrated sunlight. Later, in 1839, the French scientist A.E. Becquerel discovered that a solar collector could generate electromotive force and current when certain materials were exposed to light. This chemical and physical process converts solar energy into electricity. Solid-state semiconductor devices, such as photodiodes and solar cells, are today able to harness this effect by converting sunlight into electricity. This breakthrough has paved the way for humanity's quest to harness solar energy for sustainable power production. More than 150 years before, A. Mouchot, a visionary France instructor as well as discoverer, recognized the limitations of coal as a finite resource and began pioneering alternative energy solutions to power the world's Industrial Revolution. He explored alternative energy sources and conducted solar energy experiments (Butti et al. 1980).

11.2 What Is Energy?

In the realm of science, it exists as a measurable quality that transforms into the body or other materials. This is evident to be an accomplishment of function along with manifestation of heat and light. It lies one of the preserved measures, as per the first law of thermodynamics, which means that it can change forms but cannot be created or destroyed. The SI unit of energy is Joule (J). There are different forms of energy that are classified as force associated with the motion of an object. Objects in motion having the power to cause change or perform work are called kinetic energy. The hidden power within an object or system ready to transform into dynamic kinetic energy is called potential energy.

$$\text{Kinetic Energy} = \tfrac{1}{2} mv^2 \qquad \text{Potential Energy} = m \times g \times h$$

Forms of Energy	
1. Electrical Energy	7. Solar Energy
2. Thermal Energy	8. Wind Energy
3. Light Energy	9. Nuclear Energy
4. Wave Energy	10. Tidal Energy
5. Heat Energy	11. Biomass Energy
6. Hydroelectric Energy	12. Kinetic Energy
13. Geothermal Energy	

11.3 What Is Solar Energy?

The Sun provides sunlight, which is the primary energy received by Earth. Its low intensity at the surface is due to the dispersion of energy from the distant Sun. It exceeds our current and future energy needs. In recent years, solar energy has become increasingly popular as a renewable resource due to its unlimited availability and eco-friendly nature, unlike finite fossil fuels such as coal, petroleum, and natural gas. The potential of solar power is vast; the Earth receives around 200,000 times the total amount of electricity the world generates daily from the Sun's energy. The radiation produced by the Sun turned into heat or voltaic energy, with the former being readily accessible. Solar energy is the driving force behind numerous essential natural processes on earth. One of the most crucial of these is that photosynthesis would cease, leading to a severe shortage of food and organic matter. This would have catastrophic consequences, including the potential disappearance of many organisms and a significant reduction in oxygen levels. Furthermore, solar energy is indispensable to drive the water cycle, regulating temperatures, and influence wind formation, all of which help shape Earth's weather patterns (Ashok 2024; Shirin and Yadav 2014). ***Advantage:*** Because it does not release carbon dioxide into the atmosphere like fossil fuels, which contributes to global warming, it is considered a clean source of energy. As long as the sun shines, there will always be an abundance of energy on Earth. Unreliability does not exist because energy can be stored in batteries. ***Disadvantage:*** The drawback of this is that its output is reduced during overcast and winter days. Both the initial cost of the components and the installation are high. The consumption of space is higher.

The Sun's output energy is relatively constant, accompanied by surface intensity, i.e., 6.33×10^7 W/m^2. This intensity decreases as sunlight travels through space following an inverse square law. The average intensity reaching Earth's atmosphere is referred to as the solar constant, which varies by about 7% between January 4 (perihelion) and July 4 (aphelion) (NASA 2015). The yearly average is around 1367 W/m^2, calculating this value involves considering the Earth-Sun distance and total solar flux instead of solely solar flux per unit area. The Earth is the only region near the equator that receives sunlight at its ideal angle. In contrasting parts of these regions, sunlight arrives at an angle, resulting in a reduced average solar irradiance. This reduced angle causes the solar irradiance level to be halved, leading to an average solar irradiance of approximately 340 W/m^2. Not all solar energy reaching Earth is absorbed because of the Earth's energy budget, where some energy is reflected back into space. Approximately 70% of the solar energy in the Earth atmosphere is absorbed, amounting to 238 W/m^2. The absorbed energy differs from the energy reaching the surface, especially on clear, sunny days. The 22% decrease in solar irradiation may vary due to different Sun angles (Julian Chen 2011). AM1.5 is considered a standard representing a power density of 1000 Wm^2. The term "one sun" corresponds to 1000 Wm^2 and serves as a benchmark for evaluating solar collectors and solar panels. However, despite the significance of this amount of energy, solar power alone may not fully meet our energy needs due to challenges such as intermittent sunlight and limited energy storage.

11.3.1 Solar Cell

Solar cells were first used on the Vanguard satellite in 1958 and later Explorer 6 in 1959. These cells extended the mission time without major modifications to the spacecraft. By the 1960s, solar cells became the primary power source for satellites and solar system probes because of their high power-to-weight ratio. The space market led to the development of more efficient cells. Success in space applications was made possible because of the high costs of power systems, limited power options, and willingness to invest in ideal solar cells. The development of more effective solar cells was driven by demand in the space power industry until the National Science Foundation's program focused on solar cells for terrestrial uses. Around the early 1990s, space solar cell technology shifted from silicon to GaAs III-V semiconductor materials, resulting in the contemporary III-V multijunction photovoltaic cells used in spacecraft. Current analysis has been directed toward creating the most efficient, malleable, and feather-shaped solar cells. Sun-Earth cell technology typically uses a glass-covered solar collector to provide power and shield. Space implementation preferred a highly efficient, lightweight accumulator. Multijunction photovoltaic cells with varying bandgaps are used on satellites to capture one broad scope of solar energy. Vast spacecraft use vast solar farms, which must fit geometric launch vehicle constraints. Solar batteries on spacecraft used to be compact sheets pleated jointly and then unwinded into a large sheet succeeding distribution. New satellites are incorporating pliant spinnable solar farms that are light and can be packed in one package. These advanced arrays significantly reduce launch costs by making satellites lighter, directly impacting vehicle cost (Garcia 2017). In 2020, the US Naval Research Laboratory successfully tested energy generation in space using the Solar Collector Radiofrequency Antenna Module (PRAM) aboard the Boeing X-37 satellite (David 2021).

A solar cell, also known as a solar collector, converts light energy into electricity through the photovoltaic effect. They are types of electric eyes that change their electrical feature in response to light. These cells are the basic components of solar panels. Most solar collectors are made of crystal-clear silicon, accompanied by a market rule of 95%. The other 5% consists of thin film solar cells. A typical single-juction silicon solar cell can generate a free voltage of around 0.5–0.6 V. The solar collector can function covered by both natural and artificial light. Aside from generating energy, they can also serve as image scanners, identifying sunlight or magnetic discharge within the seeable extent, or evaluating the intensity of light.

To run a PV cell effectively, it requires three fundamental attributes:

1. It needs to absorb light to create excitons (which are bound electron-hole pairs), unbound electron-hole pairs (via excitons), or plasmons.
2. It must be able to separate charge carriers of opposite types.
3. The cell should be able to extract these carriers individually and transfer them to an external circuit.

The solar cell consists of semiconducting materials, such as silicon, arranged in a p-n junction. The junction is created by doping the device in a p-type way on one side and n-type on the other side, typically using boron or phosphorus in silicon (Schwarz 2013). When sunlight hits the solar cell, photons are engrossed by the semiconductor, forcing electrons to migrate from the valence band to the conduction band, resulting in electron-hole pairs. Near the junction, an electric field separates these pairs, causing electrons to accumulate on one side and holes on the other. In the absence of a link or large external electrical load, electrons and holes will return to equilibrium by diffusing back across the junction and recombining, releasing heat. However, with a small load, excess electrons can restore equilibrium by flowing through the external circuit and performing work. Solar cells take solar energy and convert it into direct current electricity, which can then be changed to alternating current with the help of an inverter. The silicon-based large-area p-n junction is the most prevalent type of solar cell. Alternative types include biological, Gratzel cell, perovskite (tin halide-based material), and quantum dot. The lit side of a solar cell is equipped with a see-through conducting film to let sunlight in and gather charge carriers (Kumar 2017).

11.3.2 Solar Panel

In 1839, the French physicist Edmond Becquerel made the astonishing discovery that some materials could generate an electrical charge when exposed to light. Although his early solar panels were far too meager to power anything more than the simplest of devices, they did serve as an important instrument for measuring illumination. Nearly four decades would pass before Becquerel's findings saw replication. In 1873, an English electrical engineer, Willoughby Smith, discovered that illumination could induce charge in selenium while experimenting with the substance. Heartened by that breakthrough, William Grylls Adams and Richard Evans Day conducted their own trials to duplicate Smith's effects the following year, meticulously documenting their work in the published paper "The way light interacts with selenium" in 1876 (Adams and Day 1877).

By 1881, the American inventor Charles Fritts had finally devised a practical use for the photovoltaic reaction. The solar array he assembled that year marked the first true photovoltaic cell capable of powering real applications, as Fritts boasted of its ability to generate "continuous" solar electricity. Though crude by today's standards, Fritts' solar panel represented the nascent dawn of harnessing the sun's radiance to fuel human devices and dreams. However, those solar panels during that era converted sunlight into electricity at extremely low rates compared to the large coal power plants that dominated the energy landscape. In 1939, Russell Ohl unveiled the pioneering solar cell blueprint that still underpins countless photovoltaic panels today. He protected this design with a patent filing in 1941. Bell Labs was the first to commercially exploit this innovation approximately a decade later by developing the first viable silicon solar cell. Installation companies specializing in residential solar energy systems saw an extraordinary expansion from 2008 to 2013. Due to such rapid growth, many installers encountered "non-ideal" roofs that posed integration challenges. Roofs obstructed by tall trees or facing away from optimal southwest angles created hardship. The previous obstacle was initially overcome by microwaved reconstruction, which subsequently transformed the notion of power optimizers. The manufacturers of these panels soon started to collaborate with microinverter companies to overcome such obstacles through technology that could optimally harness sunlight falling on each individual solar module regardless of its specific position or exposure to shadows. Solar panels use photovoltaic electricity (PV) to convert sunlight into electrical energy and produce direct current (DC) electricity. They are arranged in arrays or systems with an inverter that converts DC to AC. Solar panels are often used with batteries in co-residential, marketable, industrial, and space applications. They offer renewable resources and a lower emission of greenhouse gases along with a lower electricity cost. However, they depend on the availability of sunlight, require maintenance, and have higher starting prices.

11.4 Future Innovation in Solar Technology

With its unmatched potential to completely transform the production of renewable energy, solar cell technology is poised for significant breakthroughs in the future. Technologies based on quantum dots, organic photovoltaics, and perovskite solar cells are examples of emerging materials that show remarkable increases in efficiency. The production of solar cells is scalable and affordable due to advanced manufacturing processes such as roll-to-roll printing. To maximize light absorption, new light management techniques such as photon upconversion and nanostructured coatings are being used. Solar cell applications in smart grids and building integration are very promising for the future. The spotlight in the realm of renewable energy now shines on solar energy harnessed by photovoltaic (PV) systems, capturing the fascination of researchers. Solar PV systems are heralded for their carbon-free essence, eco-friendliness, abundance in nature without associated costs, and ease of access. Although they boast remarkable energy conversion efficiency and require substantial capital investment, these characteristics are also the primary limitations of this energy source. Historically, a plethora of semiconductor materials and technologies have been explored to craft photovoltaic cells (PV cells) that balance affordability with high efficiency. Conventional solar panels fashioned from silicon crystalline wafer modules tend to be bulky, posing challenges in transportation. Shiny silicon photovoltaic technologies have been steadily enhancing their efficiency and have become the preferred choice for PV installations in various applications. Through optimized manufacturing processes, technological advancements throughout the supply and value chain, and the establishment of large-scale factories, photovoltaic technology has significantly reduced costs. This transformation has made crystalline PV as one of the most cost-effective energy sources worldwide. With a 22% energy conversion efficiency, the latest generation of silicon solar panels represents notable advancements in solar efficiency. Emerging photovoltaic technology, known as perovskite cells, has a strong efficiency of around 25%, making it a competitive player in the renewable energy market. India wants to achieve 450 GW of renewable energy capacity by 2030. Fenice Energy is at the forefront of this movement, having made significant advances in photovoltaic panel manufacturing. They use modern green energy technology to integrate renewable energy sources into the national grid. The use of half-cell technology has resulted in significantly improved solar panel efficiency. This approach employs 120 half-cells instead of the standard 60 cells. It considerably increases the output of the panels. These solar panels work well in hot regions and are extremely durable. This makes them ideal for India's diverse climate. Building integrated photovoltaics (BIPV) panels are another invention. They combine efficiency and modern design, increasing urban solar power without sacrificing aesthetic appeal.

Tuticorin, Tamil Nadu, has a revolutionary floating solar facility. It shows green energy technology in a variety of environments. The floating panels use cold water to stay cool. This improves the efficiency of electricity generation. Fenice Energy was switched from lead-acid to lithium-ion solar batteries. This move improves performance while reducing the environmental effect and promotes independence from the grid. Modern microinverters improve these solar systems. They ensure long-term, high-efficiency energy generation and increased safety. Solar energy should play an important role in the transportation sector, especially with the introduction of solar-powered vehicles and the integration of solar panels into power electric vehicles. The use of artificial intelligence (AI) is expected to change solar energy management by predicting energy production, optimizing solar panel positioning, and increasing system efficiency. This development is expected to cause a significant increase in the use of AI-powered technology in solar systems, resulting in improved performance and more efficient system management.

Chemists, engineers, physicists, and materials scientists have long worked together to develop artificial photosynthesis, which uses sunlight energy to produce liquid fuels from water and waste gases. To realize this objective, these researchers are discovering new catalysts, speeding up chemical processes, creating novel materials capable of transporting electrical charges during these reactions, and devising techniques to streamline the conversion process.

Acknowledgments The authors express their heartfelt gratitude to all those who have supported and guided us in the completion of this work on future sustainability through solar energy. We are grateful to the readers who are passionate about making a positive impact on our planet.

References

Adams WG, Day RE (1877) IX. The action of light on selenium. Philos Trans R Soc Lond 167:313–316. https://doi.org/10.1098/rstl.1877.0009
Ashok S (18 June 2024) Solar energy. Encyclopedia Britannica
Bithas K, Kalimeris P (2016) A brief history of energy use in human societies. In: Revisiting the energy-development link. SpringerBriefs in Economics. Springer, Cham. https://doi.org/10.1007/978-3-319-20732-2_2
Butti K, & Perlin J (1980) Solar water heaters in California, 1891--1930. The Coevolution Quarterly; (United States)
David L (4 Oct 2021) Air Force's X-37B robotic space plane wings past 500 days in Earth orbit. LiveScience. Accessed 6 Nov 2021
Garcia M (31 July 2017) International Space Station solar arrays. NASA. Archived from the original on 17 June 2019. Accessed 10 May 2019
Hefner III RA (2009) The grand energy transition: the rise of energy gases, sustainable life and growth, and the next great economic expansion. John Wiley & Sons
Julian Chen C (2011) Physics of solar energy, 1st edn. John Wiley & Sons, Hoboken, NJ
Kumar A (2017) Predicting efficiency of solar cells based on transparent conducting electrodes. J Appl Phys 121(1):014502. https://doi.org/10.1063/1.4973117
NASA's page. Accessed 20 April 2015
Ngô C, Natowitz J (2016) Our energy future: resources, alternatives and the environment. John Wiley & Sons
Perez R, Zweibel K, Hoff TE (2011) Solar power generation in the US: too expensive, or a bargain? Energy Policy 39(11):7290–7297. https://doi.org/10.1016/j.enpol.2011.08.052
Perlin J (2022) Let it shine: the 6,000-year story of solar energy. New World Library
Roberts P (2006) HSC Ancient History. Pascal Press.
Robyns B, Davigny A, François B, Henneton A, Sprooten J (eds) (2012) Electricity production from renewable energies. John Wiley & Sons
Schwarz US, Safran SA (2013) Physics of adherent cells. Rev Mod Phys 85(3):1327–1381
Shirin S, Yadav AK (2014) Physico-chemical analysis of municipal wastewater discharge in Ganga River, Haridwar district of Uttarakhand, India. Curr World Environ 9(2):536–543. https://doi.org/10.12944/CWE.9.2.39
Smil V (2004) World history and energy. Encyclopedia of Energy 6:549–561
Sorensen B, Doble M, Broussely M, Maegaard P, Barbir F, Pistoia G, ... & Roy S (2008) Renewable energy focus e-Mega handbook. Academic Press.

Part IV
Solar Technology

Chapter 12
Efficiency, Materials, and Cost-Effectiveness of Photovoltaic Solar Panels

Dhruvit R. Parmar, Dev R. Merchant, and Sachin V. Jadhav

Abstract Photovoltaic (PV) solar panels have played a pivotal role in harnessing the energy from the Sun, helping to satisfy the global energy requirements while facilitating a transition towards a renewable and cleaner energy source. The solar PV market is constantly growing, and efforts are being made to reduce panel costs and improve their efficiency. Solar PV works on the principle of photovoltaic effect, which states that when a material is exposed to light, it absorbs the light energy, leading to excitation and flow of the charge carriers (electrons and holes), further leading to the development of current in a closed circuit. Therefore, the majority of the PV panels are p-n junction-based. Solar panels were originally selenium-based, but there were concerns related to their efficiency, toxicity, and stability. The development of Czochralski process enabled the production of high-purity crystalline Silicon, leading to the adoption of Silicon as the primary material for PV cells, providing better efficiency. Power conversion efficiency (PCE) is a key feature of solar PV, influenced by factors such as the material used in the PV cell, insolation, temperature, and shading. Ongoing developments aim to improve PV panel efficiency and reduce losses. This chapter explores the manufacturing process of solar PV panels and discusses sustainable materials to minimize environmental impact and ensure a long-term, responsible solar future.

Keywords Solar energy · Efficiency · Photovoltaic system · PV manufacturing

D. R. Parmar · D. R. Merchant · S. V. Jadhav (✉)
Department of Chemical Engineering, Institute of Chemical Technology, Mumbai, Maharashtra, India
e-mail: sv.jadhav@ictmumbai.edu.in

A. K. Yadav (ed.), *Powering Future: Clean Energy Solutions for a Sustainable Planet*, Clean Energy Production Technologies, https://doi.org/10.1007/978-981-96-5055-2_12

12.1 Introduction

The PV cell is the heart of solar energy production, helping us to make a shift from non-renewable to clean, sustainable, and renewable energy sources. The nuclear fusion reaction between two protons, continuously occurring in the Sun's core due to high temperature and pressure conditions, produces about 3.828×10^{26} W power (NASA 2024), and it is radiated in the form of electromagnetic radiation in all directions. PV cells help us to capture this abundant sunlight (radiations) and convert it into electrical energy. This reduces the dependence on energy derived from depleting fossil fuels and reduces their negative impact on the environment. Global electricity consumption is increasing every year, and the demand is predicted to rise in the future with the growing population, economies, and due to data centers, artificial intelligence, and cryptocurrency sectors (International Energy Agency 2024). Solar power, along with other renewables, has been on the rise, but their contribution to the total global energy requirement is still less than the traditional sources of electricity based on coal, oil, and gas. The two ways to increase their contribution are by installing a greater number of solar panels and by improving their efficiency. Hence, there is still some room for the development of this technology. The manufacturing process of photovoltaic cells begins with the acquisition of raw materials. From the working process of PV cells, we can understand the importance of using semiconductors as the material for constructing solar panels. Silicon, a semiconductor and a metalloid discovered in 1824 is one of the most abundant elements in the Earth's crust, making it widely available for various technological applications. Metalloid refers to a class of materials whose physical and chemical properties fall between metals and non-metals. This unique characteristic makes it an excellent semiconductor capable of conducting electricity under specific conditions. Moreover, it is easy to control the conduction properties of Silicon by employing a large variety of impurity atoms, also known as doping agents, and it can be used over an expanded temperature range without losing its properties. Consequently, Silicon's semiconducting nature is crucial in solar panel manufacturing, where it can efficiently convert sunlight into electrical energy.

12.2 Working of PV Cell

12.2.1 Principle

Figure 12.1 illustrates the solar cell's working principle of photovoltaic effect, discovered by Henri Becquerel in 1839 (Becquerel 1839). It involves generation of electric potential at the junction of two dissimilar materials due to exposure to light energy. This photovoltaic effect is linked to the photoelectric effect given by Albert Einstein, i.e., when quanta of light, known as a photon, is absorbed by a material, it leads to the emission of electrons leading to the generation of electron and hole pair

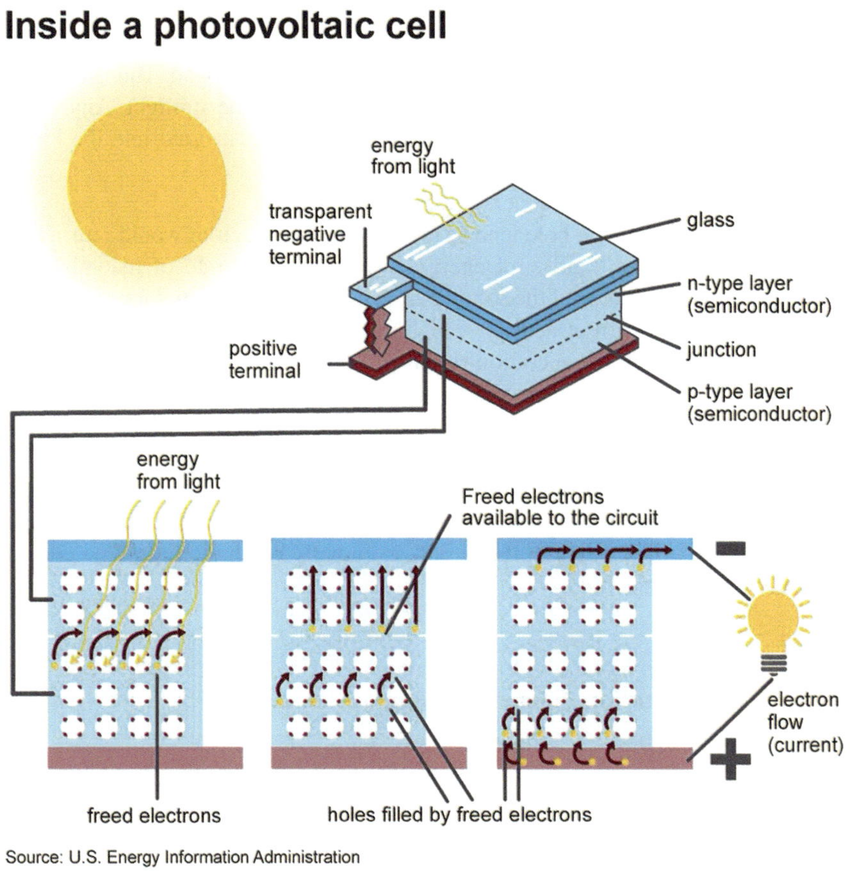

Fig. 12.1 Generation of electron-hole pairs [reprinted from U.S. Energy Information Administration (2023) {CC BY}]

when the frequency of light is greater than the threshold frequency of the material (Einstein 1905). A PV cell mainly comprises a p-n junction semiconductor and the electric potential is developed across the junction. The three main steps of the photovoltaic effect are as follows (Smets et al. 2016):

1. Absorption of the light photon giving rise to the charge carriers

   The energy of a photon (E_{photon}) can be expressed by the following equation,

$$E_{\text{photon}} = hv \tag{12.1}$$

   where h = Planck's constant, v = frequency of the incident light.

   As the electron present in the semiconductor absorbs the photon, having energy more than the bandgap energy (E_g), i.e., $E_{\text{photon}} \geq E_g$, the electron gets excited from the valence band to the conduction band, leaving behind a void (hole) in the valence band. The current generation needs the movement of electrons. Therefore, we need to knock the bonded electrons from their lattice to ensure that they are free to move around in the conduction band. Thus, the chemical energy of the electron-hole pair is obtained from the sunlight. Any excess energy beyond what is needed to free the electron is converted into the kinetic energy of the ejected electron.

2. Separation of the charge carriers (electron-hole pair)

   The electron and the hole tend to recombine. The electron would return to its original energy state and emit energy as photons. The rate of recombination affects the performance of the solar cells. A high recombination rate causes a rise in the saturation current density, which negatively affects the efficiency of the solar cell. The carrier generation rate is greater than the recombination rate by several orders of magnitude. The p- and n-type materials form the semipermeable membranes in the semiconductor, separating the holes and the electrons. The electron moves to the n-type material and the holes to the p-type material.

3. Passage of electrons through the electrical circuit

   Now, the electrons can move through the electrical circuit, leading to the generation of electric current in the circuit, eventually reaching the p-type end and combining with the hole.

12.2.2 p-n Junction Semiconductor

Silicon is the most preferred element when it comes to the materials for semiconductor in PV (Smets et al. 2016). It is doped with heterogeneous moieties which are either of electron-donating, i.e., pentavalent, nature, (P, As, Sb), or electron accepting, i.e., trivalent, nature (B, Ga, Al) to increase the concentration of the charge carriers in the semiconductor. These dopants get ionized and generate their respective charge carriers using the thermal energy at room temperature. The n-type semiconductor material has electrons as the majority charge carriers, and p-type has holes as the majority charge carriers. When these two types of material are brought in contact, due to the difference in concentration, there is movement of the charge carriers against the concentration gradient to combine at the point of contact, leading to the formation of a depletion layer between the n-type and p-type layers (Smets et al. 2016).

12.2.3 Solar Cell Under Applied Voltage

The application of voltage across the diode changes the existing potential difference across the semiconductor. If the external voltage is in forward bias, then the potential difference between the p-n junction will decrease, and the depletion layer will further widen. Whereas, if the external voltage is in reverse bias, then the potential difference across the p-n junction will increase, and the depletion layer will get narrower. Now, due to this applied voltage, the equilibrium between drift and diffusion is also disturbed (Smets et al. 2016).

12.2.4 Solar Cell Under Illumination

When light is illuminated on the semiconductor, there is a generation of electron-hole pair. The electrons generated move across the junction into the n-type region and the hole in the opposite direction. This flow generates photogeneration current density, and when the circuit is closed, only a portion of this current will flow through the circuit containing the load (Smets et al. 2016).

$$J(V_a) = J_{rec}(V_a) - J_{gen}(V_a) - J_{ph} \tag{12.2}$$

where J = current density, J_{rec} = recombination current density, J_{gen} = thermal generation current density, J_{ph} = photogeneration current density.

12.2.5 I-V and P-V Characteristics

Figure 12.2 has the I-V and P-V plots for a solar cell. These curves are influenced by irradiation and module temperature. V_{OC} is defined as the voltage when no current is flowing through the circuit and I_{SC} as the short-circuit current. The maximum power point (MPP) refers to a particular point on the power versus voltage curve which gives maximum power output from the modules for that particular irradiance and temperature condition. To achieve this, we need to make sure that the module operates at the voltage same as that at the MPP (V_m). Solar cells can be connected

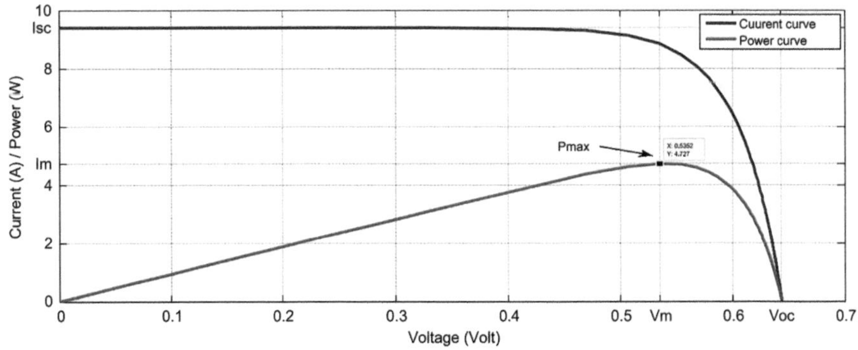

Fig. 12.2 I-V and P-V characteristics of solar cell [reprinted from Verma et al. (2016) with permission from Renewable and Sustainable Energy Reviews. Copyright © 2015 Elsevier Ltd.]

either in series or parallel configurations. The configuration does not affect the nature of the curve, but it does influence the total generated voltage or current.

12.3 PV System

It all begins with the basic solar cells, which form a solar module together, and a collection of such modules result in a solar panel. Multiple connected solar panels form a solar/PV array. The comprehensive layout of a PV system is illustrated in Fig. 12.3. The size of the PV system is dependent on the energy requirement. For example, solar panels installed on the rooftop of households produce electricity that can meet the electricity requirement of that household whereas there are solar farms where solar panels are installed in an area of several sq. km., producing electricity for commercial purposes and having capacities measured in megawatts (MW). Such purely commercial farms supply their electricity in the electricity market to the government, retailers, and companies. The Xinjiang Solar Park in China is the largest in terms of capacity, with a staggering 3.5 GW capacity. The Bhadla Solar Park, which has a capacity of 2.245 GW and is situated in Rajasthan's Thar Desert, is the largest solar park in India (Das Energie Private Limited 2024). In 2023, India became the third largest solar power-producing nation.

12.3.1 Classification of PV Systems

Depending on the connection of the PV system to the electrical grid, it can be classified as stand-alone, grid-connected, or hybrid.

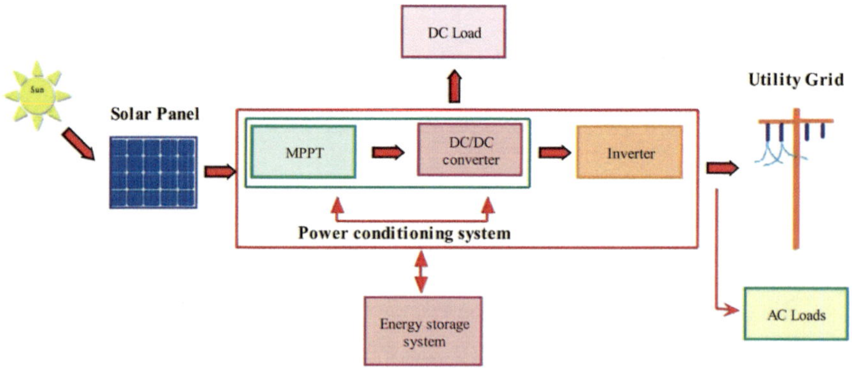

Fig. 12.3 Structure of a PV system [reprinted from Venkateswari and Sreejith (2019) with permission from Renewable and Sustainable Energy Reviews. Copyright © 2018 Elsevier Ltd.]

12.3.1.1 Stand-alone PV System

In a stand-alone PV system, there is no connection to the electrical grid, and hence, it is also known as an off-grid PV system. The energy produced can directly be supplied to the load without storing, but we can store the electrical energy in batteries and use it when needed. Charge controllers are used to connect/disconnect the battery with the source/load when maximum/minimum levels of the energy in the battery are reached. An inverter is also needed in the system to power AC loads. Such systems are suited for remote locations where there is no connectivity to the electrical grid (Goetzberger and Hoffmann 2005; Smets et al. 2016).

12.3.1.2 Grid Connected PV System

As the name suggests, grid-connected PV systems are connected to the local electricity grid. In such a system, the electrical energy generated by the PV system is used to power the facility, but when it is not enough or not available, like during unfavorable weather conditions, power is drawn from the electricity grid. In case of excess electricity production, it is supplied to the grid. Before the system is connected to the utility grid, the generated DC electricity is converted to AC via an inverter. During the daytime, when sunlight is available, the electricity generated by the PV system is enough for the house, but during the night-time when sunlight is unavailable, drawing electricity from the grid is an option, but storage of the excess electricity generated during the daytime in batteries and utilizing it at such times is also an option even though it increases the number of components in the system and hence the cost (Goetzberger and Hoffmann 2005; Smets et al. 2016).

12.3.1.3 Hybrid PV System

In a hybrid PV system, there is a supporting/secondary source of electricity that can be used in case the PV system is not able to deliver electricity. After the batteries are also drained, the system resorts to diesel/gas/wind/hydroelectric/biomass generators. Hence, such systems are not solely dependent on PV systems but can tackle situations of bad weather or time when solar radiation is not enough. Wind turbine generators would be reliable if the location is windy. Otherwise, fossil fuel-powered generators are reliable to supply electricity while simultaneously charging the batteries. However, hybrid PV systems are expensive from the capital and operating point of view (Goetzberger and Hoffmann 2005; Smets et al. 2016).

Table 12.1 Components of PV system and their uses (Ginsberg 2019; Smets et al. 2016)

Component	Use
Mounting structure	To support the solar panels and orient them in the direction of sunlight
DC-DC converter	To convert the variable output voltage from the panels to the required voltage
Inverter	To convert the produced DC power into AC power
Batteries	To store the produced electrical energy
Charge controllers	To regulate the amount of energy stored in the batteries by maintaining a minimum amount and preventing overcharging
Cables	To establish the connection between all the components of the PV system and to the load and the electrical grid
Tracker	To ensure the panel faces the Sun and receives vertical irradiance for maximizing yield
Meters and instrumentation	To keep a check on energy stored in the battery and the amount of power supplied to or drawn from the electricity grid
Safety equipment	Surge protection, overcurrent protection devices, grounding of equipment, arch fault circuit interrupters

12.3.2 Components of PV System

The entire photovoltaic system comprises numerous components, ranging from the essential elements required to capture sunlight to the supporting components and infrastructure that delivers electricity to our homes. At the heart of the system is the solar cell, but all the various components, apart from the solar cell, come under the balance of the system (BOS). These components are given in Table 12.1.

12.4 Efficiency

12.4.1 Standard Testing Condition

A module's power is estimated under Standard Test Conditions (STC), i.e., irradiance of 1000 W/m^2, constant module temperature of 25 °C, and AM1.5 spectrum. These values facilitate the comparison of performance among various solar modules (Goetzberger and Hoffmann 2005).

12.4.2 Power Conversion Efficiency

The definition of Power Conversion Efficiency (PCE) is the ratio of the PV power that the cell generates to the input solar power (PV power) that the cell receives (P_{light}).

$$\eta = \frac{P_{\text{max}}}{P_{\text{light}}} = \frac{J_{\text{m}} V_{\text{m}}}{P_{\text{light}}} = \frac{FF\, J_{\text{SC}} V_{\text{OC}}}{P_{\text{light}}} \tag{12.3}$$

Where V_m and I_m is the voltage and current at MPP (i.e., at P_m). J_{SC} is the short-circuit current density, and FF, Fill Factor, is defined as the ratio of P_{max} to J_{SC} V_{OC}. P_{light} is 1000 W/m^2 as per STC.

It is not possible to convert the entire incident energy to electrical energy. There are limits imposed due to various factors like the thermodynamic limit, Shockley-Queisser limit (for a single p-n junction cell), detailed balance efficiency, etc. We will look into the thermodynamic efficiency limit (Smets et al. 2016).

12.4.3 Thermodynamic Efficiency Limit

We assume the solar module to be a thermodynamic heat engine. There are two heat flows, Q_H flowing from a hot reservoir (T_H) to the engine and Q_C flowing from the engine to a cold reservoir (T_C) and the work done by the engine is W. Q_C is energy loss. The efficiency of the heat engine is given by:

$$\eta = \frac{W}{Q_{\text{H}}} \tag{12.4}$$

To find the efficiency limit, we need to consider this engine as an ideal engine with maximum efficiency, i.e., a Carnot engine. Entropy is conserved as work is done, and the efficiency is given by:

$$\eta_{\text{carnot}} = 1 - \frac{T_{\text{C}}}{T_{\text{H}}} \tag{12.5}$$

Since our engine is operating between the module (absorber) of temperature (hot reservoir) T_A and the surrounding (cold reservoir) at a temperature T_C and we assume that all the incident energy is being converted to electrical energy, the efficiency is given by:

$$\eta_{\text{TD}} = 1 - \frac{T_{\text{C}}}{T_{\text{A}}} \tag{12.6}$$

Using this, we derive the efficiency for an ideal solar cell:

$$\eta_{\text{SC}} = \left(1 - \frac{T_{\text{A}}^4}{T_{\text{S}}^4}\right)\left(1 - \frac{T_{\text{C}}}{T_{\text{A}}}\right) \tag{12.7}$$

Maximum efficiency of 85% is achieved when the absorber temperature is 2480 K (Smets et al. 2016).

12.4.4 Maximum Power Point Tracking

At a given time, the operating point is the voltage and current at which the modules operate. I and V of the modules are dependent on the irradiance and temperature of the module. Module output power is given by $P = I \times V$. Variations in the irradiance and temperature are reflected on the I-V and P-V graph, thus causing a change in the MPP, making it necessary to check the I-V graph and make the necessary changes to operate at MPP. This is known as Maximum Power Point Tracking (MPPT). Trackers help in estimating the MPP and are integrated into the inverter or the charge controller. MPPT can either be direct or indirect. In direct MPPT, the MPP is located using real I-V data and thus is more accurate compared to indirect MPPT, where the algorithm is hard-coded in the tracker, and MPP is determined by some simple measurements (Smets et al. 2016). After this, a DC-DC converter (buck/boost) is used to adjust the output voltage. Therefore, MPPT is a dynamic method of making solar PV systems more and more efficient.

12.4.5 Losses

Understanding the various losses in solar cells helps improve efficiency and optimize performance. The energy exceeding the band gap energy is lost in the lattice as heat (thermalization), and the photons whose energy is less than the band gap energy is not absorbed. Part of the incident light also gets reflected, and then there are transmission losses where the light propagates through the semiconductor without getting absorbed. If only a part of the total area is available for absorption, it results in shading losses. Inadequate absorption of the incident light is caused by the absorber layer's thickness constraint, which is explained by the internal optical quantum efficiency. The losses resulting from the created electron-hole pair recombining and being unable to be collected at the electrodes are represented by external quantum efficiency. Voltage drop occurs because of the series resistance of the solar cell. Then, there is a leakage current that results from the shunt resistance of the cell (Smets et al. 2016).

12.4.6 *Solar Cell Design*

The choice of materials has a significant impact on the solar cell's efficiency. The Shockley-Queisser limit indicates that the band gap of Silicon, gallium arsenide, and cadmium telluride is close to the optimum value. Therefore, these materials are preferred due to their ability to maximize energy conversion from the solar spectrum. We need to maximize the portion of the incident energy absorbed. For 100% light absorbance, the material should be optically thick and have almost 0 transmissivity. Materials with indirect band gaps have lower absorption coefficients compared to those with direct band gaps. In order to reduce reflective losses, we can use an anti-reflective coating (ARC). ARC and textured surfaces help trap the light and hence improve the chances of the photon getting absorbed. The absorber layer should be optically thick. The panels must be placed with proper spacing and inclination to reduce shading losses. Further, trackers should be used so that the panels continuously face the Sun.

12.4.7 *Efficiency Improvement*

The NREL chart of best research-cell efficiencies (Fig. 12.4) provides comparative advancements of different PV technologies. The efficiency of the solar cell is greatly dependent on the irradiation and temperature of the module. As discussed earlier, MPPT is used to direct the panels in the direction of the Sun and maximize radiation. Cooling techniques are used to reduce heat-related losses and hence improve efficiency (Manisha et al. 2022). Silicon has been the most widely used material due to its established technologies even before the photovoltaic technology. It is also the

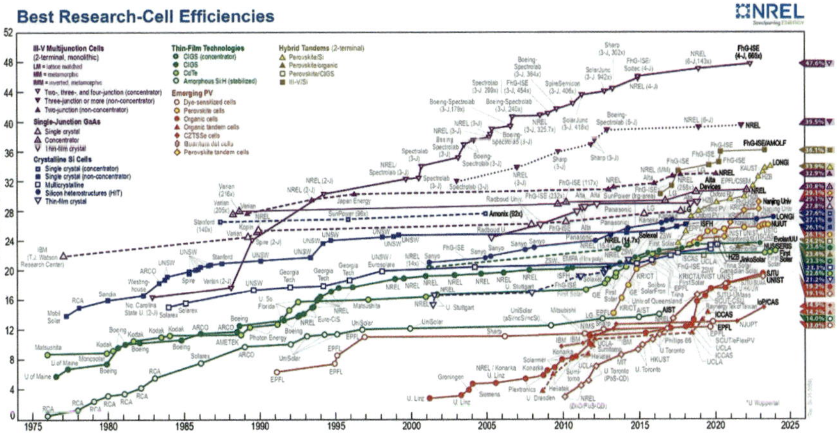

Fig. 12.4 Best research-cell efficiency chart (National Renewable Energy Laboratory 2024)

second most abundant element in the Earth's crust. Crystalline Silicon makes up 90% of the global PV market.

Bell Labs produced the earliest solar cell in 1954 with an efficiency of 6% (Chapin et al. 1954). Various companies were developing solar cells for space missions at that time (Goetzberger and Hoffmann 2005). Polycrystalline silicon cells are quite inexpensive to produce compared to monocrystalline silicon cells but have lower efficiency (Deshpande 2021; Lee and Ebong 2017). Now, the efficiency has reached around 16% (Lee and Ebong 2017).

The concept of PERC, Passivated Emitter Read Cell, was developed by a research group lead by Professor Martin Green at the University of New South Wales. It reduced optical losses by introducing textured top surfaces with inverted pyramid structures and anti-reflective coating with the rear side having a dielectric passivation, achieving an efficiency of 22.8% (Chowdhury et al. 2019), a then world record. It was further improved to 25% (Green et al. 2014).

Integrated back contact (IBC) solar cells were developed with the aim of eliminating shading losses (Smets et al. 2016). Maxeon Solar Technologies reported an efficiency of 24.9% for their IBC solar cell (Maxeon Solar Technologies 2024).

Moving on to the second generation of solar cells, thin-film solar cells are made of a-Si, CIGS, CdTe, and GaAs. Except for GaAs, these solar cells have efficiency lower than that of crystalline Si. Here, a semiconductor layer is used between a transparent conductive oxide and electric back contact (Smets et al. 2016). Hanergy's GaAs thin-film single junction solar cell has achieved a world record efficiency of 29.1% (Yun et al. 2022). CdTe is an ideal material for thin-film solar panels. CdTe-based thin solar cell from First Solar has a record efficiency of 18.6% for modules and a lab efficiency of over 22% (Lee and Ebong 2017). Perovskite solar cells are a promising technology because of their high efficiency and low manufacturing cost. Over the past decade, their efficiency has made significant advancements, reaching a peak efficiency of 25.2% (Kim et al. 2020).

Multi-junction solar cell is an advanced technology that uses the multi-junction concept, helping it to reach very high efficiencies. They have multiple p-n semiconductors in series, which enable them to absorb photons of different energies. Hence, there is more than one bandgap present, which enables it to utilize a bigger part of the solar spectrum (Smets et al. 2016). The highest efficiency recorded for multi-junction solar cells is 46% (Dimroth et al. 2016). Tandem solar cells, consisting of two cells, can deal with a larger solar spectrum compared to single junction solar cells. LONGi achieved a record 33.9% efficiency for crystalline silicon-perovskite tandem solar cells (LONGi 2023). The highest recorded efficiency for a solar cell is 47.6%, by a quadruple junction solar cell using concentrating photovoltaics technology (Schygulla et al. 2022).

12.5 Manufacturing Process

PV modules made of crystalline silicon are made in a number of steps. In an arc furnace, silicon dioxide (SiO_2) from quartz sand is reduced to metallurgical-grade silicon (MG-Si). Furthermore, impurities affecting the efficiency during operation necessitate the purification of silicon into solar-grade silicon (>99.99% purity), typically through a variety of methods, such as modified Siemens, Czochralski, and Siemens processes (Peng et al. 2013). Because it produces silicon suitable for solar or electronic use at high working temperatures (usually between 1100 and 1200 °C) in the ingot or reaction chamber, this process uses a lot of energy. In order to create monocrystalline and polycrystalline wafers, silicon ingots of either solar-grade polycrystalline or monocrystalline crystal are cut by hand or wire saw (Peng et al. 2013). To make solar cells, wafer sawing is followed by etching, doping, screen printing, coating, and checking. In the module stage, cells are encased in two layers of glass and plastic, then strung together and inserted into the system (Wong et al. 2016). After providing an overview of the manufacturing process of solar cells, let us delve into the detailed steps involved in turning high-purity Silicon into efficient photovoltaic wafers.

12.5.1 Producing Metallurgical Grade Silicon from Silica

Silicon exists in the form of its oxides and silicanes; the carbothermic reduction of silicon dioxide produces it. The metallurgical-grade Silicon is made in a graphite crucible containing raw materials, which include lumpy quartz, metallurgical-grade coal, and wood chips for better ventilation. The carbothermic reduction consists of reducing this silicon dioxide to Silicon, whereas the carbon gets oxidized to carbon monoxide. This reduction occurs between temperatures ranging from 1800 to 2100 °C. The Silicon obtained is approximately 98% pure and collected at the crucible's bottom. To produce electronic-grade Silicon, which is 99.9999% pure, the impurities (Fe, Cu, O, Al, and others) in the metallurgical-grade Silicon must be eliminated. There are many methods that can be employed to produce solar-grade Silicon. The two most important methods are the Siemens method, thc Union Carbide Monosilane process, and the New Silane Process (Chigondo 2018).

12.5.2 Producing Solar-Grade Silicon

12.5.2.1 Siemens Process

In the initial stages of this method, the crushed metallurgical-grade Silicon obtained from the crucible is reacted with hydrogen chloride at 300 °C, producing Trichlorosilane (HCl_3Si) and Hydrogen (H_2). The reaction is as follows:

$$Si_{(s)} + 3HCl_{(g)} \rightarrow SiHCl_{3(g)} + H_{2(g)}$$

The reaction also produces Silicone Tetrachloride ($SiCl_4$), Hexachlorodisilane (Si_2Cl_6), and Dichlorosilane (H_2SiCl_2). The reaction generally produces a yield of around 80–90%. Our desired product, trichlorosilane, can be obtained from this mixture by distillation. Trichlorosilane is chosen over the other silicon derivatives due to its low boiling point (31.8 °C) and high volatility, implying that it is separable from the by-products in an energy-efficient manner. In the next step, the residual impurities associated with trichlorosilane are eliminated by fractional distillation. These purification methods ensure that the impurity levels are beneath the required values for electronic-grade Silicon. In order to create polycrystalline silicon, trichlorosilane is reduced with hydrogen on a silicon wire that has been heated to almost 1100 °C. The reaction is as follows:

$$SiHCl_{3(g)} + H_{2(g)} \rightarrow Si_{(s)} + 3HCl_{(g)}$$

Polycrystalline Silicon formed via this method is virtually free of impurities, making it suitable for applications in microelectronic technology. The disadvantage of this process is that it creates hazardous and corrosive compounds known as chlorosilanes, which are produced through reactions with hydrochloric acid at temperatures that are somewhat high.

12.5.2.2 Union Carbide Monosilane Process

This process is a modification to the Siemens method. It aims to correct certain drawbacks by reducing energy consumption, thereby making the process more economical and environmentally friendly since the by-products of the reaction are recycled in this process. Unlike the Siemens method, here, the Silicon is reacted with Hydrogen and Silicon Tetrachloride ($SiCl_4$) to form trichlorosilane at a temperature of 500 °C and pressure corresponding to 30 MPa in a continuous fluidized bed reactor. The reaction is as follows:

$$3SiCl_{4(g)} + 2H_{2(g)} + Si_{(s)} \rightarrow 4SiHCl_{3(g)}$$

Post purification and rectification, the crude trichlorosilane transforms into high-purity trichlorosilane, which undergoes disproportionation to form silicon tetrachloride and silane at a temperature of 60 °C and pressure of 0.3 MPa. The reaction is as follows:

$$4SiHCl_{3(g)} \rightarrow 3SiCl_{4(g)} + SiH_{4(g)}$$

The silane obtained is purified by fractional distillation and undergoes decomposition to form solar-grade Silicon at temperatures ranging from 800 to 850 °C. The reaction is as follows:

$$SiH_{4(g)} \rightarrow Si_{(s)} + 2H_{2(g)}$$

Though the by-products, like trichlorosilane, are recycled, for every mole of Silicon formed, three moles of trichlorosilane are regenerated, inferring the inefficiency of this process. Moreover, the equipment for the reactor needs to be chemically inert to heat and pressure, increasing the costs of solar-grade silicon production. The large number of steps and production of toxic intermediates also contribute to the inefficiency of this method.

12.5.2.3 The New Silane Process

A new method has been developed aiming to make solar-grade silicon production cheaper, less toxic, and environmentally friendly compared to the Siemens method and Union Carbide process. However, it is in the research and development phase. In the first step, ethanol and powdered silicon of metallurgical grade react at 180 °C and atmospheric pressure with the help of a copper-based catalyst. The reaction yields roughly 85–90% triethoxysilane at optimum conditions (Zadde et al. 2010). The reaction is as follows:

$$Si_{(s)} + 3C_2H_5OH_{(l)} \rightarrow SiH\left(OC_2H_5\right)_{3(l)} + H_{2(g)}$$

One of the major issues faced in this step is the selectivity towards the desired intermediate triethoxysilane over the undesired tetraethoxysilane, which forms via the reaction as follows:

$$4SiH\left(OC_2H_5\right)_{3(l)} \rightarrow SiH_{4(g)} + Si\left(OC_2H_5\right)_{4(l)}$$

Under low temperatures, the triethoxysilane progressively disproportionates to form silane, and tetraethoxysilane under the influence of a catalyst, whose chemical formula is not mentioned in the literature. The purified silane undergoes decomposition at 800–850 °C to form solar-grade Silicon and hydrogen.

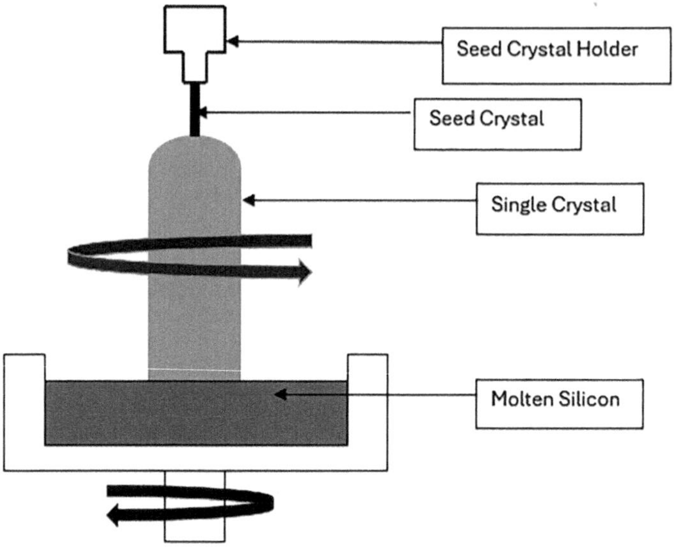

Fig. 12.5 Czochralski crystallization method [reprinted from Letcher and Fthenakis (2018) with permission from A Comprehensive Guide to Solar Energy Systems. Copyright © 2018 Elsevier Inc.]

$$SiH_{4(g)} \rightarrow Si_{(s)} + 2H_{2(g)}$$

The advantages possessed by this method include the lower temperatures and pressures at which the reactions are carried out, decreasing the overall cost of production, and the recyclability of ethanol produced, which decreases the additional requirement of ethanol. The ambient reaction conditions also imply that the process is cheaper than the conventional Siemens and Union Carbide processes. Moreover, this method boasts a better silicon yield than the other processes.

12.5.3 Crystalline Silicon Wafer Fabrication

From the working principle of solar panels, we can understand that Silicon has an indirect band structure, and the absorption coefficient falls in the infrared spectrum; therefore, a broader layer of c-Si is needed to efficiently absorb the incoming photons of energy beyond the bandgap. These layers are fabricated as wafers, cut from an ingot with a defined cross-section, which is either rectangular or square. The most common method used for ingot fabrication is Czochralski Crystallization for single-crystal c-Si and the Bridgman method for multi-crystal c-Si.

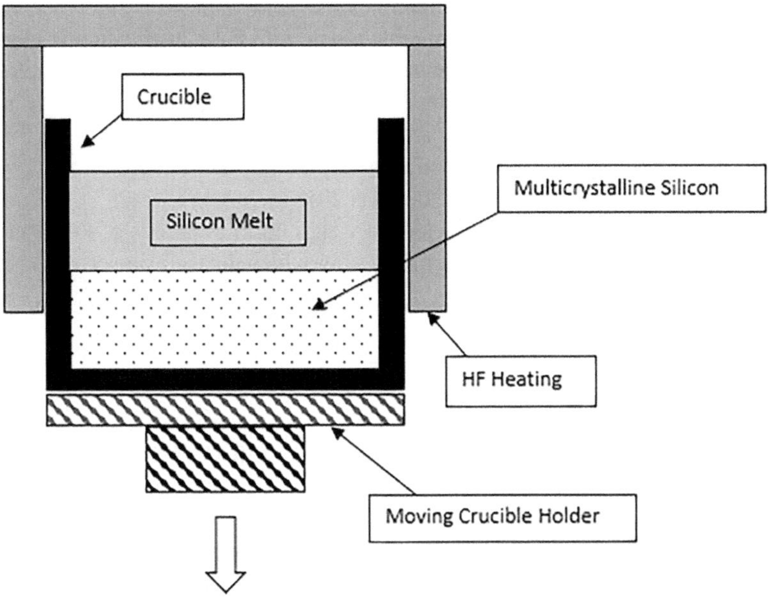

Fig. 12.6 Bridgman Process [reprinted from Ferrazza et al. (2018) with permission from McEvoy's Handbook of Photovoltaics. Copyright © 2018, Elsevier]

12.5.3.1 Czochralski Crystallization

Poly c-Si pieces are first melted in a quartz melting pot and doped with an appropriate concentration of acceptors (P-type Silicon) or donors (N-type Silicon). Figure 12.5 is a pictorial representation of the overall process. In the melted feedstock material, a seed of a defined crystal with a known crystallographic orientation is first inserted. The liquid starts to crystallize on the seed surface when it is pulled perpendicular to the melt surface. Because the pulling velocity determines the diameter of the crystals that are formed, it is only increased to a certain point in order to prevent abnormalities in the size of the crystals. Along with the upward movement, the seed crystal is rotated so the crystallized product can attain a cylindrical shape.

Phosphorous is commonly used as the doping element to prepare N-type silicon by adjusting the P_2O_5 concentration, and boron is typically used as the doping agent to prepare P-type silicon by adding the necessary amount of B_2O_3 to the raw material before melting.

12.5.3.2 Bridgman Process

Another method that can be used in the manufacture of silicon wafers is multicrystalline Silicon as the raw material. The advantage that mc-Si has is that the fabricating costs are much lower, and the module region is superior to that utilized for wafer generation. Figure 12.6 illustrates the Bridgman Process, which begins with melting poly c-Si in a quartz crucible along with B_2O_3 as the doping element. Si_3N_4 is also added to prevent the molten Silicon from sticking to the crucible. The crucible is moved downwards to induce the crystallization process, which begins at the bottom of the crucible at temperatures below the melting point of Silicon. Column-like crystal grains grow spontaneously and do not need the same crystallographic orientation (Letcher and Fthenakis 2018).

The multi-crystalline silicon blocks can weigh up to hundreds of kilograms, so they are cut into the shape of a square prism, after which the wafering process begins.

12.5.4 Wafering Process

The silicon ingots that are cut are of a standard size, 156×156 mm^{2,} and the largest cross-sectional area can reach up to 210×210 mm^2. The wafering process basically means cutting the silicon ingot obtained in the previous stage into smaller equal square prisms that are suitable for the further stage of production. The Multiwiring cutting method is the technique used in the wafering process. As the name suggests, this method involves the use of a series of wires to cut the ingot. To create a web of wires that are regularly spaced and parallel to one another, steel wire is first wound around wire guides. To produce hundreds of wafers at once, the silicon ingot is fastened to a glass plate that is adhered to a steel plate and then pressed up against the moving wires. The most common thickness of wafers is 180 μm.

12.5.5 PV Cell Fabrication Technology

The fabrication method for both monocrystalline and multi-crystalline is almost the same. Figure 12.7 portrays the overall process involved in the fabrication of photovoltaic cells. Figure 12.7a represents the c-Si wafer in its raw form. The wafers are first etched in an isotropic chemical polishing etchant, and then the surface layers damaged during the cutting process are removed by washing them in deionized water. Wafers are anisotropically etched to prepare the surface texturing after the damaged layer has been removed (Fig. 12.7b). For monocrystalline wafers, texturing is done in an alkaline medium, resulting in pyramidal structures on the surface, and for multi-crystalline wafers, an acidic medium is used to form a haywire texture on the surface (Fig. 12.7c). Following texturing, the surface is cleansed, and phosphorus is applied to a depth of almost 0.4 μm (Fig. 12.7d). The surface will then be passivated using a dielectric layer in the following step. A positively charged dielectric on an N-type surface or a negatively charged dielectric on a P-type surface is preferred because it reduces the surface recombination effect by forming an accumulation layer beneath the silicon surface. This is paired with the application of a silicon nitride layer and an antireflection coating to the surface via the PECVD

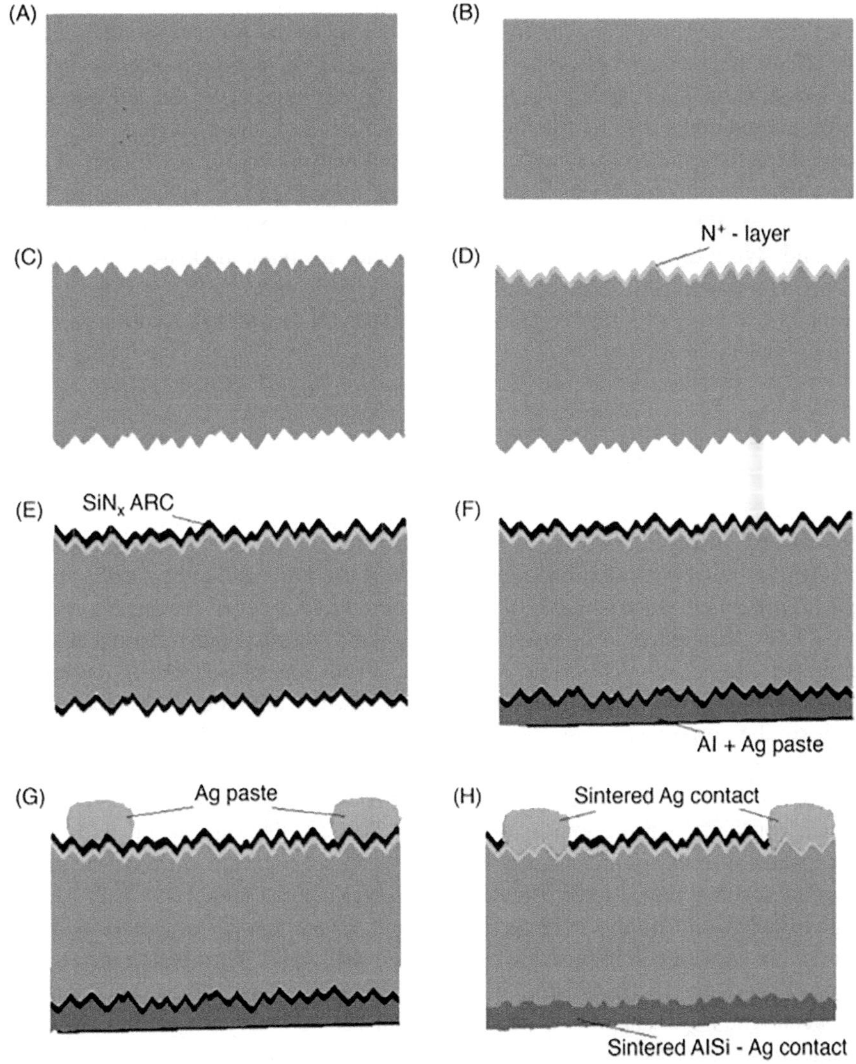

Fig. 12.7 Fabrication of photovoltaic cells. (**A**) Plain c-Si wafer. (**B**) c-Si wafer after isotropic etching. (**C**) c-Si wafer after texturing. (**D**) Addition of N^+ layer on the surface. (**E**) Addition of SiN_x anti-reflective coating (ARC). (**F**) Addition of Al and Ag paste layer. (**G**) Application of Ag paste on the top. (**H**) Sintered Ag contact and sintered Al contact layers. [reprinted from Ferrazza et al. (2018) with permission from McEvoy's Handbook of Photovoltaics. Copyright © 2018 Elsevier Ltd.]

(plasma-enhanced chemical vapor deposition) process. The silicon nitride layer's thickness, which is controllable during layer fabrication, is roughly 75 nm. (Fig. 12.7e). The reaction is as follows:

$$3SiH_4 + 4NH_3 \rightarrow Si_3N_4 + 12H_2$$

The back side is then covered with an additional aluminum paste by screen printing, with each print being followed by a 200 °C drying of the wafer (Fig. 12.7f). Over the silicon nitride layer, phosphorous-doped silver paste is screen-printed as thick, narrow metallic lines. Seventy percent of the paste is made up of the active powder, with the remaining 10% consisting of an organic binder and an organic solvent to keep the active powder attached to the layer below it. This process, known as front metallization, is a crucial step in the fabrication process (Fig. 12.7g). Both the layers are sintered on their respective sides at 800 °C to make good electrical contact with the underlying layer to increase the efficiency and decrease the losses (Fig. 12.7h). The I-V characteristics are measured for the cells using a solar simulator, which simulates sunlight of known parameters, and the cells are sorted according to their throughput.

12.5.6 Module Design and Fabrication

PV cells are susceptible to corrosion from diffusion of moisture, bending, mechanical damage during installation, and vibrations caused by wind after installation. To ensure the panels work properly, they are covered by a layer of toughened glass on both sides. This glass layer provides the required stability and is coated with an antireflection layer, ensuring the panel works efficiently by decreasing the losses. As mentioned in the previous section, solar cells, which are grouped together due to their similar properties, are used for module assembly. As a standard practice, tinned copper wires are soldered to cells in an alternate manner to connect the front of one cell to the back of the next. These ribbons provide a flexible link between cells, which considers thermal expansion as well. Beneath the glass layer, there is an encapsulant present to prevent any moisture from seeping in to corrode the panel. This encapsulant needs to have low light absorption, high thermal conductivity, and high resistivity to ensure less leakage of current. Ethylene-vinyl-acetate is regularly used as an encapsulant in the form of a 0.5 mm thick layer. Further, lamination and curing are done in a laminator, which has two chambers with a diaphragm in between. The lower chamber is heated to about 120 °C before the module is placed inside; this is done to melt the EVA. The melted EVA then fills the volume between the glass and the module on both sides. The upper chamber is filled with air to ensure the diaphragm presses the laminate. Next, the curing agents are added as the temperature is increased to 150 °C to induce the cross-linking of EVA for it to acquire rubber-like properties. For modules with glass-cell-plastic foil structure, EVA is used as the encapsulant, whereas for glass-cell-glass structure, Polyvinyl butyral (PVB) is used. Post-lamination, the encapsulant is removed from the edges, which are then sealed with silicone rubber. Hot spots caused by local shading are eliminated by putting a diode in parallel but opposite polarity with a group of cells. These diodes, also known as Schottky diodes, are connected in the junction box for each substring. A power optimizer may also be added, which is a circuit that keeps the PV module operating at the maximum power point. The power output of the module is provided by cables connected to the junction box, which are provided with connectors that keep the wiring weatherproof.

12.6 Sustainable Materials

In today's scenario, there is a lot of scope for technology optimization by using alternative materials for solar cell manufacturing. Though Silicon has always dominated the market in terms of the leading choice for construction materials, there are other non-silicon materials as well, which are used in smaller proportions in PV cells. The materials discussed are used in the working photovoltaic layer.

12.6.1 Cadmium-Tellurium-Based Cells

Cadmium Tellurium is a promising material for the construction of photovoltaic cells. CdTe films are relatively cheap to manufacture, and they allow the use of alloying elements in the CdTe films, for example, Zn, Hg, and other elements, to create layers with the desired properties. These cells have a very high radiation absorption capacity along with an ideal band gap. The cells made from CdTe films have charge carriers that have high mobility, adding to their efficiency, which varies between 10% and 16% depending upon the volume of charge carriers (Papež et al. 2021). Cadmium and Tellurium are rare earth elements, making their availability very low compared to Silicon, the second-most abundant element on Earth.

Moreover, Cd and Te are also highly toxic, which adds to their drawbacks. The production of these cells is declining due to the issues related to the disposal of Cd and its compounds. The plan is to abandon the use of CdTe-based cells completely due to the lack of recyclability of these materials.

12.6.2 Gallium Arsenide-Based Cells

To convert sunlight directly into electricity, a semiconductor material known as gallium arsenide (GaAs) is used in the absorbing layer of these high-efficiency photovoltaic cells. The main reason why these cells are the frontrunners behind silicon-based solar cells is because of their high efficiency. GaAs have a bandgap of 1.42 eV, whereas Silicon has a bandgap of 1.12 eV, which means that GaAs retain their semiconductive nature even at higher temperatures due to the wider bandgap. With an increase in temperature, the generation of carriers also increases, which reflects the use of these cells in challenging temperature circumstances. Their performance can be enhanced by increasing the number of layers in the solar cell, where each layer absorbs a specific wavelength of light and has a different composition. Such cells are known as multi-junction cells. By covering a wide range of wavelengths for absorption, multi-junction cells have higher efficiency, which can reach up to 47.1% (Papež et al. 2021). GaAs cells have applications in aviation and the military due to their flexibility and adaptive nature, and they can be used in unmanned aerial vehicles. However, the crystals of GaAs are brittle upon crystallization, which makes the manufacturing process expensive. Therefore, they are not used in common applications due to their increased cost though they have high efficiencies.

12.6.3 Copper-Indium-Gallium Selenide-Based Cells

The raw materials used to manufacture CIGS cells are less abundant. The absorption coefficient for these cells is higher than Silicon, meaning for the same amount of material, CIGS cells can absorb more light as compared to Silicon. Therefore, the thickness of wafers of CIGS cells is less, which brings down the cost of manufacturing both energy and material compared to c-Si cells. Along with these advantages, there are a few challenges present as well. The technology used in the production of these cells is lab-scale, and scaling it to the industrial level decreases the efficiency of the cells. Indium is obtained as a by-product from the production of zinc; consequently, its availability is dependent on it. CIGS is a quaternary compound, meaning it has four chemical elements; therefore, uniformity, in terms of film thickness and composition, over a large area presents an issue. So, at present, nearly 2% of solar panels are made from CIGS (Ramanujam and Singh 2017).

12.6.4 Silicon-Based Cells

The abundant nature of Silicon, along with its semiconducting properties, makes it the go-to material for manufacturing solar cell wafers. In the current scenario, nearly 92% of solar panels are made from Silicon (Ramanujam and Singh 2017). These can further be classified into monocrystalline and polycrystalline silicon solar panels. Monocrystalline solar cells are made from single-crystal Silicon; they offer higher efficiency and longevity due to their uniform structure. Polycrystalline solar cells, made from silicon crystals melted together, provide a cost-effective pathway with lower efficiency. The losses in these cells include a high reflection coefficient of bare Silicon and the absence of an antireflection coating. These can be corrected by adding a thin layer of dielectric, which acts as an antireflection coating, and by changing the texture from smooth to randomly arranged, it encourages light trapping and reduces losses, thereby increasing efficiency. Silicon is an environmentally benign material that possesses minimal hazards compared to other materials of construction. It can also be recycled by reprocessing the silicon wafers and purifying them for use in panels. Silicon can also be disposed of safely, as compared to other materials, and is less toxic.

12.7 Cost-Effectiveness

Along with the efficiency, the cost of the panel also needs to be considered. Presently, residential solar panels cost around 3 dollars/watt, and commercial solar panels cost around 2.25 dollars/watt (Szlufcik et al. 1997). There are a lot of factors that contribute to the cost of a solar panel system. Of the total module cost, the silicon wafer accounts for 46%, the cell fabrication for 24%, and the module fabrication for 30% (Szlufcik et al. 1997). The material of construction chosen for the panel contributes the maximum percentage towards the total cost, so it is essential to avoid rare elements, which have high material costs, and prefer a cheaper material like Silicon. Since solar cells generate electricity, they save some amount of money every month in the electricity bill. This amount gradually adds up to the initial investment made to install the solar system. The amount of time needed to equal the initial investment is known as the payback period. Out of the materials discussed in this chapter, Silicon has the lowest payback period, which makes it the most cost-effective material.

12.8 Conclusions

Solar cells harvest energy from the sunlight, which makes them a renewable and dependable source of electricity in regions where sunlight is prevalent. Even though the laboratory efficiencies of the cells can produce impressive numbers, the real-world efficiency of PV modules is around 15–23%, and this considerable difference is important to address. Solar cells must be carefully cut, cleaned, and coated to make solar panels. Clean energy for a future that is greener is made possible by efficient and long-lasting panels, which are ensured by modern machinery and stringent inspections. The materials of construction have evolved over the years, and ultimately, after reviewing the chemical, physical, and structural properties, Silicon stands out due to its abundant availability, long-term stability, and excellent semiconductor properties in photovoltaic applications. Although other materials, such as CIGS, CdTe, and GaAs, offer higher efficiencies at lower costs, their toxic nature, disposal issues, and recyclability prevent them from being the pioneers in materials of construction. The limited availability of individual elements compared to the surplus of Silicon present motivates the use of Silicon in solar cell wafer production. The future of photovoltaic solar panels lies in the continuous development of technology and materials to improve efficiency, lower the cost, and address environmental concerns, and this will have a significant role in making solar energy a dependable source of energy as we aim to move towards renewable and green energy.

Acknowledgments The authors declare no conflict of interest. This work did not receive any specific grant from funding agencies in the public, commercial, or not-for-profit sectors. The authors acknowledge the Institute of Chemical Technology, Mumbai, for the resources provided in writing this chapter.

References

Becquerel E (1839) On electron effects under the influence of solar radiation. Comptes Rendus de 'l' Academie Sciences Paris 9:561
Chapin DM, Fuller CS, Pearson GL (1954) A new silicon p-n junction photocell for converting solar radiation into electrical power. J Appl Phys 25(5):676–677. https://doi.org/10.1063/1.1721711
Chigondo F (2018) From metallurgical-grade to solar-grade silicon: an overview. SILICON 10(3):789–798. https://doi.org/10.1007/s12633-016-9532-7
Chowdhury S, Kumar M, Dutta S, Park J, Kim J, Kim S, Ju M, Kim Y, Cho Y, Cho E-C, Yi J (2019) High-efficiency crystalline silicon solar cells: a review. New RenewEnergy 15(3):36–45. https://doi.org/10.7849/ksnre.2019.3.15.3.036
Das Energie Private Limited (2 April 2024) Why Bhadla Solar Park is important for India. https://dasenergie.com/blog/bhadla-solar-park-largest-solar-power-plant-in-india/
Deshpande RA (2021) Advances in solar cell technology: an overview. J Sci Res 65(02):72–75. https://doi.org/10.37398/jsr.2021.650214
Dimroth F, Tibbits TND, Niemeyer M, Predan F, Beutel P, Karcher C, Oliva E, Siefer G, Lackner D, Fus-Kailuweit P, Bett AW, Krause R, Drazek C, Guiot E, Wasselin J, Tauzin A, Signamarcheix T (2016) Four-junction wafer-bonded concentrator solar cells. IEEE J Photovolt 6(1):343–349. https://doi.org/10.1109/JPHOTOV.2015.2501729
Einstein A (1905) Über einen die Erzeugung und Verwandlung des Lichtes betreffenden heuristischen Gesichtspunkt. Ann Phys 322(6):132–148. https://doi.org/10.1002/andp.19053220607
Ferrazza F (2018) Crystalline silicon: manufacture and properties. In Elsevier eBooks (pp. 75–93). https://doi.org/10.1016/b978-0-12-809921-6.00004-5
Ginsberg M (2019) Solar photovoltaic power optimization; enhancing system performance through operations, measurement, and verification. Routledge
Goetzberger A, Hoffmann VU (2005) Photovoltaic solar energy generation. Springer
Green MA, Emery K, Hishikawa Y, Warta W, Dunlop ED (2014) Solar cell efficiency tables (version 43). Prog Photovolt Res Appl 22(1):1–9. https://doi.org/10.1002/pip.2452
International Energy Agency (2024) Electricity 2024 - analysis and forecast to 2026. www.iea.org
Kim JY, Lee JW, Jung HS, Shin H, Park NG (2020) High-efficiency perovskite solar cells. Chem Rev 120(15):7867–7918. https://doi.org/10.1021/acs.chemrev.0c00107
Lee TD, Ebong AU (2017) A review of thin film solar cell technologies and challenges. Renew Sustain Energy Rev 70:1286–1297. https://doi.org/10.1016/j.rser.2016.12.028
Letcher T, Fthenakis VM (2018) A comprehensive guide to solar energy systems: with special focus on photovoltaic systems. Elsevier
LONGi (1 Nov 2023) LONGi sets a new world record of 33.9% for the efficiency of crystalline silicon-perovskite tandem solar cells.
Manisha, Pinkey, Kumari M, Sahdev RK, Tiwari S (2022) A review on solar photovoltaic system efficiency improving technologies. Appl Sol Energy (English Translation of Geliotekhnika) 58(1):54–75. https://doi.org/10.3103/S0003701X22010108
Maxeon Solar Technologies, Ltd. (21 March 2024) Maxeon sets another solar panel efficiency benchmark and achieves leading reliability certification.
NASA (9 May 2024) Sun fact sheet. https://nssdc.gsfc.nasa.gov/planetary/factsheet/sunfact.html
National Renewable Energy Laboratory (NREL) (2024) Best research-cell efficiency chart. https://www.nrel.gov/pv/cell-efficiency.html. Accessed 11 Nov 2024

Papež N, Dallaev R, Ţălu Ş, Kaštyl J (2021) Overview of the current state of gallium arsenide-based solar cells. Materials 14(11):3075. https://doi.org/10.3390/ma14113075
Peng J, Lu L, Yang H (2013) Review on life cycle assessment of energy payback and greenhouse gas emission of solar photovoltaic systems. Renew Sustain Energy Rev 19:255–274. https://doi.org/10.1016/j.rser.2012.11.035
Ramanujam J, Singh UP (2017) Copper indium gallium selenide based solar cells - a review. Energy Environ Sci 10(6):1306–1319. https://doi.org/10.1039/c7ee00826k
Schygulla P, Beutel P, Heckelmann S, Höhn O, Klitzke M, Schön J, Oliva E, Predan F, Schachtner M, & Siefer G (2022) Quadruple junction solar cell with 47.6% conversion efficiency under concentration. In: Presentation held at the 20th international conference on metal organic vapor phase epitaxy (ICMOVPE XX), Stuttgart, Germany
Smets A, Jäger K, Isabella O, van Swaaij R, Zeman M (2016) Solar energy: the physics and engineering of photovoltaic conversion, technologies and systems. Bloomsbury
Szlufcik J, Sivoththaman S, Nijs JF, Mertens RP, Van Overstraeten R (1997) Low-cost industrial technologies of crystalline silicon solar cells. Proc IEEE 85:711
U.S. Energy Information Administration (2023) Solar explained photovoltaics and electricity. https://www.eia.gov/energyexplained/solar/photovoltaics-and-electricity.php
Venkateswari R, Sreejith S (2019) Factors influencing the efficiency of photovoltaic system. Renew Sustain Energy Rev 101:376–394. https://doi.org/10.1016/j.rser.2018.11.012
Verma D, Nema S, Shandilya AM, Dash SK (2016) Maximum power point tracking (MPPT) techniques: recapitulation in solar photovoltaic systems. In. Renew Sust Energ Rev 54:1018–1034. https://doi.org/10.1016/j.rser.2015.10.068
Wong JH, Royapoor M, Chan CW (2016) Review of life cycle analyses and embodied energy requirements of single-crystalline and multi-crystalline silicon photovoltaic systems. Renew Sustain Energy Rev 58:608–618. https://doi.org/10.1016/j.rser.2015.12.241
Yun Y, Moon S, Kim S, Lee J (2022) Flexible fabric-based GaAs thin-film solar cell for wearable energy harvesting applications. Sol Energy Mater Sol Cells 246:111930. https://doi.org/10.1016/j.solmat.2022.111930
Zadde VV, Pinov AB, Strebkov DS, Belov EP, Efimov NK, Lebedev EN, Korobkov EI, Blake D, Touryan K (2010) New method of solar grade silicon production. In: 12th Workshop on crystalline silicon solar cell materials and processes

Chapter 13
Harnessing Generative AI for Photovoltaic Systems to Enhance Safety Features of Tailored Industrial Solar Solutions

Citlaly Perez Briceño, Mario Rojas, Isabel Mendez, Pedro Ponce, and Qipei Mei

Abstract This chapter explores the transformative potential of Generative Artificial Intelligence (GAI) in tailoring solar photovoltaic (PV) systems for industrial needs, emphasizing safety features. With fossil fuel depletion and rising emissions, a shift to renewable energy is crucial, positioning solar PV as a key sustainable alternative. Traditional solar installations often fail to optimize performance, but GAI can create customized solutions considering environmental parameters, energy consumption, and financial constraints. AI techniques like feature extraction, multi-objective optimization, and predictive modeling enhance system efficiency.

The *Tailoring PV Systems Solution* API, developed with ChatGPT, exemplifies AI's role in providing risk assessments, safety protocols, and real-time support, ensuring international safety compliance. AI improves equipment protection, operator safety, and community well-being, driving innovation and sustainability in renewable energy.

However, challenges like data privacy, algorithmic biases, technical barriers, and economic viability must be addressed. Robust data protection, unbiased AI models, and supportive policies are essential. Future advancements in GAI, integrated with smart grids and the Internet of Things (IoT), will enhance forecasting and real-time adaptation, optimizing solar performance. Collaborative efforts are needed to overcome obstacles and fully harness AI's potential for a cleaner, equitable energy future.

Keywords Generative AI · Photovoltaic system · Social security

---

C. P. Briceño · M. Rojas · I. Mendez · P. Ponce (✉)
Institute of Advanced Materials for Sustainable Manufacturing, Tecnologico de Monterrey, Monterrey, NL, Mexico
e-mail: pedro.ponce@tec.mx

Q. Mei
Civil and Environmental Engineering Department, University of Alberta, 6-263 Donadeo Innovation Centre for Engineering, Edmonton, Canada

A. K. Yadav (ed.), *Powering Future: Clean Energy Solutions for a Sustainable Planet*, Clean Energy Production Technologies, https://doi.org/10.1007/978-981-96-5055-2_13

13.1 Introduction

The depletion of fossil fuels and escalating carbon emissions are driving a shift towards renewable energy systems, notably solar photovoltaic (PV) installations. Historically, fossil fuels—coal, oil, and gas—have dominated global energy production, but their reserves are projected to run out by 2083 without a transition to renewable sources (Chen et al. 2024). Solar PV technology has seen rapid growth due to cost reductions and efficiency improvements. It is anticipated that by 2050, PV installations could satisfy 25% of the world's electricity demand (Nain and Anctil 2024).

Consequently, the global adoption of solar energy is accelerating, with nations investing heavily to lessen dependence on fossil fuels and combat environmental pollution. In 2023, global renewable energy capacity increased by 507 GW, with solar PV accounting for 75% of this rise increase (International Energy Agency (IEA) 2024a). China led this expansion, installing approximately 260 GW of PV capacity in 2023, supported by supportive government policies and substantial investments in renewable energy infrastructure (Office of Energy Efficiency and Renewable Energy 2024).

The increasing adoption of solar energy systems, driven by technological advancements and the imperative for sustainable energy sources, highlights the need for customized PV systems tailored to specific environmental conditions and user requirements (Schulte et al. 2022). This customization not only enhances the effectiveness of solar energy adoption but also ensures cost-effectiveness and a positive return on investment by aligning the system with a company's consumption patterns (Allouhi 2020).

Furthermore, tailored PV systems can address regulatory requirements and leverage incentive programs, helping organizations achieve sustainability goals while optimizing performance and achieving cost savings. Energy consumption plays a crucial role in advancing several Sustainable Development Goals (SDGs), including SDG7 (Affordable and Clean Energy), SDG13 (Climate Action), and SDG9 (Industry, Innovation, and Infrastructure). SDG9 emphasizes the need to upgrade infrastructure and industrial processes to enhance resource-use efficiency and promote the adoption of clean energy technologies (United Nations 2023). Therefore, the objective of upgrading infrastructure and retrofitting industries is to make them sustainable by enhancing energy use efficiency and increasing the adoption of clean energy sources, thereby reducing CO_2 emissions.

AI technology plays a crucial role in customizing PV systems for companies by analyzing large datasets to optimize system design and performance based on specific business needs and site conditions. AI employs techniques such as feature extraction, multi-objective optimization, and predictive modeling to improve the efficiency and performance of solar installations (Alam et al. 2023). Recent research has leveraged deep learning and machine learning algorithms for energy management and prediction. These methods utilize historical data, weather patterns, energy consumption profiles, and financial constraints to determine the optimal system configuration (Tien et al. 2022). AI also enhances predictive maintenance by identifying potential issues preemptively, thereby ensuring reliability and efficiency through real-time monitoring and adaptive control strategies.

As the adoption of tailored PV systems continues to rise, addressing safety and security aspects is essential to protect equipment, the environment, the community, and operators while mitigating risks (Duroha and Macht 2023). Proper installation, secure mounting, and high-quality components ensure equipment durability. Environmental considerations include selecting resilient sites and minimizing ecological impact (Li et al. 2024). Community's safety involves preventing fire and electrical hazards through proper wiring, grounding, and overcurrent protection devices, alongside clear signage and barriers (Juarez-Lopez et al. 2023). Operators' safety focuses on comprehensive training, adherence to safety protocols, use of personal protective equipment (PPE), and establishing emergency response plans (Duroha and Macht 2023). Risk management requires assessing and mitigating potential hazards, including cybersecurity measures to protect against unauthorized access and data breaches, ensuring the overall safety and security of the PV system (Harrou et al. 2023).

This chapter focuses on applying a Generative AI tool and framework designed to enhance safety features in installing industrial-tailored PV systems, addressing safety related to equipment, environment, operators, community, and risk management. This tool, implemented as an API in the ChatGPT environment, helps stakeholders to identify hazards, perform comprehensive risk assessments, provide recommendations for risk mitigation, generate customized safety solutions, ensure compliance with international safety standards, provide safety protocols and secure practices, etc.

13.2 AI and Its Potential to Enhance the Design and Implementation of Tailored Photovoltaic Systems

AI-based models can evaluate environmental parameters, user energy consumption, and financial constraints to generate optimal PV system configurations. This involves AI techniques like feature extraction, multi-objective optimization, and predictive modeling to enhance solar installation performance and efficiency (Alam et al. 2023).

Using AI to tailor solar PV systems offers several benefits, such as optimizing the design and performance of installations through advanced data analysis.

The integration of advanced data acquisition and monitoring systems further supports the security aspects of solar PV systems, ensuring optimal performance and reliability even under challenging conditions (Abdulla et al. 2024). AI algorithms analyze and interpret complex datasets, enabling the creation of optimized and personalized designs by learning from existing data to produce new and efficient designs tailored to specific user requirements (Abdelaal 2024).

13.2.1 AI for Equipment Protection and Maintenance

AI can enhance equipment protection in solar PV systems by predicting potential faults and optimizing maintenance schedules. Systems designed to monitor climatic parameters and dust accumulation can predict and optimize PV performance, thereby improving overall efficiency and reliability (Touati et al. 2016). AI models can monitor and analyze data from sensors placed on solar panels and associated equipment to detect anomalies and predict failures before they occur. This predictive maintenance approach helps in reducing downtime and extending the lifespan of solar equipment. For example, advance AI algorithms can optimize the placement of panels and predict shading effects, thereby preventing physical damage and ensuring consistent performance (Abubakar et al. 2021).

13.2.2 AI for Operator Safety and Community Well-being

Regarding operator safety, AI plays a critical role by designing systems that minimize human intervention and exposure to hazardous conditions. AI-driven automation can handle routine maintenance tasks and system monitoring, reducing the need for operators to work in potentially dangerous environments (Whitley 2024). Additionally, AI algorithms can simulate various scenarios to optimize safety protocols and emergency response strategies, ensuring that operators are well-prepared to handle any incidents that may arise (Ghaffarian et al. 2023).

Besides, AI can contribute to community well-being by designing solar systems that are not only efficient but also environmentally friendly and socially acceptable. AI algorithms can optimize the integration of solar panels into urban landscapes, ensuring minimal visual impacts and maximizing aesthetic appeal (Etukudoh et al. 2024). Furthermore, AI can help in designing systems that reduce noise pollution and other environmental impacts, promoting a healthier living environment for the community. The use of AI in energy systems also supports the development of sustainable and resilient energy infrastructures, contributing to overall community resilience and well-being (Koubaa Turki and Ben Saci 2023).

13.2.3 Generative Artificial Intelligence in the Design and Implementation of PV Systems

Recently, there have been significant improvements in the development of Large Language Models (LLMs) and GAI. GAI refers to a class of AI that uses algorithms to generate new data or content based on learned patterns from existing data (Lv 2023). Furthermore, GAI is considered in (Kanbach et al. 2024) as a groundbreaking technology because it creates new, original content and solutions by learning from existing data, surpassing traditional AI's focus on analysis and prediction. Additionally, GAI has enhanced human-computer interaction through Natural Language Processing (NLP), allowing computers to understand and interact with people conversationally. This assists users in automating tasks, retrieving information, and providing customer services, among other functions. (Saka et al. 2024).

Generative Pre-trained Transformer (GPT) models from OpenAI have gained significant attention through multiple iterations, culminating in the release of GPT-4 in 2023, which further enhanced performance and capabilities. Each version has built upon the previous ones, highlighting the rapid evolution and potential of AI in understanding and generating human-like text. As a result, ChatGPT have found a widespread endorsement, with significant implementations in healthcare (38.6%), computer science/IT (18.6%), and education/research (17.3%) (Khan et al. 2024). Moreover, organizations like Amazon, Google, IBM, and Microsoft are actively developing various GAI solutions aimed at enhancing productivity and efficiency across individuals, businesses, and organizations (Kalota 2024).

GAI holds immense potential in customizing solar PV systems for companies, offering unmatched benefits in their design and implementation phases, for instance by providing assistance in advanced data analysis, predictive maintenance, performance monitoring, risk management, sustainability optimization, economic analysis, and enhanced human-machine collaboration. Thus, GAI can play a crucial role across other stages of their life cycle such as predictive maintenance, customer support, operational optimization, energy forecasting, resource allocation, among others. Nevertheless, while powerful and transformative, GAI has several limitations that can impact its application in various fields, including solar PV systems, such as data requirements (Zhang et al. 2021), model complexity (Bandi et al. 2023), scalability (Avci 2023), interpretability (Kuzlu et al. 2020), and ethical and social concerns (Qadir 2023).

Therefore, by leveraging GAI across these stages, stakeholders in the solar energy sector can optimize operations, improve decision-making processes, enhance system reliability, and accelerate the transition to sustainable energy solutions.

13.3 Identified Safety and Security Aspects Related to the Installation of Solar PV Systems

When installing tailored PV systems, several safety aspects must be considered to ensure the protection of the equipment, the environment, the community, and operators' security and effective risk management.

- For the equipment, proper installation requires ensuring that the mounting structures and roofs can support panels' weight and withstand environmental conditions like wind and snow. Also, consistent management and maintenance of large-scale solar plants are essential for achieving maximum performance and long-term profitability. However, as presented by (Abdulla et al. 2024) the prevalent "install and forget" mentality in the PV industry results in minimal maintenance beyond periodic cleaning, relying on component reliability. However, this approach can lead to unexpected failures, production losses, higher costs, and compromised power quality.

- Environmental considerations include minimizing ecological impact and durability against harsh conditions. The increasing deployment raises end-of-life (EoL) management concerns, translated into an estimated global PV waste of 1.8–8 Mtons by the end of 2030 and up to 60–78 Mtons cumulative by 2050 (Tsanakas et al. 2020). Therefore, it is essential to understand the approaches, mechanisms, and regulations necessary for the sustainable management of EoL solar PVs. (Nain and Anctil 2024) shows that the US lacks federal PV waste management regulations, resulting in varying state requirements. Conversely, European countries have adopted the Waste Electrical and Electronic Equipment Directive and country-specific PV compliance schemes. Without regulations, 20 MT of PV waste is expected to end up in US landfills. Moreover, PV modules contain small amounts of toxic metals, and EoL assessment and regulation of these metals vary widely across nations. Li et al. (2024) overviews the metal composition of PV modules and common toxicity assessment procedures based on extensive research and review of technical literature and legislative documents.
- Community safety involves preventing electrical hazards by including proper wiring, grounding, and installations of overcurrent protection devices to prevent electric shock and fires. Fires in PV installations can cause severe economic, environmental, and safety risks. Furthermore, PV installation fires may emit more harmful substances than other fires, producing carcinogenic and toxic compounds. These emissions pose health risks to firefighters and raise concerns about the impact on their protective clothing and the safe handling and cleaning after such fires (Szmytke et al. 2022). Most fires due to electrical issues like arcing, short circuits, and overloads, exacerbated by extreme weather, poor lightning protection, and insufficient insulation.
- Operators' safety focuses on comprehensive training, adherence to safety protocols, use of personal protective equipment, and establishing emergency response plans. With the rise in PV installations, developing health and safety knowledge in the sector is essential. Duroha and Macht (2023) identified four major safety risk categories associated with PV installations: (1) electrical and fire risks, (2) heat stress, (3) manual handling risks, and (4) fall risks. Further, data on occupational accidents reveal that human errors and unsafe behaviors are primary causes, highlighting the importance of trained employees, safety culture awareness, and workplace precautions (Erten et al. 2022).
- Risk management requires incorporating cybersecurity measures to protect against unauthorized access and data breaches, ensuring the overall safety and security of the PV system. As PV system adoption rises, their vulnerability to cybersecurity threats increases. Cyberattacks on PV installations can cause operational disruptions, financial losses, and compromise broader energy infrastructure reliability (Harrou et al. 2023). Cyber-physical security has been addressed

from the perspectives of hardware, firmware, communications, and networks (Ye et al. 2022).

Therefore, a tailored approach to safety is critical for the successful installation and long-term operation of tailored PV systems, encompassing equipment durability, environmental sustainability, community protection, operator safety, and robust risk management strategies.

13.3.1 Understanding Company Requirements and Constraints for Designing Solar PV Systems

Traditional one-size-fits-all approaches to solar panel deployment often fall short in addressing the unique needs and constraints of different users and environments. For example, a traditional approach might involve installing standard solar panels on rooftops without considering specific site conditions or user energy requirements. These generic solutions may not optimize energy production or efficiency, leading to suboptimal performance and low adoption rates. Standardized solar solutions do not account for the specific conditions of each installation site, such as local weather patterns (Awad and Gül 2018) shading from nearby buildings or trees, and varying energy consumption patterns. This lack of customization can result in significant inefficiencies.

On the contrary, personalized solar solutions tailor the design and deployment of solar systems to the specific requirements of each user. This tailoring involves optimizing panel placement, orientation, and advanced tracking systems to maximize energy capture (Schulte et al. 2022). The advantages of personalized design over traditional approaches include higher energy efficiency, better return on investment, and improved user satisfaction due to the tailored fit to their specific needs. Additionally, integrating storage systems and smart grid technologies can further enhance the efficiency and reliability of solar power. To create a customized solar solution aiming safety and security, several comprehensive steps need to be followed:

- Conduct a Comprehensive Risk Assessment: Identify potential hazards and suggest mitigation strategies to ensure the safety and reliability of the solar installation.
- Implement Safety Protocols for Equipment: Develop safety protocols and training materials for the proper handling and maintenance of solar equipment.
- Enhance Community Safety: Facilitate community engagement and develop emergency response plans to ensure the safety of the surrounding community.
- Protect the Environment: Assess the environmental impacts of the solar installation and recommend eco-friendly practices to minimize negative effects.
- Integrate Monitoring and Reporting Systems: Set up monitoring systems and generate reports to track the performance and condition of the solar installation.

- Promote Continuous Improvement: Collect feedback and suggest improvements to enhance the performance and efficiency of the solar system over time.
- Compliance and Certification: Ensure compliance with relevant standards and assist in obtaining necessary certifications to validate the quality and safety of the solar installation.

The design and deployment of tailored solar panel systems are influenced by a variety of factors critical in determining the optimal configuration and efficiency of solar installations. These factors include:

- Building Type: The structure and layout of the building can influence the design of the solar system.
- Available Space: The physical space available for installing solar panels is one of the primary constraints. In urban environments, the limited space necessitates efficiently using every square meter to maximize energy generation.
- Energy Needs: The specific energy requirements of a household or facility play a significant role in the design of solar systems. Larger energy needs may require more extensive solar arrays and potentially integrated storage solutions to ensure a stable energy supply (International Energy Agency (IEA) 2024b).
- Budget Constraints: Financial considerations are crucial when planning solar installations. The initial investment cost, maintenance expenses, and potential saving on energy bills must all be factored into the decision-making process (Pérez et al. 2022).
- Operation and Maintenance: Proper operation and maintenance planning are essential to ensure the long-term performance of the solar system.
- Retrofitting: Assessing the feasibility and benefits of retrofitting existing structures with solar panels.

Generative AI employs several key principles and methodologies to customize solar solutions effectively. These include:

- **Modularity and Reuse:** AI systems often use modular and reusable components to create flexible and scalable designs. This approach allows to AI to combine different elements to meet specific requirements (Abdelaal 2024).
- **Multi-Objective Optimization:** AI algorithms optimize multiple objectives simultaneously, such as maximizing energy output, minimizing costs, and ensuring reliability. Techniques like genetic algorithms and neural networks are commonly used for this purpose (Harrington and Ross 2013).
- **Predictive Analytics:** AI models analyze historical data to predict future conditions and adjust system designs accordingly. This helps in creating robust and adaptive solutions that perform well under varying conditions (Aldoseri et al. 2023).

13.3.2 Addressing Environmental and Regulatory Considerations and Ethical and Social Implications

When designing and implementing tailored PV systems, it is essential to consider a range of environmental, regulatory, ethical, and social factors. Adhering to international standards ensures the safety, efficiency, and reliability of PV installations and supports sustainable development goals. Therefore, Table 13.1 outlines key international standards related to various aspects of PV system installation and management.

13.3.3 Environmental and Regulatory Considerations

Customization of solar solutions takes into account various social factors, including operator safety and community well-being. Advanced AI-driven predictive maintenance and real-time monitoring systems to detect and mitigate potential hazards before they pose risks can significantly reduce the likelihood of accidents and ensure a safer working environment for operators (Abubakar et al. 2021). Additionally, community well-being is enhanced by designing solar installations that minimize visual and noise pollution, ensuring that the deployment of solar panels does not adversely affect the local population.

The deployment of solar panel systems is governed by a variety of regulatory frameworks and standards, which significantly influence the customization of solar solutions. Regulatory standards ensure solar PV systems' safety, efficiency, and reliability, including requirements for grid connection, safety, and performance metrics. For example, the National Electric Code in the United States sets forth regulations for the installation of solar systems, including safety standards for electrical connections and system integrity (Mundada et al. 2016).

Additionally, renewable energy policies, such as feed-in tariffs and renewable portfolio standards, play a crucial role in promoting the adoption of solar technologies. These policies often include incentives for the deployment of customized solar solutions that can adapt to specific local conditions and user needs.

13.3.4 Ethical and Social Implications

AI plays a pivotal role in achieving energy equity and reducing environmental impacts associated with solar energy deployment. By optimizing the placement and performance of solar PV systems, AI ensures that solar energy is accessible and beneficial to a broader range of communities, including underserved and low-income areas. AI-driven customization allows for the efficient use of resources and minimizes environmental impacts by optimizing energy production and reducing waste (Kurukuru et al. 2021; Shirin et al. 2019; Shirin and Yadav 2021).

Table 13.1 International standards for addressing safety and security related to PV systems

Equipment security	Environmental protection	Operator safety	Risk management	Community safety	General safety and compliance
1. IEC 61730—PV module safety qualification 2. IEC 62446—Grid Connected PV Systems—Minimum Requirements for System Documentation, Commissioning Tests, and Inspection 3. UL 1703—Standard for Safety of Flat-Plate PV Modules and Panels 4. IEC 62109—Safety of Power Converters for Use in PV Power Systems 5. NFPA 70 (NEC)—National Electrical Code 6. UL 9540—Standard for Energy Storage Systems and Equipment	1. ISO 14001—Environmental Management Systems 2. ISO 14040—Environmental Management—Life Cycle Assessment	1. ISO 45001—Occupational Health and Safety Management Systems 2. OSHA 1926.28—Personal Protective Equipment	1. ISO 31000—Risk Management 2. ISO 55001—Asset Management 3. IEC 61853—PV Module Performance Testing and Energy Rating 4. IEC 62938—PV Modules—Non-uniform Snow Load Testing	1. ANSI/ASSP Z10—Occupational Health and Safety Management Systems. 2. ISO 26000—Guidance on Social Responsibility	1. IEC 61215—Terrestrial PV Modules—Design Qualification and Type Approval 2. IEC 62941—Terrestrial PV Modules—Quality System for PV Module Manufacturing 3. IEEE 1547—Standard for Interconnection and Interoperability of Distributed Energy Resources with Associated Electric Power Systems Interfaces

AI can also enhance the integration of renewable energy into the grid, reducing the reliance on fossil fuels and lowering greenhouse gas emissions. By improving the accuracy of solar energy forecasts and optimizing energy storage and distribution, AI contributes to a more sustainable and resilient energy systems. This reduces the overall environmental footprint of energy production and supports global efforts to combat climate change (Yang et al. 2021a).

For instance, AI-driven energy management systems can dynamically adjust energy production and consumption patterns to match real-time demand, reducing energy wastage and enhancing the overall sustainability of energy systems. Moreover, AI can facilitate the integration of solar energy with other renewable sources, creating a more resilient and sustainable energy infrastructure (Kromer et al. 2018).

13.4 Scaling Generative AI for Large-Scale Commercial and Industrial PV Installations

The scalability of AI in managing large-scale commercial and industrial PV installations is crucial for optimizing energy production, reducing operational costs, and enhancing system reliability. AI technologies, such as ML and generative AI, can handle the complexities and scale of industrial PV systems by processing large datasets, predicting performance, and optimizing operational parameters in real time.

Scalable AI frameworks, such as the AI Model Factory, provide scalable platforms for deploying AI applications across multiple assets, facilitating joint execution of multiple AI tasks such as anomaly detection, failure pattern analysis, and asset health forecasting. These frameworks ensure that the benefits of AI, including improved predictive maintenance and operational efficiency, can be realized across large-scale PV installations (Patel et al. 2023).

Distributed optimization frameworks allow large-scale aggregation of prosumers with residential PV-battery systems, supporting both day-ahead and real-time optimization, and enhancing scalability. Such frameworks ensure efficient operation and energy management, essential for large-scale commercial PV systems (Dolatabadi and Siano 2020).

Integrating the Internet of Things (IoT) with AI enhances the scalability of control algorithms used in commercial building HVAC systems, demonstrating energy savings and improved system efficiency. The decentralized control architecture enabled by IoT can be adapted for PV systems to optimize performance and scalability (Png et al. 2019).

13.4.1 *Impact of Customized Solar Solutions on Companies*

Using AI and advanced analytics in customizing solar solutions tailored to specific commercial and industrial needs can maximize energy production while operational costs are minimized. Some of the potential impacts of these solutions can include:

- **Improved Sustainability:** Customized solar solutions enable companies to optimize their energy production, reducing reliance on fossil fuels and lowering carbon emissions. AI-driven models can predict energy production and optimize the placement and configuration of solar panels to maximize exposure to sunlight, thus increasing overall efficiency and supporting regulatory compliance and corporate social responsibility goals (Kurukuru et al. 2021).
- **Enhanced Operational Efficiency:** AI can customize PV system designs to meet specific operational requirements, optimizing both performance and cost-efficiency.
- **Economic Benefits:** The economic benefits of customized solar solutions include reduced energy costs and improved return of investment (ROI). AI algorithms can forecast energy production and consumption, enabling better energy management and cost savings. For example, optimized configurations of PV systems that consider component types and arrangements can significantly lower the levelized cost of electricity (LCOE), maximizing economic benefits over the system's operational lifetime (Kerekes et al. 2013).
- **Operational Resilience:** AI-enhanced solar solutions provide greater operational resilience by predicting and mitigating potential issues before they impact performance. AI can analyze environmental data to optimize PV system configurations, ensuring consistent energy production even under variable conditions (Abubakar et al. 2021).

13.5 Case Study: Leveraging Generative AI for Comprehensive Safety and Security Protocols in Solar PV Systems

This case study explores how Generative AI (GAI) can be leveraged to develop a comprehensive safety and security protocol tailored to the specific needs and historical data of an industrial company deploying solar PV systems. By focusing on key areas such as equipment security, environmental protection, operator safety, risk management, and community safety, this case study demonstrates the practical application and benefits of GAI in enhancing the overall safety and security framework.

An API named *Tailoring PV Systems Solution* was created using ChatGPT. This API is designed to support the management of PV system safety and security protocols, adhering to global standards, and aids in risk management, safety training, community engagement, environmental protection, system monitoring, improvement, and compliance.

13.5.1 API Overview

The instructions given for creating the API were the following:

(a) **Name:** Tailoring PV Systems Solution
(b) **Description:** This API is tailored to support the management of PV system safety and security protocols, adhering to global standards, and aids in risk management, safety training, community engagement, environmental protection, system monitoring, improvement, and compliance.
(c) **Instructions for the API:**

- What does this GPT do? This GPT assists organizations in managing the safety and security of PV systems by providing guidance on international standards, generating safety and risk management documentation, developing training materials, and offering real-time troubleshooting support.
- How does this GPT behave? This GPT behaves as an interactive and responsive AI tool that processes and analyzes large volumes of data, tailors recommendations to specific contexts, and engages users with adaptive learning experiences. It operates with a focus on enhancing user understanding and ensuring adherence to the highest safety and environmental standards.

  – Who are you?

  – Response: I am a GPT developed by the Enabling Technologies Unit from the Institute of Advanced Materials for Sustainable Manufacturing at Tecnologico de Monterrey.

- What should this GPT avoid doing? This GPT should avoid:

  – Providing outdated or unverified information.
  – Making decisions or providing advice without referencing the appropriate standards and compliance guidelines.
  – Overstepping in areas that require human judgment, particularly in complex safety evaluations and high-stakes decision-making processes.
  – Handling personal data without strict adherence to data privacy laws and ethical guidelines.

13.5.2 API Implementation Steps

The API follows seven main steps, considering the global standards shown in Table 13.1:

1. Conduct a Comprehensive Risk Assessment: Utilize ChatGPT to identify potential hazards and suggest mitigation strategies.
2. Implement Safety Protocols for Equipment: Use of ChatGPT to develop safety protocols and training materials for equipment handling and maintenance.

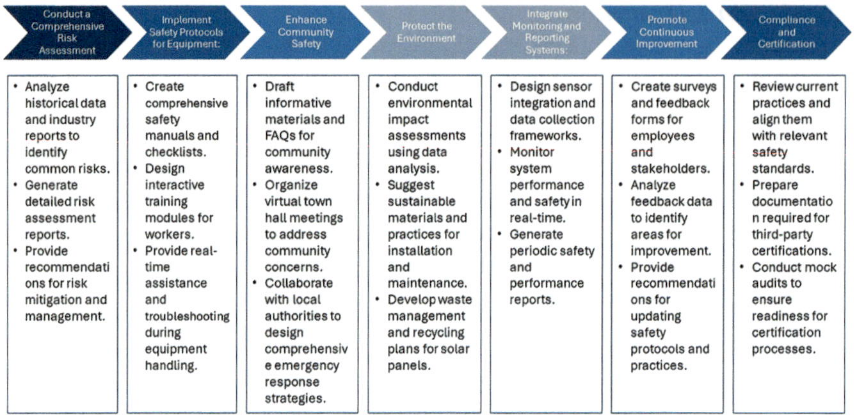

Fig. 13.1 Steps and tasks for the API *Tailoring PV Systems Solution*

3. Enhance Community Safety: ChatGPT will facilitate community engagement and develop emergency response plans.
4. Protect the Environment: ChatGPT will assess environmental impacts and recommend eco-friendly practices.
5. Integrate Monitoring and Reporting Systems: Use ChatGPT to assist in setting up monitoring systems and generating reports.
6. Promote Continuous Improvement: Use ChatGPT to facilitate feedback collection and suggest improvements.
7. Compliance and Certification: Utilize ChatGPT to ensure compliance with standards and assist in obtaining certifications.

Figure 13.1 illustrates a step-by-step process that the API follows to develop a comprehensive security protocol for PV systems. Each step in the process is designed to address specific aspects of safety and security, leveraging GAI's capabilities to provide tailored recommendations based on historical data and user needs. The diagram also outlines detailed tasks for each step to give a clear understanding of the API's functionality.

13.5.3 Interaction with the API

To facilitate user interaction with the API, several phrases can be used to start conversations:

- Can I provide you historical data or statistics on incidents in PV systems that can help me understand common risks and conductor a risk analysis according to international standards?
- What are the steps for enhancing Safety PV Systems?
- What are the common risks based on standards?

13.5.4 Contribution of ChatGPT in Each Step

For each step, the contributions of ChatGPT are the following:

- It processes and analyzes large volumes of unstructured data from various sources in real time, provides insights that static systems or traditional methods might miss.
- It offers interactive and adaptive learning experiences, tailored to individual user needs and providing instant feedback, which static training programs cannot match.
- It can engage with the community in multiple languages and provide personalized responses, enhancing accessibility and understanding more effectively than standard communication methods.
- It continuously updates suggestions based on the latest research and environmental data, offering dynamic and current advice that static systems cannot provide.
- It synthesizes data from multiple sources and generates comprehensive, easy-to-understand reports, making it easier for decision-makers to act on real-time insights.
- It analyzes qualitative feedback, recognizing patterns and sentiments that automated survey tools might miss, providing deeper insights into areas for improvement.
- It provides ongoing, adaptive compliance checks and updates, ensuring that the latest regulatory changes are always incorporated, unlike static compliance checklists.

This case study highlights the significant potential of Generative AI, specifically ChatGPT, in enhancing the efficiency and customization of safety and security protocols for solar PV systems. By leveraging AI, organizations can ensure their PV installations are safe, compliant, and optimized for performance, ultimately contributing to the broader goals of sustainability and social responsibility. Thus, Fig. 13.2 shows an example of interaction between user and the API developed in this work.

13.6 Other GAI Applications

GAI can significantly enhance the analysis of energy data. Processing large datasets, it identifies patterns and trends that are not immediately apparent. Key points where generative AI aids in data analysis and forecasting include:

- Data Preprocessing and Cleaning: AI assists in preprocessing the raw data, handling missing values, and ensuring thc data is clean and ready for analysis (Yang et al. 2021b).

Fig. 13.2 Exemplification of the interaction between the user and the API

- Pattern Recognition: It identifies patterns in energy production, such as peak generation times and the impact of weather conditions on solar output (Avci 2023).
- Production Forecasting: AI models can forecast future energy production based on historical data, helping to predict solar energy availability and integrate it efficiently with other energy sources (Alam et al. 2023).

13.6.1 Optimization of Energy Solutions

Optimization involves determining the best configuration and operational strategy for the solar energy systems to maximize efficiency and reliability.

Generative AI contributes to optimization through:

- **Simulation of Scenarios:** AI can simulate various operational scenarios to identify the most efficient configurations and strategies for solar PV systems (Rosa de Jesús et al. 2021).
- **Resource Allocation:** It helps in optimizing the allocation of resources, ensuring that solar energy is used efficiently and in combination with other energy sources (Alam et al. 2023).
- **Cost-Benefit Analysis:** AI can perform cost-benefit analyses to evaluate the economic viability of different solar energy solutions, ensuring that the chosen solution is both efficient and cost-effective (Ismail et al. 2014).

13.6.2 Decision-Making and Customization

Generative AI supports informed decision-making by providing insights based on data analysis and optimization results. This is crucial for customizing solar PV solutions to meet specific needs, particularly in the context of enhancing social security.

AI aids decision-making through:

- **Customized Solutions:** It enables the customization of solar PV systems based in the specific energy needs and conditions of different regions, enhancing social security by ensuring reliable energy access.
- **Predictive Maintenance:** AI can predict potential issues with solar PV systems, allowing for proactive maintenance and reducing downtime.
- **Policy and Regulatory Compliance:** AI helps ensure that the solutions comply with local policies and regulations, which is essential for sustainable and socially secure energy systems.

13.6.3 Adaptability to Different Scenarios and Countries

The methodology presented in this work can be adapted to larger databases and varied scenarios, making it applicable in different countries and regions. This adaptability ensure that the benefits of AI-driven solar PV optimization can be realized globally, catering to the specific energy needs and social security goals and diverse populations.

Generative AI supports this adaptability through:

- **Scalability:** AI models can be scaled to handle larger datasets, ensuring that they remain effective even as the volume of data grows.
- **Versatility:** The AI-driven approach can be customized to account for different climatic conditions, regulatory environments, and energy consumption patterns across various countries.
- **Localization:** By incorporating local data and conditions, AI can provide tailored solutions that address the unique challenges and opportunities of each region, thereby enhancing the relevance and impact of the solutions.

13.6.4 Ensuring Operator Safety and Equipment Reliability

In addition to optimizing energy solutions, ensuring operator safety and equipment reliability are critical components of any solar PV system. AI plays a crucial role in enhancing these safety features through:

- Safety Protocols: AI helps develop comprehensive safety protocols for the installation and maintenance of solar PV systems, ensuring that all procedures adhere to international safety standards.
- Risk Assessment: AI conducts thorough risk assessments to identify potential hazards and suggest mitigation strategies, enhancing the overall safety of solar PV installations.
- Real-Time Monitoring: AI-enabled real-time monitoring systems can detect anomalies and potential safety issues, allowing for immediate action to prevent accidents and equipment failure.
- Training and Support: AI can generate training materials and provide real-time support for operators, ensuring they are well-equipped to handle the equipment safely and effectively.

This exemplification demonstrates the significant potential of GAI in enhancing the efficiency and customization of solar PV solutions. AI plays a crucial role in improving social security through reliable and sustainable energy solutions by analyzing energy data, forecasting production, optimizing solutions, and aiding decision-making. The integration of AI in energy management not only maximizes the adoption of solar energy but also ensures that it is done in a socially responsible and economically viable manner. The adaptability of these methods to larger databases and different geographic contexts further underscores their value in achieving global energy sustainability and social security goals.

13.7 Discussion

Integrating GAI in designing and implementing solar PV systems presents a transformative potential to meet the growing demand for renewable energy. The global shift towards renewable energy is driven by the depletion of fossil fuels and rising carbon emissions (Chen et al. 2024). The rapid adoption of solar technologies, supported by favorable government policies and investments (Office of Energy Efficiency and Renewable Energy 2024), necessitates tailored PV solutions to enhance efficiency and sustainability.

GAI aids in this endeavor by leveraging complex data inputs to optimize system designs. AI-based models evaluate environmental parameters, user energy consumption, and financial constraints to generate optimal PV system configurations (Alam et al. 2023). Optimizing energy production and reducing reliance on fossil fuels enhances performance and supports the broader goals of sustainability.

Furthermore, AI-driven tools like the *Tailoring PV Systems Solution* API, developed using ChatGPT, provide comprehensive risk assessments, safety protocols, and real-time troubleshooting support. This API exemplifies how AI can facilitate the development of customized safety solutions, ensuring compliance with international safety standards and promoting continuous improvement.

The stages of AI adoption for enhancing safety and security in PV systems are outlined in Table 13.2, which categorizes the maturity levels of GAI integration

Table 13.2 Stages of GAI adoption for enhancing safety and security in PV systems

Category	Behind	On track	Ahead
Belief	AI is a toy, AI is all hype	AI is a tool, AI helps me be more efficient and secure PV systems	AI is a system, AI is a shift in how we enhance safety and security in PV systems
Policy on AI usage	Bans all versions and forms of generative AI for PV systems (or lacks clear policies)	Allows limited AI tools for PV system security with minimal policies	Comprehensive AI policies supporting diverse AI tools for PV security, including real-time monitoring, predictive analytics, and continuous improvement
Tool Choice	No adoption of AI tools for PV security	Selective adoption of proven AI security tools for PV systems	Broad adoption of AI tools, including experimental technologies for advanced PV system security, working with multiple AI providers to enhance safety
Internal Training	No training on AI for PV system security	AI training programs for select teams involved in PV system management and security	Extensive AI training programs for all employees, with specialized tracks on PV system security, risk management, and emergency response strategies
Budget	No budget for AI in PV system security	Allocates an experimental budget for AI tools in PV system security (often a percent of revenue or R&D re-allocation)	Comprehensive budget for AI integration in PV security, covering training, tool acquisition, testing, and development of AI-driven security and safety projects
Use Case Progress	No use cases for AI in PV security, awaiting perfect proof of AI's efficacy before starting	Limited use cases for AI in PV security: initial trials of AI tools for monitoring, risk assessment, and predictive maintenance	Numerous AI-driven security use cases in production for PV systems, universal employee access to AI tools for security, and integration of AI into PV security products for top-line growth
Privacy and Security	No formal policies on AI-related data privacy and security for PV systems, no training or incident response plan	Basic data privacy policies for AI in PV systems, including encryption, access controls, regular training, compliance with key regulations, anonymizing data where possible	Dedicated privacy team for AI in PV systems, privacy by design, managing third-party risks, continuous monitoring and audits, detailed incident response plan, transparent communication
Talent	No changes in hiring for AI roles related to PV security	Hiring a few data scientists or machine learning engineers for PV security; added AI-related skills to job descriptions	Reallocated headcount across the company to include AI talent for PV security, upskilling current employees, integrating AI security qualifications into job descriptions

from initial skepticism to advanced, comprehensive adoption. This structured approach helps organizations identify their current standing and the necessary steps to progress towards utilizing AI for improved safety and efficiency in PV systems.

13.8 Conclusions: Future Directions and Challenges

The future of tailored solar solutions powered by GAI is poised to revolutionize the renewable energy sector. Emerging trends suggest significant advancements in predictive analytics, enabling precise forecasting and real-time adaptive systems that optimize solar panel performance under varying conditions. Integrating GAI with smart grids and the Internet of Things (IoT) will create intelligent, adaptive energy solutions, enhancing grid stability and resilience. Personalized energy solutions tailored to specific consumption patterns and environmental conditions will drive higher adoption rates and customer satisfaction. Additionally, future developments will focus on sustainability, including advancements in recycling technologies for end-of-life solar panels and eco-friendly materials.

However, several challenges must be addressed to achieve widespread adoption of AI-powered solar solutions. Data privacy concerns necessitate robust protection frameworks to safeguard-sensitive information. Algorithmic biases must be identified and mitigated to ensure equitable outcomes. Technical barriers, such as interoperability and high computational requirements, need resolution to realize AI benefits fully. Economic viability, especially for small and medium-sized enterprises, requires making these solutions more affordable. Navigating the complex regulatory landscape and ensuring compliance with standards is crucial for acceptance and implementation. Ethical and social implications, including job displacement and equitable distribution of benefits, must be addressed through thoughtful policymaking and stakeholder engagement. Thus, the integration of GAI in tailored solar solutions offers transformative potential for the renewable energy sector. AI-powered solar PV systems can significantly contribute to a cleaner, more efficient, and equitable energy future by driving innovation, enhancing efficiency, and promoting sustainability. The adaptability of these AI-powered solutions to various geographic and socio-economic contexts will drive global implementation, addressing diverse energy needs and contributing to worldwide sustainable efforts. Addressing the associated challenges through collaborative efforts and continued advancements will pave the way for widespread adoption, making the vision of a solar-powered future well within reach.

Acknowledgment The authors acknowledge the technical and financial support of the Institute of Advanced Materials for Sustainable Manufacturing, Tecnologico de Monterrey, Mexico in producing this work.

References

Abdelaal O (12 Jan 2024) Design a modular AI: from theory to practice in machine learning code architecture. Medium. https://medium.com/@osama1339669/embracing-modularity-in-ai-part-1-from-theory-to-practice-in-machine-learning-code-architecture-5ab1257dfe69

Abdulla H, Sleptchenko A, Nayfeh A (2024) Photovoltaic systems operation and maintenance: A review and future directions. Renew Sust Energ Rev 195:114342. https://doi.org/10.1016/j.rser.2024.114342

Abubakar A, Almeida CFM, Gemignani M (2021) Review of artificial intelligence-based failure detection and diagnosis methods for solar photovoltaic systems. Mach Des 9(12):Article 12. https://doi.org/10.3390/machines9120328

Alam MM, Alshahrani T, Khan F, Hakami J, Shinde SM, Azim R (2023) AI-based efficiency analysis technique for photovoltaic renewable energy system. Phys Scr 98(12):126006. https://doi.org/10.1088/1402-4896/ad0bb4

Aldoseri A, Al-Khalifa KN, Hamouda AM (2023) Re-thinking data strategy and integration for artificial intelligence: concepts, opportunities, and challenges. Appl Sci 13(12):Article 12. https://doi.org/10.3390/app13127082

Allouhi A (2020) Solar PV integration in commercial buildings for self-consumption based on life-cycle economic/environmental multi-objective optimization. J Clean Prod 270:122375. https://doi.org/10.1016/j.jclepro.2020.122375

Avci E (2023) Generative AI in electricity distribution: a qualitative exploration. PressAcademia Procedia 17:208–211. https://doi.org/10.17261/Pressacademia.2023.1788

Awad H, Gül M (2018) Load-match-driven design of solar PV systems at high latitudes in the northern hemisphere and its impact on the grid. Sol Energy 173:377–397. https://doi.org/10.1016/j.solener.2018.07.010

Bandi A, Adapa PVSR, Kuchi YEVPK (2023) The power of generative AI: a review of requirements, models, input–output formats, evaluation metrics, and challenges. Future Internet 15(8):Article 8. https://doi.org/10.3390/fi15080260

Chen Y, Li W, Wang X (2024) An investigation into the risk assessment of building-integrated photovoltaic residential project development utilizing the DEMATEL-ANP methodology: a Chinese case study. Buildings 14(3):Article 3. https://doi.org/10.3390/buildings14030623

Dolatabadi M, Siano P (2020) A scalable privacy preserving distributed parallel optimization for a large-scale aggregation of prosumers with residential PV-battery systems. IEEE Access 8:210950–210960. https://doi.org/10.1109/ACCESS.2020.3035432

Duroha JC, Macht GA (2023) Solar installation occupational risks: a systematic review. Saf Sci 160:106048. https://doi.org/10.1016/j.ssci.2022.106048

Erten B, Oral B, Yakut MZ (2022) The role of virtual and augmented reality in occupational health and safety training of employees in PV power systems and evaluation with a sustainability perspective. J Clean Prod 379:134499. https://doi.org/10.1016/j.jclepro.2022.134499

Etukudoh EA, Nwokediegwu ZQS, Umoh AA, Ibekwe KI, Ilojianya VI, Adefemi A, Etukudoh EA, Nwokediegwu ZQS, Umoh AA, Ibekwe KI, Ilojianya VI, Adefemi A (2024) Solar power integration in urban areas: a review of design innovations and efficiency enhancements. World J Adv Res Rev 21(1):Article 1. https://doi.org/10.30574/wjarr.2024.21.1.0168

Ghaffarian S, Taghikhah FR, Maier HR (2023) Explainable artificial intelligence in disaster risk management: achievements and prospective futures. Int J Disaster Risk Reduct 98:104123. https://doi.org/10.1016/j.ijdrr.2023.104123

Harrington A, Ross BJ (2013) Generative representations for artificial architecture and passive solar performance. In: 2013 IEEE Congress on evolutionary computation. IEEE, pp 537–545. https://doi.org/10.1109/CEC.2013.6557615

Harrou F, Taghezouit B, Bouyeddou B, Sun Y (2023) Cybersecurity of photovoltaic systems: challenges, threats, and mitigation strategies: a short survey. Front Energy Res 11. https://doi.org/10.3389/fenrg.2023.1274451

International Energy Agency (IEA) (2024a) Electricity – renewables 2023 – analysis. IEA. https://www.iea.org/reports/renewables-2023/electricity

International Energy Agency (IEA) (2024b) Grid-scale storage. IEA. https://www.iea.org/energy-system/electricity/grid-scale-storage

Ismail MS, Moghavvemi M, Mahlia TMI (2014) Genetic algorithm based optimization on modeling and design of hybrid renewable energy systems. Energy Convers Manag 85:120–130. https://doi.org/10.1016/j.enconman.2014.05.064

Juarez-Lopez JM, Franco JA, Hernandez-Escobedo Q, Muñoz-Rodríguez D, Perea-Moreno A-J (2023) Analysis of a novel proposal using temperature and efficiency to prevent fires in photovoltaic energy systems. Fire 6(5):196. https://doi.org/10.3390/fire6050196

Kalota F (2024) A primer on generative artificial intelligence. Educ Sci 14(2). https://doi.org/10.3390/educsci14020172

Kanbach DK, Heiduk L, Blueher G, Schreiter M, Lahmann A (2024) The GenAI is out of the bottle: generative artificial intelligence from a business model innovation perspective. Rev Manag Sci 18(4):1189–1220. https://doi.org/10.1007/s11846-023-00696-z

Kerekes T, Koutroulis E, Séra D, Teodorescu R, Katsanevakis M (2013) An optimization method for designing large PV plants. IEEE J Photovolt 3(2):814–822. https://doi.org/10.1109/JPHOTOV.2012.2230684

Khan N, Khan Z, Koubaa A, Khan MK, Salleh RB (2024) Global insights and the impact of generative AI-ChatGPT on multidisciplinary: A systematic review and bibliometric analysis. Connection Science 36(1). https://doi.org/10.1080/09540091.2024.2353630

Koubaa Turki L, Ben Saci A (2023) Generative design for a sustainable urban morphology. Archit Plan J (APJ) 28:Article 15. https://doi.org/10.54729/2789-8547.1210

Kromer M, Zeifman M, Roth K, Yip T (2018) The SunDial framework: enabling high penetration solar through the integration of energy storage, demand management, and forecasting. In: 2018 IEEE 7th world conference on photovoltaic energy conversion (WCPEC) (A joint conference of 45th IEEE PVSC, 28th PVSEC & 34th EU PVSEC). IEEE, pp 1436–1439. https://doi.org/10.1109/PVSC.2018.8547884

Kurukuru VS, Haque A, Khan MA, Sahoo SS, Malik A, Blaabjerg F (2021) A review on artificial intelligence applications for grid-connected solar photovoltaic systems. Energies. https://doi.org/10.3390/en14154690

Kuzlu M, Cali U, Sharma V, Güler Ö (2020) Gaining insight into solar photovoltaic power generation forecasting utilizing explainable artificial intelligence tools. IEEE Access 8:187814–187823. https://doi.org/10.1109/ACCESS.2020.3031477

Li F, Shaw S, Libby C, Preciado N, Bicer B, Tamizhmani G (2024) A review of toxicity assessment procedures of solar photovoltaic modules. Waste Manag 174:646–665. https://doi.org/10.1016/j.wasman.2023.12.034

Lv Z (2023) Generative artificial intelligence in the metaverse era. Cogn Robot 3:208–217. https://doi.org/10.1016/j.cogr.2023.06.001

Mundada AS, Nilsiam Y, Pearce JM (2016) A review of technical requirements for plug-and-play solar photovoltaic microinverter systems in the United States. Sol Energy 135:455–470. https://doi.org/10.1016/J.SOLENER.2016.06.002

Nain P, Anctil A (2024) End-of-life solar photovoltaic waste management: a comparison as per European Union and United States regulatory approaches. Res Conserv Recycl Adv 21:200212. https://doi.org/10.1016/j.rcradv.2024.200212

Office of Energy Efficiency & Renewable Energy (2024) Quarterly solar industry update. Energy.Gov. https://www.energy.gov/eere/solar/quarterly-solar-industry-update

Patel D, Lin S, Shah D, Jayaraman S, Ploennigs J, Bhamidipati A, Kalagnanam J (2023) AI model factory: scaling AI for Industry 4.0 applications. Proc AAAI Conf AI 37(13):Article 13. https://doi.org/10.1609/aaai.v37i13.27081

Pérez C, Ponce P, Meier A, Dorantes L, Sandoval JO, Palma J, Molina A (2022) S4 Framework for the integration of solar energy systems in small and medium-sized manufacturing companies in Mexico. Energies 15(19):Article 19. https://doi.org/10.3390/en15196882

Png E, Srinivasan S, Bekiroglu K, Chaoyang J, Su R, Poolla K (2019) An internet of things upgrade for smart and scalable heating, ventilation and air-conditioning control in commercial buildings. Appl Energy 239:408–424. https://doi.org/10.1016/j.apenergy.2019.01.229

Qadir J (2023) Engineering education in the era of ChatGPT: promise and pitfalls of generative AI for education. In: 2023 IEEE global engineering education conference (EDUCON). IEEE, pp 1–9. https://doi.org/10.1109/EDUCON54358.2023.10125121

Rosa de Jesús DA, Mandal P, Senjyu T, Kamalasadan S (2021) Unsupervised hybrid deep generative models for photovoltaic synthetic data generation. In: 2021 IEEE power & energy society general meeting (PESGM). IEEE, pp 1–5. https://doi.org/10.1109/PESGM46819.2021.9637844

Saka A, Taiwo R, Saka N, Salami BA, Ajayi S, Akande K, Kazemi H (2024) GPT models in construction industry: opportunities, limitations, and a use case validation. Dev Built Environ 17:100300. https://doi.org/10.1016/j.dibe.2023.100300

Schulte E, Scheller F, Sloot D, Bruckner T (2022) A meta-analysis of residential PV adoption: the important role of perceived benefits, intentions and antecedents in solar energy acceptance. Energy Res Soc Sci 84:102339. https://doi.org/10.1016/j.erss.2021.102339

Shirin S, Jamal A, Ranjan P, Yadav AK (2019) Study on assessment of slope stability and mixed disposal of overburden in voids of Singrauli Coalfield. Environ Qual Manag 28(3):131–139. https://doi.org/10.1002/tqem.21616

Shirin S, Yadav AK (2021) Assessment and suitability for the dumping of flyash in an abandoned mine. Environ Chall 4:100111. https://doi.org/10.1016/j.envc.2021.100111

Szmytke E, Brzezińska D, Machnowski W, Kokot S (2022) Firefighters' clothing contamination in fires of electric vehicle batteries and photovoltaic modules—literature review and pilot tests results. Int J Environ Res Public Health 19(19). https://doi.org/10.3390/ijerph191912442

Tien PW, Wei S, Darkwa J, Wood C, Calautit JK (2022) Machine learning and deep learning methods for enhancing building energy efficiency and indoor environmental quality – a review. Energy AI 10:100198. https://doi.org/10.1016/j.egyai.2022.100198

Tsanakas JA, van der Heide A, Radavičius T, Denafas J, Lemaire E, Wang K, Poortmans J, Voroshazi E (2020) Towards a circular supply chain for PV modules: review of today's challenges in PV recycling, refurbishment and re-certification. Prog Photovolt Res Appl 28(6):454–464. https://doi.org/10.1002/pip.3193

United Nations (2023) The sustainable development goals report. Divisiôn de Estadística de las Naciones Unidas para los Objetivos de Desarrollo Sostenible. https://unstats.un.org/sdgs/report/2023/

Whitley BE (22 Jan 2024) Integrating AI in maintenance protocols for safer industrial environments. Occupational Health & Safety. https://ohsonline.com/Articles/2024/01/22/Integrating-AI-in-Maintenance-Protocols-for-Safer-Industrial-Environments.aspx

Yang D, Li W, Yagli GM, Srinivasan D (2021a) Operational solar forecasting for grid integration: standards, challenges, and outlook. Sol Energy 224:930–937. https://doi.org/10.1016/j.solener.2021.04.002

Yang L, Zhang H, Mu Y (2021b) Data-driven method of renewable energy based on generative adversarial networks and energyPLAN. In: 2021 33rd Chinese control and decision conference (CCDC), pp 938–943. https://doi.org/10.1109/CCDC52312.2021.9601975

Ye J, Giani A, Elasser A, Mazumder SK, Farnell C, Mantooth HA, Kim T, Liu J, Chen B, Seo G-S, Song W, Greidanus MDR, Sahoo S, Blaabjerg F, Zhang J, Guo L, Ahn B, Shadmand MB, Gajanur NR, Abbaszada MA (2022) A review of cyber–physical security for photovoltaic systems. IEEE J Emerg Sel Top Power Electron 10(4):4879–4901. https://doi.org/10.1109/JESTPE.2021.3111728

Zhang W, Luo Y, Zhang Y, Srinivasan D (2021) SolarGAN: multivariate solar data imputation using generative adversarial network. IEEE Trans Sustain Energy 12(1):743–746. https://doi.org/10.1109/TSTE.2020.3004751

Chapter 14
Assessing the Environmental Impact of Solar Panel Manufacturing and Disposal

R. A. Zerubabel Michael and Sourav Maity

Abstract Energy harnessed from the sun is a fundamental element of renewable energy plans; nevertheless, considerable environmental issues must be resolved to achieve genuine sustainability. The findings of this review give a top-level view of the present state of the investigation in evaluating these dangers linked with solar energy making. It reviews the environmental effects of solar thermal structures, solar power production, and photovoltaic (PV) panels life cycle assessment. Vital issues include the power and assets necessary for producing, conveyance, setup, operation, and solar panel recycling or disposal at the end of life. Moreover, the impact of land usage and its modification, environment disruption, and biodiversity loss because of huge-scale solar installations are addressed. Additionally, the overview elaborates on the safety and dangers related to solar power manufacturing during the production, installation, preservation, and decommissioning of solar power systems. It studies the effects of lethal substances, including lead, cadmium, and silicon, on production, as well as the dangers of natural disasters, electrical hazards, etc., during installation and maintenance activities. Furthermore, the review focused on the increasing advances in technologies and excellent practices to mitigate environmental health and safety dangers in solar energy production. These include developments in PV panel recycling technology and enhancements in production tactics to reduce environmental impacts and ensure the sustainable and secure deployment of solar power technologies within the transition to create a low carbon future.

Keywords Photovoltaic (PV) panels · Life cycle assessment · Environmental disruption · Biodiversity loss · Low carbon

R. A. Zerubabel Michael
Department of Biotechnology, Andhra University, Visakhapatnam, Andhra Pradesh, India

S. Maity (✉)
Centre for the Environment, Indian Institute of Technology Guwahati, Guwahati, India

A. K. Yadav (ed.), *Powering Future: Clean Energy Solutions for a Sustainable Planet*, Clean Energy Production Technologies, https://doi.org/10.1007/978-981-96-5055-2_14

14.1 Introduction

Energy is a crucial parameter for socio-economic development in today's era. Humans have been using fossil fuels as their primary energy source since the advent of the Industrial Revolution. With an increasing demand for electricity and the reduction in fossil fuels, harnessing electricity from renewable resources is becoming inevitable (Ahuja and Tatsutani 2009). India is the fourth largest power consumer after America, China, and Russia, but it is not bequeathed with bountiful nonconventional energy resources. Therefore, to meet its development needs, it must utilize all domestically available resources of oil, coal, uranium, and renewable resources, including hydro and even import them. The increase in problems without passports, such as increasing GHG emissions, energy scarcity, climate variability, and development through sustainable approaches, has made it necessary to accelerate the evolution of renewable technologies. Being an inexpensive, clean, and limitless energy source, solar energy is appropriate (Xu et al. 2018).

Among renewable energy resources, the sun is the universal and readily available energy supply to all life. With industrialization, the power necessities have expanded notably. Solar photovoltaic technology can be exploited to convert sun radiation into electricity using solar cells. A sun cellular is a photovoltaic tool that uses the photovoltaic effect to convert mild strength into electric energy. There has been a large deployment of solar photovoltaics (PV) worldwide (Brooks 2014). Advancements in the installed capacity of photovoltaic panels have invariably resulted in an increase in decommissioned panels yearly. The world photovoltaic potential by 2050 is anticipated to be 4500 GW (Brunisholz 2015). Solar PVs have an average operational validity of 25–30 years; those mounted in the early 1990s are close to expiring, and the early 2030s will produce massive waste. These increasing wastes not only result in environmental challenges, but also open up new economic avenues and unprecedented opportunities, such as the recovery of raw materials, and will also result in the growth of new industries focused on end-of-life photovoltaic panels (Artaş et al. 2023). If a framework for recycling photovoltaic waste is not established in time, one will miss the bus to take advantage of the benefits of such industries that are coming to be established.

14.2 Monitoring Solar Panel End-of-Life

The technique to assess the environmental resources and consequences used during the period of a product, taking into account all characteristics, resources, and human health of the ecosystem, is known as life cycle assessment or LCA. It is also a procedure for assessing and analyzing a system's, product's, or service's ecological impacts over its lifetime. LCA examines the environmental impact of products from the acquisition of raw materials to manufacturing, using, and maintaining, and finally, life cycle management, whether through reuse, recycling, or reduction (Latunussa et al. 2016).

IRENA attempted a projection for future PV panel waste volumes until 2050 by assuming an average 30-year lifetime for solar panels ("Regular loss scenario") and accounting for infant failures (light-induced degradation, poor support constructions, improper mounting, and poor planning), mid-life failures within first 12 years (potential induced degradation, delamination, blotching of ethylene vinyl acetate (discoloration), deterioration of the antireflective coating on the surface of the glass, cracked cell isolation, cell interconnect breakages, glass breakage, contact failures in the loose frames, junction box, and diode defects) and wear out failures (most of the mid-life failures and also severe corrosion of cells and interconnector) before the end-of-life cycle ("Early Loss scenario"). A study estimated that 40% of new modules manufactured after 2008 have thinner cells and are frequented by micro-cracks in at least one cell (Brunisholz 2015). Figure 14.1 shows the estimated waste generation from 2016 to 2050 (Stolz et al. 2017). In 2050, waste generation is expected to reach 78 million tons in an early loss scenario. In the subsequent figure (Fig. 14.2), China, Germany, and Japan will account for higher waste generation by 2030. By 2050, China, India, and Japan will be the top producers of PVV waste.

Compared to nuclear power plants, toxic waste per unit of energy created by solar panels is 300 times more. A 250 MW (DC) solar plant shall contain 7,81,250 solar panels. Two hundred and thirty nine MNRE estimated that 75 metric tons of modules are required to install 1 MW capacity (260 Wp modules of 20 kg each). India has a current installed capacity of 31696.24 MW, indicating ~10.56 crores of panels of 300 Wp each. India intends to install 100 GW of solar capacity in the next 3–4 years, which would result in ~33.33 crores of panels of 300 Wp each, indicating the potential waste that India would generate after 25 years for these installations over the next 2 years. These numbers are discussed at the end of the life period. Natural calamities such as tornadoes, storms, earthquakes, and hurricanes can significantly increase the number and, to top it, damage sustained during the construction and transit phase. Therefore, sustained policy action is required to counter the challenges posed by the generation of module waste and allow the framework to be deployed according to the requirements and environment of each region or country.

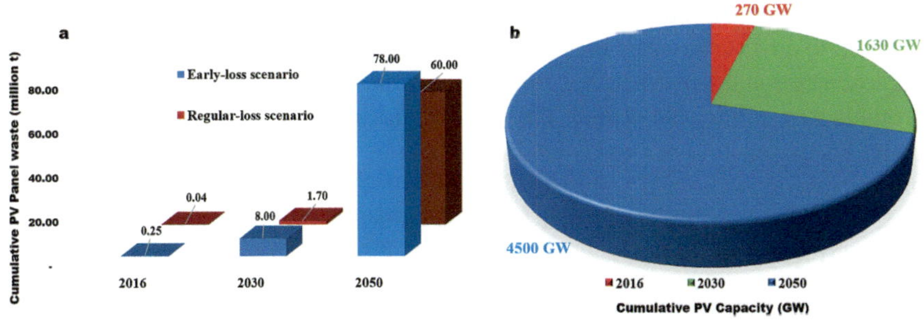

Fig. 14.1 Global solar panel waste projection: Overview, 2016–2050, (**a**) cumulative waste (in million t) and (**b**) cumulative capacity (in GW)

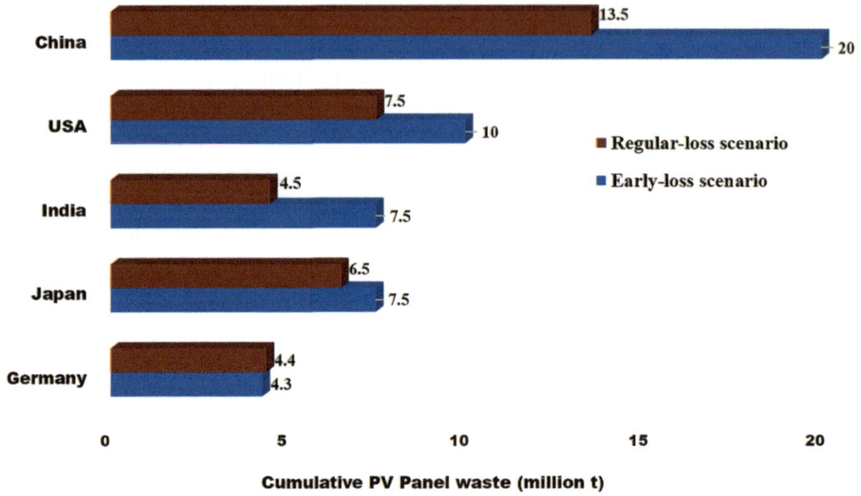

Fig. 14.2 Estimated photovoltaic waste generation from leading countries

14.3 Composition of Photovoltaic Module and Waste Classification

Photovoltaics are monolithic devices that are produced to attain longevity and sturdiness with high-strength encapsulation, limiting the potential exposure of module components to the environment. Suppose that poorly managed broken modules are likely to be landfilled or exposed to other countries (Zimmermann et al. 2013). They may be considered environmentally safe throughout the operational segment, as all layers are sealed and using encapsulants or laminated glass protects them from humidity, excessive warmth, and vicious weather conditions. Dangerous toxic elements, i.e., Cd, Ga, In, U, Te can threaten environmental risks (Espinosa et al. 2016; Yadav et al. 2014). If these damaged metallic modules are dumped in habitats by lack of adequate control actions, acidic/rainwater that falls at the overwhelmed or damaged modules can reach the encapsulated semiconductor layer and might leach metals/substances (Abdollahi and Rahimi 2020). Suppose that a damaged module is exposed to adverse external conditions, such as heavy rain. In that case, the resulting leachate can contaminate the surrounding soil, surface water in our bodies, and groundwater, ultimately posing a health risk to the environment and the human body. The population of nearby landfills is a more significant threat to exposure to infected groundwater and polluted soil.

Consequences of the used solar PV cells on the leakage of PV chemicals are as follows (Fig. 14.3):

- Improper disposal by fire accidents, unrestrained dumping, and spreading from landfills into soil and water systems.
- A pile-up in storerooms, renewing panels that have slight maintenance.

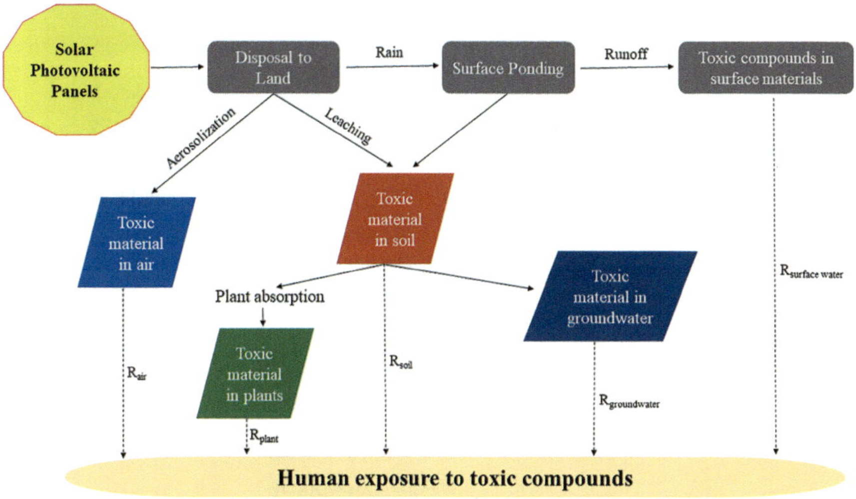

Fig. 14.3 Probable social contact to solar-linked elements: A schematic overview

- Recycling leads to the recovery of glass and valued components.
- Appropriate disposal and energy recovery through incineration, organized landfilling, and assortment until an effective recycling method is concluded.

14.4 Categorization of Solar Cells in Different Timelines and Their Panel Composition

Solar photovoltaic solar cells are categorized according to the materials of which they are made, as represented in Fig. 14.4.

First-Generation PV Cells Silicon wafers are used to produce these solar cells. It is a famous technology for its extreme primitiveness and excellent efficiency. These could be further classified into (a) a monocrystalline silicon solar cell made of a single silicon crystal, and its efficiency ranges from 17% to 18%. (b) Multi crystalline silicon solar cell is a cell that combines several crystals (Kim et al. 2014). It is currently the most popular. In addition, they are inexpensive to produce than monocrystalline solar cells and are only about 12% to 14% efficient. Table 14.1 provides an overview of the environmental effect of mc-Si PV as reported by Yang et al. (2015).

Second-Generation Photovoltaic Cells The thin-film solar PV modules have extra economic aspects compared to wafer-based silicon cells because a 350-μm layer is present to absorb the incident photons. On the contrary, a wafer-based cell has a layer that is only 1 μm thick. Thin-film solar cells are additionally divided into

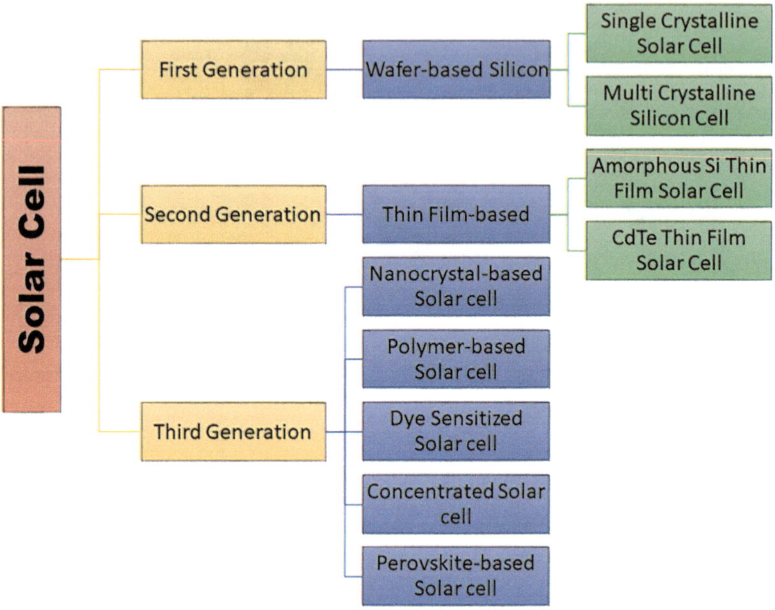

Fig. 14.4 Categorization of photovoltaic solar cells

Table 14.1 A summary of the first-generation solar cells impact on the environment [reprinted from Yang et al. (2015) with permission from Journal of Cleaner Production. Copyright © 2015 Elsevier Ltd.]

S. no	Impact on the environment	Values
1	Acidification potential	17.6 kg SO_2eq
2	Eutrophication potential	1.80 kg PO_4eq
3	Abiotic depletion potential	3.40 kg SBeq
4	Ozone layer depletion potential	4.51×10^{-4} kg CFC-11 eq
5	Terrestrial eco-toxicity potential	7.10×10^{-2} kg 1.4DBeq
6	Marine aquatic eco-toxicity potential	$2.69 \times 10^{+2}$ kg 1.4DBeq
7	Freshwater aquatic eco-toxicity	15.2 kg 1.4DBeq
8	Human toxicity potential	62.5 kg 1.4DBeq
9	Photochemical ozone creation potential	0.61 kg C2H4eq

- *Amorphous silicon thin-film (a-Si) Solar module*: Silicon particles are randomly placed in the crystal lattice. It was the first primary solar cell created on an industrial scale. It has a comparatively low price charge and is widely available. Its efficacy is little, from 4% to 8%.
- *Cadmium telluride (CdTe) thin-film solar module*: The most widely used form of a thin-film solar module is cadmium telluride. These are also economically feasible. Compared to other solar cells, their price is significantly more affordable. Their productivity ranges from 9% to 11%. Despite the advantages of using extra

CdTe solar cells, the CD found in them has a negative impact on existing organisms in the habitat, including humans, animals, and plants. Cadmium reclamation is a time- and financially demanding operation with a low rate of viability. In addition, there are few transitional elements in nature. Another area of concern is the dangerous impact of cadmium on the environment.

- *Copper Indium Gallium Di-Selenide (CIGS) solar cell*: A solar module of this kind consists of tetra-components that contain representative elements and transition elements. These cells have an efficiency of 10–12% higher than CdTe solar cells. Additionally, these solar photovoltaic cells are also reasonably inexpensive compared to other thin-film solar photovoltaic module equivalents.

A summary of the environmental outcome of the Collier et al. (2014) investigation is shown in Table 14.2.

Third-Generation Solar Photovoltaic Cells They have recently been developing in business and research. This category includes the following types:

- *Polymer-Based Solar Module (PSC)*: A skinny polymer layer is enveloped around the absorber layer in this form of solar module.
- *Concentrated solar cells (CSC)*: The latest solar technique. This device concentrates the incident photon energy in a narrow area.
- *Dye-Sensitized Solar Cells (DSCs):* This technique involves dye molecules in the space between the collection electrodes.
- *Perovskite Solar Cells*: In this form of the solar cell, halogen ions such as chloride, bromide, and iodide are present, as well as two different cations with different radii. Chemicals of the formula ABX3 are usually included. This Perovskite Solar Technology is a relatively new addition to the photovoltaic industry.
- *Nanocrystal Solar Cells*: This type of solar module comprises nanometer-sized semiconductor material. There are also transition group elements. Quantum dot cells are solar cells made of nanocrystals.

The general intention for the enhancement of solar photovoltaics was to increase the performance of the PV module to lower the estimated price of this generation and, at the same time, improve the associated production era. Solar modules are enhanced from one or more coatings of materials to absorb sun rays. The waste from solar modules harms the environment, people, the ecosystem, and biodiversity. Among the main side effects on humans were the effects of indium compounds on

Table 14.2 An outline of the ecological effect of second-generation solar cells (Collier et al. 2014)

S. no	Environmental impact	Production phase	End of life	Life cycle
1	Global warming potential	45.3	6.0	86.1
2	Eutrophication potential	0.018	8.03×10^{-4}	0.029
3	Acidification potential (kg SO_2eq)	0.21	9.10×10^{-3}	0.36
4	Primary energy	750	81	1270
5	Photochemical ozone creation potential	0.015	6.63×10^{-4}	0.032

the intestinal tract, the effects of germanium, the cancer-causing effects of Cd, Te, Se, etc. (Strachala et al. 2017; Wu et al. 2017).

14.5 Ecological Effect of Solar Panel Manufacturing and Disposal

The ecological impressions of energy derived from the sun for manufacturing systems from the producing level (Fthenakis et al. 2005) to setup and processing (Turney and Fthenakis 2011), decommission and dumping or re-utilization of solar PV panels (Hernandez et al. 2014; Fthenakis et al. 2008), which involves the construction of a PV facility, include the use of heavy equipment which leads to noise and visible disturbances, therefore, annoying the herbal habitation and the surroundings (Guerin 2017; Soliño et al. 2009). Numerous impacts could be associated with human fitness (Aman et al. 2015), weather (Alsema 2012), the natural world (Pimentel Da Silva and Branco 2018), groundwater and soil (Tammaro et al. 2016; Shirin and Yadav 2014), and land usage (Denholm and Margolis 2008). Turney and Fthenakis (2011) recognized the ecological impact of using solar electricity instead of traditional power resources. Dimond and Webb (2017) demonstrated that essential components of a habitat with excessive environmental impacts include biodiversity, temperature, precipitation, solar insulation, humidity, and biomass density (Tsang et al. 2016). Furthermore, it was stated that the spectrum of localities where it was installed in a varieties, of variety from forests to deserts, at some point of grasslands, farmlands, could affect the surroundings.

14.5.1 Impacts on Soil

In particular, the building of solar power plants is technologically expensive compared to various power plant building projects due to the unpredictable or negligible influence of ground factors on the consistency of the system, its efficacy, and its operation during its life cycle (Bukhary et al. 2018). Therefore, significant landscaping is required, including clearing vegetation, leveling, compacting the soil, removing unnecessary roads, and building critical accessible roads, which can lead to more profound surface erosion, increased turbidity or sedimentation in local water bodies, reduced infiltration of stormwater and air pollutants, reduced underground water and increased risk of drowning (Turney and Fthenakis 2011; Yadav 2015, 2021). For instance, access streets between solar modules have a high potential to cause erosion, transpiration, infiltration, and runoff rate (Mao and Cherkauer 2009). For instance, deforestation resulted in a total loss of organic carbon of 25–30% in the United States and Spain without actual field measurements or hypothetical scenarios that support the need to quantify and screen such influences during the solar plant operating procedure (Dhar et al. 2020).

14.5.2 Impact on Air Quality and Climate Change

The green energy source solar power has less of an impact on climate change and air quality than conventional energy generation systems. Therefore, it can help alleviate some environmental problems related to fossil fuel use (Avril et al. 2012). During production, the photovoltaic system has no or less CO_2, CH_4, SOX, and NOX emissions and does not contribute to air contamination and global warming (Fthenakis et al. 2008). This depends on the performance of the PV system and fails to account for all stages of the life cycle stages. Therefore, for a more comprehensive assessment, emissions must be considered during entire phases of the PV system life cycle from production to transport, installation, operation, and disposal/removal. The emissions associated with the transport of photovoltaic units remain negligible with respect to the waste emissions during and after manufacturing. Vehicle emissions are only 0.1–1% of production emissions (Tsoutsos et al. 2005). Construction-related emissions occur mainly at the production point of aluminum and steel production point for building poles and frames, glass manufacturing, and the reduction of silicon from solar cells (Alsema 2012). Additionally, the type of fuel or fuel blend used to power the manufacture of photovoltaic modules directly affects the sort and intensity of emissions. Numerous gas emissions are generally transformed into equivalents of sulfur dioxide and carbon dioxide, as these are the two critical gases released from photovoltaic energy systems. The Kyoto Protocol suggested that six greenhouse gases' capacity to cause global warming: hydrofluorocarbons (HFCs), sulfur fluoride (SF6), methane, carbon dioxide nitrous oxide, and perfluorocarbons (PFCs), as CO_2 scavengers, is determined primarily using carbon dioxide equivalent (CO_2) (Alami et al. 2020a; Cucchiella and D'Adamo 2012). On the contrary, SO_2 is the same as SO_2 used to determine the acidification or formation potential, respectively. Sulfur dioxide is primarily mixed with hydrogen sulfide, hydrofluoric acid, hydrochloric acid, nitrogen dioxide, and ammonia. Reduced CO_2 emissions are the most significant benefit of using solar energy. However, attention should be paid to installing nuclear power plants in the forest. In these areas, florae must be removed to a height of <1 m to avoid solar panels or completely removed during installation (Swart and Hertzog 2016; Turney and Fthenakis 2011). As a result, this often results in low levels of CO_2 sequestration by plants, with around 50–75% of cut plantations being transformed into fuel wood and CO_2 emissions. (Turney and Fthenakis 2011). The genuine drop in CO_2 emissions will be lower. Researchers have explored many approaches to improve how photovoltaics reduce their carbon footprint (Nugent and Sovacool 2014). This can be attained by using the finest protocols in policy-making and implementing phases, increasing performance, and minimizing total emissions. Some properties are longevity, increased system capacity, ground installation, and added ambiguity (desert). To achieve the lowest level of gas emissions, a mixture of renewable forms of energy sources and cadmium selenide quantum dots or thin-film photovoltaic plants must be considered. Another aspect of analyzing the impact of photovoltaic energy production systems effect on climate variability is the effect of albedo.

Photovoltaic panels exhibit a low albedo of 0.18–0.23, indicating their low reflectivity and therefore convert a large part of solar radiation into heat that can affect climate (Kotak et al. 2015; Brennan et al. 2014; Andrews and Pearce 2013; Nemet 2009). Nemet (2009) reported on the effect of albedo on climate change globally and found that its influence is small and associated with a reduction in greenhouse gas emissions linked with PV use.

14.5.3 Impact of the Emission of Hazardous Elements from Photovoltaic Energy

The manufacture of photovoltaic cells includes numerous unsafe constituents during solar cell extraction or semiconductor purification and surface treatment (Üçtuğ and Azapagic 2018; Marwede et al. 2013). Photovoltaic cells are made from a variety of basic elements, including potassium, copper, cadmium, silicon, selenium, and tellurium (Alami et al. 2020b; Stamford and Azapagic 2019). Mining and processing these raw materials requires several extraction and refining steps. This substance is produced in large quantities as a by-product of mining other metals. For example, waste streams that process lead and zinc minerals produce cadmium as a by-product (Şengül and Theis 2011). The lead or zinc receives 99.5% pure cadmium that has been extracted and sent to the recovery unit. Then, using procedures that include electrolytic reduction, precipitation, oxidation, and leaching, it is concentrated to 99.99% purity. To obtain the 99.99% purity needed to produce CdTe cells, vacuum distillation is the next step (Marwede et al. 2013). Furthermore, several metal ores, including copper, tin, and gold, are mineral processed to produce tellurium (Te) as a by-product, which can limit CdTe solar cell production (Fthenakis et al. 2005). Recycle and recovery are key to maintaining current cell growth rates. Minerals containing silicon silicate, primarily silicon dioxide (SiO_2), are used to produce silicon for solar cells. Silicon is the second most prevalent material, with siliceous minerals comprising more than 90% of the crust, following oxygen (Hamasha et al. 2015). The primary process to produce pure silica involves reducing silicon at high temperatures with coke present (Okutani 2009), followed by blending the effluent with hydrochloric acid in the presence of copper. High-temperature processing of high-purity silicon results in monocrystalline or polycrystalline silicon crystals. Remarkably, high-temperature processing causes 80% of high-purity silicon to disintegrate (Okutani 2009). Moreover, different solar cells are separated, extracted, processed, and processed using a variety of chemicals and solvents. These substances include selenium hydride, hydrogen, ammonia, nitric acid, hydrochloric acid, and isopropanol. Many of these substances are hazardous, carcinogenic, combustible, and corrosive, so they must be handled carefully. The releases of these chemical solvents and harmful gases vary depending on the specific type of photovoltaic cell material. Chemical compounds used in the production of photovoltaic cells have many health and environmental impacts (Aman et al. 2015): acetone for cleaning, ammonia for reflective coatings, arsenic for producing potassium arsenide solar photovoltaic cells, cadmium for the production of cadmium telluride solar cells, hexavalent chromium for solar panel coatings, hydrochloric acid for the production of electrical grade silicon, hydrogen for manufacturing amorphous-Si solar cells, isopropanol for cleaning microscopic impurities from chips, lead for wires and soldering photovoltaic components, nitric acid for cleaning, silicon for PV semiconductor materials, Toulon dust for cleaning, xylene for cleaning, etc.

14.5.4 Impacts Regarding Water Resources

The review shows that photovoltaic systems have negligible water consumption during operation. Water is used mainly to cool and clean the panels during the procedure. Water consumption for cooling can be significantly reduced by using dry or hybrid cooling systems and recirculating water with cooling wind (Hasanuzzaman et al. 2016; Johst and Rothstein 2014). In contrast, flow cooling systems use less energy than recycling or dry systems (Salameh et al. 2020a). Photovoltaic panels also become more efficient when cleaned although the cost of cleaning them is usually higher due to the water used (Sahm et al. 2005). However, the volume of water used and the degree of cleaning, however, are depending on several variables, including humidity, temperature, wind direction and speed, air pollution, rainfall, dust, vegetation, glass qualities, tilt angle, and panel orientation (Al-Qodah et al. 2020). The water from the rear of the panel evaporates to cool floating photovoltaic arrays, increasing energy output without using water (Choi 2014). The use of manufacturing and processing water is much higher than the use of operating water usage. Mineral extraction, processing, and chemical purification are all steps in the manufacturing process. For instance, water consumption in the production of silicon is about 180 kg/kg, and in polycrystalline, about 470 kg/kg; accordingly, 3.7–5.2 tons of water will be needed per kW (Du et al. 2015). However, the wastewater produced must be treated and reused. Different techniques, such as biological processes, have been used to treat wastewater (Ashkanani et al. 2019), adsorption (Al-Bsoul et al. 2021, 2020; Hoslett et al. 2020), advanced oxidation technique (Almomani et al. 2019a, b), membrane fusion and fission (Tawalbeh et al. 2018). Fthenakis and Kim (2010) researched recent studies on water use in both conventional and renewable energy technology, looking at life-consumption factors. They indicated that switching to photovoltaic technology could be the top choice to save water supplies. They investigate the use of water in fuel extraction, preparation, and construction of fuel facilities/facilities. The fuel cycle of renewable energy systems involves the construction of power plants or the manufacture of machinery, excluding the biomass fuel cycle, which requires a lot of irrigation water. Accurate use of water for renewable technologies is lacking, as there are no adequate statistics on water treatment in power plants. Thin-film CdTe, mono-Si, and bulk Si are the three popular PV types that they included. The cleaning and cooling of wafers, cells, and modules involves the use of water. However, the majority of the water was attributable to the formation of single crystals and the synthesis of silica (Frisvold and Marquez 2013).

14.5.5 *Noise and Visual Impacts*

The effects of sound interference are caused by differences in power intensity, which are responsible for generating different voltages. Dehra (2018) studied noise characterization of solar energy conversion and wind photovoltaic devices. All sources of sound waves are defined according to the level of sound interference (Dehra 2018). The noise of the solar system is caused by the voltage difference between two PV arrays. PV modules have no moving or rotating parts, so their operation does not generate significant noise (Tsoutsos et al. 2005). However, in the building process, there were numerous heavy machines and operational vehicles on site, causing noise pollution to residents, tourists, and wildlife (Fernandez-Jimenez et al. 2015). Among the latest ideas is the development of a sound barrier (NB) out of a photovoltaic system. It is often located close to the highway and offers power and sound control. The visual impression of PV systems is adverse because of the scale of PV installations and projects (Poe et al. 2017). The visual pollution is probably the most frequently mentioned problem by social and environmental activists. Subjective to the grade of impact, the public viewpoint can be strong against the construction of PV plants and significantly impede its implementation. Visibility impacts generally depend on the installation area, and particularly negative impacts are expected for massive PV projects. Numerous PV projects have been developed in rural areas, which has a negative impact on a significant landscap. A probable approach to reduce this negative effect is to install PV panels on the roof and facade of the building (Salameh et al. 2020b; Bazán et al. 2018). Integrating PV panels into building facades usually has a positive visual impact. Also, the floating structure's new design and shape look great. Another option to reduce visibility is to locate PV installations in areas remote from populated places, like desert regions (Fernandez-Jimenez et al. 2015). However, numerous approaches and knowledge procedures are available to minimize the visual impact of PV. These actions include, among others, proper site planning and selection, adopting environmentally friendly policies and regulations, implementing appropriate installation procedures, improving the PV panel's integration into building facades, and preventing the placement of PV panels in buildings. Early public involvement in the design of historical and culturally significant monuments or heritage areas is followed by popular acceptance (Fernandez-Jimenez et al. 2015; Tsoutsos et al. 2005).

14.6 Disposal Strategies for Making Solar Energy More Environmentally Friendly

The first shipment of PV setups from the beginning of 1980 has already stopped operating. The management of the generated PV waste is also urgent. There are two forms of waste that get generated at the end:

- The waste produced in the manufacturing plants
- End life generated waste

Due to the increasing installations, solar PV waste estimation must be carried out. With successive years at an interval of five, the PV installations have significantly risen since 2015 (Chowdhury et al. 2020). The installations will further increase in 2050.

The following criteria and factors must be analyzed to estimate PV waste.

- ***Production of the panel*:** Commercial and non-commercial solar panel manufacturing is increasing, i.e., Crystalline Silicon, Cadmium Telluride solar panels, etc.
- ***Panel transportation*:** After the solar panels are manufactured, they are to be transferred to the fabricator companies.
- ***Panel installation and use*:** Once the solar panels' transportation is done, they will be installed.
- ***End-of-life and disposal*:** After their life cycle, the disposal part also has a vital role in estimating the total PV waste.

For the evaluation of the PV waste, the total positioned capacity of the solar photovoltaic panels is to be taken in terms of their mass in metric tons. Further end-life cycle management also decides the overall financial requirement of the PV waste model (Fig. 14.5).

The regular losses include the waste generated after the life cycle of the solar panels is over. The solar panels' life cycle is estimated to be 30 years (Suresh and Rajanna 2019). At the same time, the early losses include the PV waste generated in the course of the operating phase of the solar panels (Fu et al. 2015). This phase includes the infancy lifetime to the middle life up to the completion of the 30 years life span (Fig. 14.6). The early loss of PV waste in 2016 was 250,000 million tons, whereas the regular loss was 43,500 million tons. In 2030, the regular loss will be 8 million tons, and the early loss will be 1.7 million tons. By the extreme of 2050, regular loss was estimated to be 78 million tons, and the early loss was 60 million tons (Weckend et al. 2016).

Many of the resources used in the production of TFPV and PV are generally lethal, very valued, and scarce, and their release into the surroundings through water and air can create severe complications. Hence, it must be used throughout the system's entire life cycle, from the building process to the discarding phase at the expiration of our system's life. Methods for appropriate waste management from the PV industry have been established and debated, mainly since manufacturing is related to the semiconductor industry. The below-mentioned subject outlines the

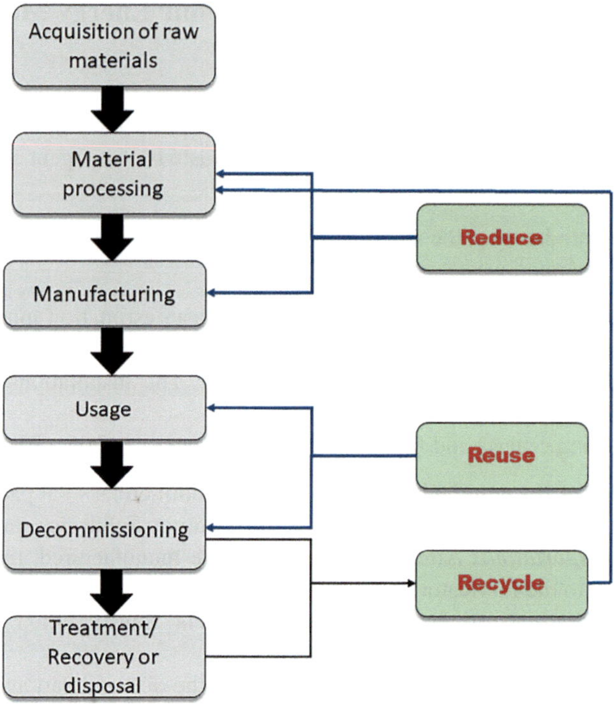

Fig. 14.5 Life cycle assessment of a product: Overview

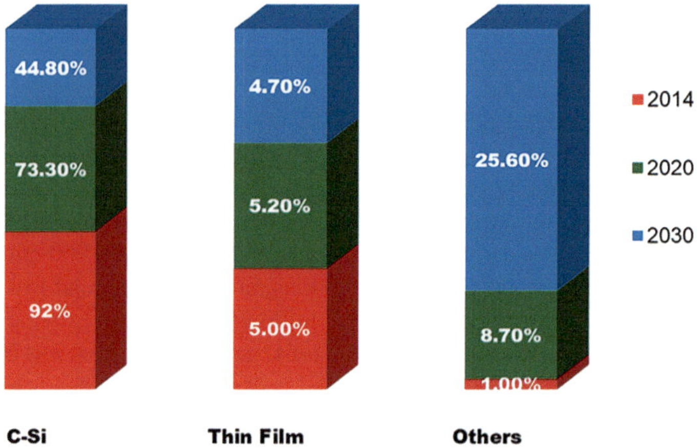

Fig. 14.6 Photovoltaic Panels market availability (2014–2030) (Dataset adopted from https://www.irena.org/-/media/Files/IRENA/Agency/Publication/2016/IRENA_RE_Statistics_2016.pdf)

finest practices for reducing the ecological impact of PV systems, such as reducing waste throughout end-of-life production and recycling (Turney and Fthenakis 2011; Tsoutsos et al. 2005).

14.7 Prospects to Reduce, Reuse, and Recycle PV Panels: The Ultimate Principle of Waste Reduction—3Rs (Reduce, Reuse, and Recycle) Also Apply to Modules

- ***Reduce*:** The preferred option is to reduce the utilization of raw materials for producing PV panels; this results in an increase in efficiency. Lack of raw materials and increased demand for panels and low-cost modules lead to mass production, less consumption of raw materials and substituting one raw material for another, continuous research and development, new technology with high efficiency, and technological progress. In the future, the production of modules will require fewer raw materials, reducing the use of hazardous and rare materials and thus increasing the recovery of panels' resources (Padoan et al. 2019). Crystalline silicon panels contain more than 90% of components classified as non-hazardous waste, and only 10% of materials such as lead, lead, and silver 247 are hazardous. Thin-film panels contain more than 98% harmless materials and 2% toxic constituents such as copper, zinc, semiconductors, and other entities (like potassium, selenium, cadmium, tellurium, lead, and indium). Valuable materials must undergo careful processing procedures depending on their geographical location. As in the manufacturing firm, waste reduction is used in the photovoltaic industry. Waste reduction strategies comprise using again treated water, substituting lethal chemicals with lesser harmless ones, and replacing acid baths with acidic sprays in cleaning operations. In addition to better regulation of spills and leakages, decreasing steam losses and using industrial procedures with less harmful materials. Few past approaches need pure materials; therefore, they are not extensively used in semiconductor waste reduction. Methods to reduce the lethal chemicals utilized in thin-film production have been discovered, particularly in CdTe and GDA TFPV cells. For instance, in GDA TFPV with CdTe buffer layer and CdS thin-film synthesis, changing the solution temperature and concentration during chemical bath deposition could decrease the density of harmful cadmium by a factor of 10. The above-mentioned diverse working approaches can significantly reduce the waste formed in manufacturing PV module industries.
- ***Reuse*:** This option follows in the 3R priority order after the reduction option and includes various reuse and repair methods. With the rapid global growth of photovoltaics, a secondary market for used modules and their accessories is expected to develop. The repaired PV panels can be further sold to countries that cannot spend massively on newer projects and have budget constraints but, at the same time, want to tap into the solar sector.

- ***Recycle*****:** This is the most popular option and can only happen after the above options have been completed. It involves the manipulation and repair of modules and leads to the release of raw materials to create innovative panels or other products. Currently, the processing of waste panels is carried out in separate groups in the existing public recycling plant, which allows the recovery of the main components (Ali and Kumar 2017). To date, 85% of the total mass of the crystal panel can be recovered. As technology advances, specialized panel recycling plants may increase turnover due to better processing and the ability to recover more recovered materials. A technical and regulatory framework must be developed to ensure that waste PV panels are profitable enough to operate at scale. After recycling, the recovered material will be used again to make new panels or traded internationally (Berger et al. 2010). According to preliminary estimates, based on current technical capabilities, by 2030, $450 million (in 2016) of raw materials can be obtained from PV panels. The recovery of this raw material is expected to lead to the production of approximately 6 million new modules, equal to approximately 18 million GW of capacity. If this figure is extrapolated to 2050, which is equal to 630 GW of energy production capacity or two billion panels, $15 billion worth of material can be recovered.

Research and development efforts and important patent work related to PV processing technology were recently more focused (Stolz et al. 2017). However, it is primarily due to the need for environmental security, the development of effectual and economical processing approaches, and the lack of essential resources for photovoltaic production (Mathur et al. 2020). Entities such as indium (In), tellurium (Te), and germanium (Ge) are becoming rarer as their availability decreases in the future timeline, exclusively with the rapid development of the photovoltaics industry. Although the lifetime of photovoltaic panels is 30 years, material reprocessing must be appropriately designed in advance from the initial stages of proposal and production. Some researchers have found a way to recycle harmful chemicals into no or less toxic forms. An effective technique of TFPV recycling is the melting of photovoltaic components of waste by utilizing metal smelters. For instance, considering the CIGS system copper indium gallium selenide solar cell, zinc and copper smelters can melt the PV components and retrieve the trace and invaluable cadmium. Few companies have designed a CdTe (Cadmium telluride cell) recovering technique that disassembles the unit, uses crystal-crushing, and metal parting by a mixture blend of chemical dissolution, electroplating, precipitation, and mechanical parting causing an 80% saving of the tellurium. A protocol for reprocessing both CIGS and CdTe module units by a mixture of the above techniques saves about 95% of the tellurium and 96% of the lead utilized for cell bonding allows for the utilization of raw materials (McDonald and Pearce 2010).

14.8 Conclusion

The synthesized references show the significance of effective processing, sustainable material sources, and lessening environmental impact in construction as critical advances in evolving technologies and finest practices in solar panel reprocessing and sustainable energy processes.

The reuse of materials emphasizes the sustainability of raw substrate utilization and minimizes pollution and surrounding ecological effects. Advances in the PV panel processing era and upgrades in manufacturing methods require improved safety practices and drill practices to address current and future industrial advancement.

In the future, it will be determined that solar electricity manufacturing will provide essential ecological and financial earnings, but now it is full of dangers. Constant research and advancement are at the vanguard of addressing ecological effects, safety issues, and human health in the current and future solar energy era. In addition, non-stop study is required to recognize rising threats and expand answers to ensure safe disposal practices in producing green energy. A practical regulatory framework plays a crucial position in coping with the EHS dangers related to the sun strength era. Governments and regulators must start and impose ethics and protocols for building, setting up, operating, and decommissioning sun power modules to shield our habitat and protect employees. A sustainable and secure future for solar power can be predicted through current studies and stakeholder coalition, paving the way to a green and neat surrounding for future generations.

Acknowledgment The authors would like to express our appreciation to the National Institute of Technology Andhra Pradesh for providing the infrastructure and instrumental support to proceed with the work.

References

Abdollahi N, Rahimi M (2020) Potential of water natural circulation coupled with nano-enhanced PCM for PV module cooling. Renew Energy 147:302–309

Ahuja D, Tatsutani M (2009) Sustainable energy for developing countries. SAPIENS Surveys and Perspectives Integrating Environment and Society, (2.1)

Al-Bsoul A, Al-Shannag M, Tawalbeh M, Al-Taani AA, Lafi WK, Al-Othman A, Alsheyab M (2020) Optimal conditions for olive mill wastewater treatment using ultrasound and advanced oxidation processes. Sci Total Environ 700:134576

Al-Bsoul A, Hailat M, Abdelhay A, Tawalbeh M, Al-Othman A, Al-Taani AA (2021) Efficient removal of phenol compounds from water environment using Ziziphus leaves adsorbent. Sci Total Environ 761:143229

Alami AH, Faraj M, Aokal K, Abu Hawili A, Tawalbeh M, Zhang D (2020b) Investigating various permutations of copper iodide/FeCu tandem materials as electrodes for dye-sensitized solar cells with a natural dye. Nanomaterials 10(4):784

Alami AH, Hawili AA, Tawalbeh M, Hasan R, Al Mahmoud L, Chibib S, Mahmood A, Aokal K, Rattanapanya P (2020a) Materials and logistics for carbon dioxide capture, storage and utilization. Sci Total Environ 717:137221

Ali B, Kumar A (2017) Development of water demand coefficients for power generation from renewable energy technologies. Energy Convers Manag 143:470–481

Almomani F, Al Ketife A, Judd S, Shurair M, Bhosale RR, Znad H, Tawalbeh M (2019a) Impact of CO2 concentration and ambient conditions on microalgal growth and nutrient removal from wastewater by a photobioreactor. Sci Total Environ 662:662–671

Almomani F, Bhosale R, Khraisheh M, Kumar A, Tawalbeh M (2019b) Photocatalytic conversion of CO_2 and H_2O to useful fuels by nanostructured composite catalysis. Appl Surf Sci 483:363–372

Al-Qodah Z, Tawalbeh M, Al-Shannag M, Al-Anber Z, Bani-Melhem K (2020) Combined electrocoagulation processes as a novel approach for enhanced pollutants removal: a state-of-the-art review. Sci Total Environ 744:140806

Alsema E (2012) Energy payback time and CO_2 emissions of PV systems. In: Practical handbook of photovoltaics. Academic Press, pp 1097–1117

Aman MM, Solangi KH, Hossain MS, Badarudin A, Jasmon GB, Mokhlis H, Bakar AHA, Kazi SN (2015) A review of safety, health and environmental (SHE) issues of solar energy system. Renew Sust Energ Rev 41:1190–1204

Andrews RW, Pearce JM (2013) The effect of spectral albedo on amorphous silicon and crystalline silicon solar photovoltaic device performance. Sol Energy 91:233–241

Artaş SB, Kocaman E, Bilgiç HH, Tutumlu H, Yağlı H, Yumrutaş R (2023) Why PV panels must be recycled at the end of their economic life span? A case study on recycling together with the global situation. Process Saf Environ Prot 174:63–78

Ashkanani A, Almomani F, Khraisheh M, Bhosale R, Tawalbeh M, AlJaml K (2019) Bio-carrier and operating temperature effect on ammonia removal from secondary wastewater effluents using moving bed biofilm reactor (MBBR). Sci Total Environ 693:133425

Avril S, Mansilla C, Busson M, Lemaire T (2012) Photovoltaic energy policy: financial estimation and performance comparison of the public support in five representative countries. Energy Policy 51:244–258

Bazán J, Rieradevall J, Gabarrell X, Vázquez-Rowe I (2018) Low-carbon electricity production through the implementation of photovoltaic panels in rooftops in urban environments: a case study for three cities in Peru. Sci Total Environ 622:1448–1462

Berger W, Simon FG, Weimann K, Alsema EA (2010) A novel approach for the recycling of thin film photovoltaic modules. Resour Conserv Recycl 54(10):711–718

Brennan MP, Abramase AL, Andrews RW, Pearce JM (2014) Effects of spectral albedo on solar photovoltaic devices. Sol Energy Mater Sol Cells 124:111–116

Brooks AE (2014) Solar energy: photovoltaics. In: Future energy. Elsevier, pp 383–404

Brunisholz G (2015) Snapshot of global photovoltaic markets (Report IEA PVPS T1-29: 2016). International Energy Agency (IEA), Paris

Bukhary S, Ahmad S, Batista J (2018) Analyzing land and water requirements for solar deployment in the Southwestern United States. Renew Sust Energ Rev 82:3288–3305

Choi YK (2014) A study on power generation analysis of floating PV system considering environmental impact. Int J Softw Eng Appl 8(1):75–84

Chowdhury MS, Rahman KS, Chowdhury T, Nuthammachot N, Techato K, Akhtaruzzaman M, Tiong SK, Sopian K, Amin N (2020) An overview of solar photovoltaic panels' end-of-life material recycling. Energ Strat Rev 27:100431

Collier J, Wu S, Apul D (2014) Life cycle environmental impacts from CZTS (copper zinc tin sulfide) and Zn3P2 (zinc phosphide) thin film PV (photovoltaic) cells. Energy 74:314–321

Cucchiella F, D'Adamo I (2012) Estimation of the energetic and environmental impacts of a roof-mounted building-integrated photovoltaic systems. Renew Sust Energ Rev 16(7):5245–5259

Dehra H (2018) Solar energy conversion and noise characterization in photovoltaic devices with ventilation. In: Recent developments in photovoltaic materials and devices. IntechOpen, Rijeka

Denholm P, Margolis RM (2008) Impacts of array configuration on land-use requirements for large-scale photovoltaic deployment in the United States (No. NREL/CP-670-42971). National Renewable Energy Lab (NREL), Golden, CO
Dhar A, Naeth MA, Jennings PD, El-Din MG (2020) Perspectives on environmental impacts and a land reclamation strategy for solar and wind energy systems. Sci Total Environ 718:134602
Dimond K, Webb A (2017) Sustainable roof selection: environmental and contextual factors to be considered in choosing a vegetated roof or rooftop solar photovoltaic system. Sustain Cities Soc 35:241–249
Du L, Hanley A, Wei C (2015) Marginal abatement costs of carbon dioxide emissions in China: a parametric analysis. Environ Resour Econ 61:191–216
Espinosa N, Zimmermann YS, dos Reis Benatto GA, Lenz M, Krebs FC (2016) Outdoor fate and environmental impact of polymer solar cells through leaching and emission to rainwater and soil. Energy Environ Sci 9(5):1674–1680
Fernandez-Jimenez LA, Mendoza-Villena M, Zorzano-Santamaria P, Garcia-Garrido E, Lara-Santillan P, Zorzano-Alba E, Falces A (2015) Site selection for new PV power plants based on their observability. Renew Energy 78:7–15
Frisvold GB, Marquez T (2013) Water requirements for large-scale solar energy projects in the West. J Contemp Water Res Educ 151(1):106–116
Fthenakis V, Kim HC (2010) Life-cycle uses of water in US electricity generation. Renew Sust Energ Rev 14(7):2039–2048
Fthenakis VM, Fuhrmann M, Heiser J, Lanzirotti A, Fitts J, Wang W (2005) Emissions and encapsulation of cadmium in CdTe PV modules during fires. Prog Photovolt Res Appl 13(8):713–723
Fthenakis VM, Kim HC, Alsema E (2008) Emissions from photovoltaic life cycles. Environ Sci Technol 42(6):2168–2174
Fu Y, Liu X, Yuan Z (2015) Life-cycle assessment of multi-crystalline photovoltaic (PV) systems in China. J Clean Prod 86:180–190
Guerin T (2017) A case study identifying and mitigating the environmental and community impacts from construction of a utility-scale solar photovoltaic power plant in eastern Australia. Sol Energy 146:94–104
Hamasha K, Hamasha E, Masadeh G, Shariah A, Hamasha MM (2015) Aluminum-induced crystallization of hydrogenated amorphous silicon thin films with assistance of electric field for solar photovoltaic applications. J Disp Technol 12(1):82–88
Hasanuzzaman M, Malek ABMA, Islam MM, Pandey AK, Rahim NA (2016) Global advancement of cooling technologies for PV systems: a review. Sol Energy 137:25–45
Hernandez RR, Easter SB, Murphy-Mariscal ML, Maestre FT, Tavassoli M, Allen EB, Barrows CW, Belnap J, Ochoa-Hueso R, Ravi S, Allen MF (2014) Environmental impacts of utility-scale solar energy. Renew Sust Energ Rev 29:766–779
Hoslett J, Ghazal H, Mohamad N, Jouhara H (2020) Removal of methylene blue from aqueous solutions by biochar prepared from the pyrolysis of mixed municipal discarded material. Sci Total Environ 714:136832
Johst M, Rothstein B (2014) Reduction of cooling water consumption due to photovoltaic and wind electricity feed-in. Renew Sust Energ Rev 35:311–317
Kim BJ, Lee JY, Kim KH, Hur T (2014) Evaluation of the environmental performance of sc-Si and mc-Si PV systems in Korea. Sol Energy 99:100–114
Kotak Y, Gul MS, Muneer T, Ivanova SM (2015) Investigating the impact of ground albedo on the performance of PV systems. In: CIBSE technical symposium 2015, London
Latunussa CE, Ardente F, Blengini GA, Mancini L (2016) Life Cycle Assessment of an innovative recycling process for crystalline silicon photovoltaic panels. Sol Energy Mater Sol Cells 156:101–111
Mao D, Cherkauer KA (2009) Impacts of land-use change on hydrologic responses in the Great Lakes region. J Hydrol 374(1–2):71–82
Marwede M, Berger W, Schlummer M, Mäurer A, Reller A (2013) Recycling paths for thin-film chalcogenide photovoltaic waste–Current feasible processes. Renew Energy 55:220–229

Mathur N, Singh S, Sutherland JW (2020) Promoting a circular economy in the solar photovoltaic industry using life cycle symbiosis. Resour Conserv Recycl 155:104649

McDonald NC, Pearce JM (2010) Producer responsibility and recycling solar photovoltaic modules. Energy Policy 38(11):7041–7047

Nemet GF (2009) Net radiative forcing from widespread deployment of photovoltaics. Environ Sci Technol 43(6):2173–2178

Nugent D, Sovacool BK (2014) Assessing the lifecycle greenhouse gas emissions from solar PV and wind energy: a critical meta-survey. Energy Policy 65:229–244

Okutani T (2009) Utilization of silica in rice hulls as raw materials for silicon semiconductors. J Metals Mater Miner 19(2)

Padoan FC, Altimari P, Pagnanelli F (2019) Recycling of end of life photovoltaic panels: a chemical prospective on process development. Sol Energy 177:746–761

Pimentel Da Silva GD, Branco DAC (2018) Is floating photovoltaic better than conventional photovoltaic? Assessing environmental impacts. Impact Assess Proj Apprais 36(5):390–400

Poe C, Plovnick A, Hodges T, Hastings A, Dresley S (2017) Highway renewable energy: photovoltaic noise barriers. US Department of Transportation

Sahm A, Gray A, Boehm R, Stone K (2005) Cleanliness maintenance for an Amonix lens system. Int Sol Energy Conf 47373:817–822

Salameh T, Assad MEH, Tawalbeh M, Ghenai C, Merabet A, Öztop HF (2020b) Analysis of cooling load on commercial building in UAE climate using building integrated photovoltaic façade system. Sol Energy 199:617–629

Salameh T, Tawalbeh M, Juaidi A, Abdallah R, Issa S, Alami AH (Feb 2020a) A novel numerical simulation model for the PVT water system in the GCC region. In: 2020 Advances in science and engineering technology international conferences (ASET). IEEE, pp 1–5

Şengül H, Theis TL (2011) An environmental impact assessment of quantum dot photovoltaics (QDPV) from raw material acquisition through use. J Clean Prod 19(1):21–31

Shirin S, Yadav AK (2014) Physico-chemical analysis of municipal wastewater discharge in Ganga river, Haridwar district of Uttarakhand, India. Curr World Environ 9(2):536–543. https://doi.org/10.12944/CWE.9.2.39

Soliño M, Prada A, Vázquez MX (2009) Green electricity externalities: forest biomass in an Atlantic European region. Biomass Bioenergy 33(3):407–414

Stamford L, Azapagic A (2019) Environmental impacts of copper-indium-gallium-selenide (CIGS) photovoltaics and the elimination of cadmium through atomic layer deposition. Sci Total Environ 688:1092–1101

Stolz P, Frischknecht R, Wambach K, Sinha P, Heath G (2017) Life cycle assessment of current photovoltaic module recycling. IEA PVPS task 12, International Energy Agency Power Systems Programme, Report IEA-PVPS T12, 13:2018

Strachala D, Hylský J, Vaněk J, Fafilek G, Jandová K (2017) Methods for recycling photovoltaic modules and their impact on environment and raw material extraction. Acta Montan Slovaca 22(3):257–269

Suresh HN, Rajanna S (2019) Performance enhancement of Hybrid interconnected Solar Photovoltaic array using shade dispersion Magic Square Puzzle Pattern technique under partial shading conditions. Sol Energy 194:602–617

Swart AJ, Hertzog PE (2016) Varying percentages of full uniform shading of a PV module in a controlled environment yields linear power reduction. J Energy South Afr 27(3):28–38

Tammaro M, Salluzzo A, Rimauro J, Schiavo S, Manzo S (2016) Experimental investigation to evaluate the potential environmental hazards of photovoltaic panels. J Hazard Mater 306:395–405

Tawalbeh M, Al Mojjly A, Al-Othman A, Hilal N (2018) Membrane separation as a pre-treatment process for oily saline water. Desalination 447:182–202

Tsang MP, Sonnemann GW, Bassani DM (2016) Life-cycle assessment of cradle-to-grave opportunities and environmental impacts of organic photovoltaic solar panels compared to conventional technologies. Sol Energy Mater Sol Cells 156:37–48

Tsoutsos T, Frantzeskaki N, Gekas V (2005) Environmental impacts from the solar energy technologies. Energy Policy 33(3):289–296
Turney D, Fthenakis V (2011) Environmental impacts from the installation and operation of large-scale solar power plants. Renew Sust Energ Rev 15(6):3261–3270
Üçtuğ FG, Azapagic A (2018) Environmental impacts of small-scale hybrid energy systems: coupling solar photovoltaics and lithium-ion batteries. Sci Total Environ 643:1579–1589
Weckend S, Wade A, Heath GA (2016) End of life management: solar photovoltaic panels (No. NREL/TP-6A20-73852). National Renewable Energy Lab (NREL), Golden, CO
Wu P, Ma X, Ji J, Ma Y (2017) Review on life cycle assessment of energy payback of solar photovoltaic systems and a case study. Energy Procedia 105:68–74
Xu L, Zhang S, Yang M, Li W, Xu J (2018) Environmental effects of China's solar photovoltaic industry during 2011–2016: a life cycle assessment approach. J Clean Prod 170:310–329
Yadav AK (2015) Elemental composition and source apportionment of suspended particulate matters and health risk assessment in mining and nonmining areas of Odisha, India. J Hazard Toxic Radioact Waste 19(3):04014037. https://doi.org/10.1061/(asce)hz.2153-5515.0000248
Yadav AK (2021) Human health risk assessment in opencast coal mines and coal-fired thermal power plants surrounding area due to inhalation. Environ Chall 3:100074. https://doi.org/10.1016/j.envc.2021.100074
Yadav AK, Sahoo SK, Mahapatra S et al (2014) Concentrations of uranium in drinking water and cumulative, age-dependent radiation doses in four districts of Uttar Pradesh, India. Toxicol Environ Chem 96(2):192–200. https://doi.org/10.1080/02772248.2014.934247
Yang D, Liu J, Yang J, Ding N (2015) Life-cycle assessment of China's multi-crystalline silicon photovoltaic modules considering international trade. J Clean Prod 94:35–45
Zimmermann YS, Schäffer A, Corvini PFX, Lenz M (2013) Thin-film photovoltaic cells: long-term metal (loid) leaching at their end-of-life. Environ Sci Technol 47(22):13151–13159

Part V
Clean Energy Solutions

Chapter 15
Leveraging the Atmosphere Renewable Energy and Climate Mitigation Strategies

Rahul Raj, Rashmi Raghav, Muskan Mayank, and A. R. Laiju

Abstract Transformative potential of harnessing air resources for sustainable development, as societies worldwide grapple with pressing environmental and energy challenges, exploring innovative solutions becomes paramount. The air, abundant and renewable, emerges as a crucial resource offering multifaceted opportunities for sustainability. Renewable energy technologies, notably wind, and solar power, epitomize the harnessing of air for clean energy generation. Wind turbines and solar panels leverage the kinetic and radiant energy of the air to produce electricity, offering scalable and eco-friendly alternatives to fossil fuels. Emerging advancements like airborne wind energy systems promise further efficiencies in harnessing air currents for power generation. Initiatives aimed at enhancing air quality underscore the diverse ways in which air resources can be leveraged for environmental well-being. Urban green spaces, green infrastructure, and innovative air purification technologies demonstrate the potential to mitigate pollution and improve public health. By fostering natural ventilation and reducing energy consumption, these approaches contribute to sustainable urban development and resilience. Further addressing climate change necessitates leveraging air-based solutions for carbon sequestration and mitigation. Afforestation projects, carbon capture technologies, and sustainable land management practices represent avenues for enhancing carbon sinks and reducing greenhouse gas emissions. By harnessing the air to combat climate change, societies can mitigate its adverse impacts and foster climate resilience. Through the utilization of atmospheric energy for the production of clean energy, enhancement of air quality, and reduction of climate change, communities can create the foundation for a future that is more robust and sustainable. This

R. Raj (✉) · M. Mayank
Department of Civil Engineering, NIT Uttarakhand, Srinagar, India
e-mail: rahul.rajphd2021@nituk.ac.in

R. Raghav
Department of Chemistry, NIT Uttarakhand, Srinagar, India

A. R. Laiju
Department of Civil Engineering, NIT Trichy, Tiruchirappalli, Tamil Nadu, India

A. K. Yadav (ed.), *Powering Future: Clean Energy Solutions for a Sustainable Planet*, Clean Energy Production Technologies, https://doi.org/10.1007/978-981-96-5055-2_15

abstract advocates for concerted efforts to explore and implement air-based solutions, thereby advancing toward a greener and more prosperous tomorrow.

Keywords Green spaces · Purification · Sequestration · Afforestation · Pollution

Environmental Significance
Promise of airborne renewable energy is its ability to get us closer to a sustainable energy future while addressing urgent environmental issues. By harnessing the force of the air, it can significantly reduce greenhouse gas emissions, mitigate global warming, conserve water, preserve ecosystems, and improve air quality. Unlike fossil fuel-based energy generation, airborne renewables produce electricity without emitting harmful pollutants, thereby reducing air pollution and mitigating health impacts associated with poor air quality. These energy systems have a smaller environmental footprint compared to conventional energy infrastructure, minimizing land use, habitat disruption, and water consumption. Wind turbines and solar panels can be deployed on existing agricultural land, reducing pressure on natural habitats and preserving biodiversity. Additionally, the integration of airborne renewables with energy storage systems and grid services enhances grid flexibility, stability, and resilience, facilitating the transition to a low-carbon energy system. It extends beyond local and regional impacts, contributing to global efforts to combat climate change and protect the planet for future generations. By prioritizing environmental sustainability in the development and deployment of airborne renewable energy technologies, we can create a cleaner, healthier, and more sustainable world for all.

15.1 Introduction

Using air power becomes a promise for a better future in an era marked by environmental problems and the pressing need for sustainable solutions. As humanity grapples with the consequences of fossil fuel dependency and the looming specter of climate change, innovative approaches to energy production have become imperative. It is within the boundless expanse of the atmosphere that we find a potent source of renewable energy waiting to be unlocked. The study of harnessing the power of the air encompasses a myriad of technologies, each offering unique opportunities to transition toward more sustainable future (Ahmad and Beheiry 2023). Wind power, perhaps the most recognizable form of airborne energy, has surged to the forefront of renewable energy solutions. From vast wind farms dotting coastlines to sleek turbines adorning urban landscapes, wind energy represents a tangible manifestation of our commitment to cleaner, greener power generation. Yet, the potential of the air extends far beyond wind alone. High-altitude platforms, such as airborne wind turbines and solar-powered drones, offer novel avenues for harvesting energy from the skies. These innovative technologies leverage the constant motion and abundant sunlight found at higher altitudes, presenting a compelling vision of sustainable energy generation without the need for vast land installations (Yan et al. 2023). Advancements in atmospheric science and engineering continue to unlock new possibilities for harnessing the power of the air. From airborne electrostatic energy harvesting to utilizing atmospheric pressure differentials for energy production, researchers are exploring the full spectrum of opportunities presented by our atmospheric environment. The journey toward a sustainable tomorrow through the harnessing of airborne energy is not without its challenges. Technological innovation must be coupled with thoughtful planning and consideration for environmental impacts. Balancing the benefits of clean energy with the preservation of ecosystems and biodiversity remains paramount. As it navigates the complexities of the modern world, the vision of harnessing the power of the air serves as a guiding light toward a more sustainable and resilient future (Karthik 2024). Through collaboration, innovation, and a steadfast commitment to stewardship of our planet. It can use the air's limitless potential to build a society where clean, plentiful energy is available to everyone.

15.1.1 Airborne Renewable Energy

Looking for environmentally friendly energy options, humanity finds itself at a pivotal juncture where innovation meets necessity. As climate change's effects intensify and the finite nature of reserves of fossil fuels grows progressively apparent, the imperative to shift to more sustainable energy sources grows ever more urgent. Among the myriad options available, there exists a realm of possibility high above the vast expanse of the atmosphere where the potential for sustainable energy generation knows virtually no bounds (Huang et al. 2012). A burgeoning field of technological innovation poised to revolutionize the way we harness power from the skies. At its core lies a simple yet profound idea of tapping into the kinetic and radiant energy abundant in the atmosphere to produce clean, renewable electricity. From the relentless gusts of wind sweeping across vast plains to the boundless expanse of sunlight bathing the earth from above, the air offers a treasure trove of energy waiting to be unlocked. The concept of airborne renewable energy encompasses a diverse array of technologies, each harnessing different aspects of atmospheric energy in unique ways (Huang et al. 2015). Wind power stands as perhaps the most iconic example, with towering turbines dotting landscapes around the world, generating electricity from the wind's kinetic energy. Yet, the potential of airborne energy extends far beyond traditional wind farms, encompassing innovative approaches such as airborne wind turbines, high-altitude platforms, and solar-powered drones. What sets airborne renewable energy apart is its inherent versatility and scalability. Unlike traditional energy sources reliant on finite reserves of fossil fuels, airborne renewables tap into a virtually inexhaustible resource the ceaseless motion of the atmosphere and the unrelenting radiance of the sun. Moreover, they offer the flexibility to deploy energy generation infrastructure in a variety of settings, from remote off-grid locations to densely populated urban centers. Realizing the full potential of airborne renewable energy is not without its challenges (Ajiere et al. 2024). Technical hurdles, regulatory considerations, and economic viability pose significant barriers to widespread adoption. Nevertheless, the promise of less carbon emissions and better air, and greater energy independence beckons us forward into uncharted territory. In this manuscript, embark on a journey to explore the vast potential of airborne renewable energy and its role in shaping a sustainable energy future. From the physics of atmospheric energy to the latest technological innovations and policy frameworks, we delve deep into the complexities and opportunities inherent in harnessing the power of the air (Platzer and Sarigul-Klijn 2021). Pioneer is a new frontier of sustainable energy generation, where the sky is not the limit, but the beginning of a brighter, cleaner tomorrow.

15.2 Environmental Benefits of Airborne Renewable Energy

The environmental advantages of airborne renewable energy are many and diverse, providing a strong argument for reducing the damaging effects of conventional energy generation on the environment. It can significantly reduce emissions of greenhouse gases, consume less water and land, protect ecosystems, and work toward a cleaner, more sustainable future by utilizing the power of the air (Rasugu Ayub et al. 2021; Shirin et al. 2023). It is an overview of the benefits to the environment that are covered in this article.

Reduced Greenhouse Gas Emissions When electricity is produced by airborne renewable energy sources like wind and solar energy, greenhouse gases like carbon dioxide (CO_2), methane (CH_4), and nitrous oxide (N_2O) are not discharged. This reduction in emissions contributes to mitigation of climate change by lowering the carbon footprint associated with energy generation. (Sacchi et al. 2022).

Minimized Land Use Unlike conventional energy infrastructure, which often requires extensive land clearance for mining, drilling, and power plant construction, airborne renewable energy systems have a smaller physical footprint. To lessen the need for habitat loss and land conversion, solar panels may be put on roofs and wind turbines can be erected on farms or in offshore locations (Anjum and Chauhan 2023).

Preservation of Ecosystems and Biodiversity Airborne renewable energy contributes to biodiversity conservation and the preservation of natural ecosystems by reducing habitat damage and land usage. Traditional energy extraction and generation activities, like coal mining and hydropower dams, can significantly harm wildlife habitats, water quality, and biodiversity (Awewomom et al. 2024; Shirin and Yadav 2014). Airborne renewables offer an environmentally preferable substitute that minimizes these impacts.

Conservation of Water Resources Conventional energy production techniques frequently call for substantial volumes of water for cooling purposes, leading to water pollution, habitat destruction, and competition with agriculture and other sectors. Airborne renewable energy technologies, such as wind and solar power, require minimal water for operation, thereby encouraging water conservation and lessening the burden on freshwater environments (Cui et al. 2020).

Improved Air Quality Airborne renewable energy contributes to better air quality and less air pollution by replacing fossil fuel-based electrical generation. Power plants that burn coal, oil, or natural gas produce pollutants such as sulfur dioxide (SO_2), nitrogen oxides (NOx), and particulate matter into the atmosphere. These pollutants lead to respiratory illnesses, the development of smog, and acid rain (Wang and Ding 2023; Yadav et al. 2022). In contrast, wind and solar power produce electricity without emitting harmful pollutants, leading to cleaner air and healthier communities.

Enhanced Resilience to Climate Change Worldwide, communities, infrastructure, and ecosystems face serious problems as a result of climate change. By encouraging sustainable energy habits and reducing greenhouse gas emissions, airborne renewable energy contributes to the mitigation of climate change. By doing this, the repercussions of global warming such as severe weather, rising sea levels, and ecological disruption are lessened (Pajek and Košir 2021).

The ability of airborne renewable energy to reduce global warming, protect natural ecosystems, maintain water supplies, and improve air quality are just a few of the benefits it offers to the environment. Its benefits will help create a more resilient and sustainable future for upcoming generations. A range of renewable energy sources can be used to create forms of energy that replace fossil fuels, which can also serve as a tactic to mitigate climate change. Figure 15.1 illustrates this. The most common energy sources are solar, wind, geothermal, hydropower, and biomass (Osman et al. 2023).

15.3 Technological Innovations and Advances

Technological innovations and advances in airborne renewable energy are driving the rapid evolution of this field, unlocking new possibilities for harnessing the power of the air in increasingly efficient, cost-effective, and sustainable ways. From advancements in turbine design to the development of novel airborne platforms, these innovations are reshaping the landscape of renewable energy generation.

Next-Generation Wind Turbines The past several years have seen incredible breakthroughs in wind turbine technology, resulting in bigger, more dependable, more efficient turbines. Innovations such as advanced blade designs, aerodynamic enhancements, and optimized control systems have increased energy capture and reduced maintenance requirements (Chen et al. 2021). Additionally, floating offshore wind turbines and vertical-axis wind turbines (VAWTs) represent promising avenues for expanding wind energy deployment in new environments and improving energy yield.

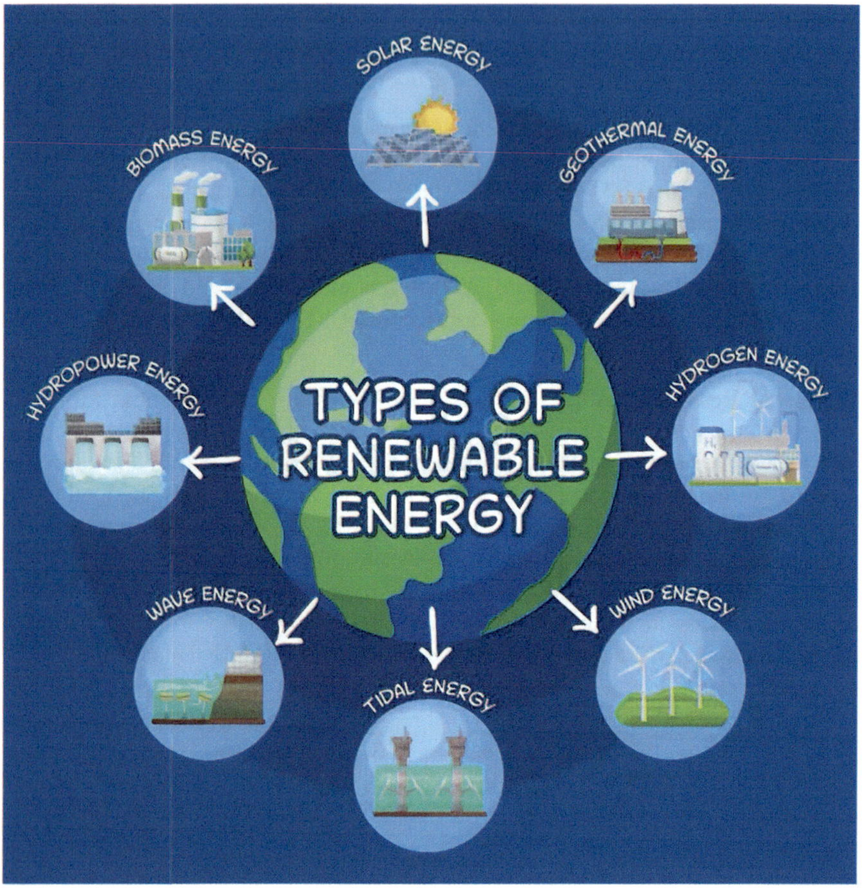

Fig. 15.1 Types of renewable energy

Airborne Wind Energy Systems AWES represent a cutting-edge approach to capturing wind energy using tethered airborne devices such as kite turbines and flying drones. These systems operate higher elevations where the prevailing winds are stronger and more consistent, enabling increased energy production with reduced material and infrastructure costs compared to traditional wind turbines (Stambouli et al. 2012). Advances in AWES technology include improved tether materials, autonomous control systems, and innovative launching and retrieval mechanisms.

High-Altitude Platforms HAPs like solar-powered drones and balloons provide unique opportunities for solar energy gathering and wireless power transmission systems to the ground. By operating above traditional airspace, these platforms may access a wealth of sunshine and reduce the requirement for land-based solar systems (Anil 2018). Technological advancements in HAPs include lightweight materials, advanced photovoltaic cells, and efficient energy storage and transmission systems.

Advanced Materials and Manufacturing Techniques With developments in materials science and manufacturing techniques, aerial renewable energy systems are becoming more effective, durable, and reasonably priced. Reducing material needs and environmental impact, wind turbine blades, solar panels, and aerial platforms are made more efficient and long-lasting via the use of lightweight composite materials, additive manufacturing, and improved coatings.

Grid Integration and Smart Grid Technologies As airborne renewable energy systems become increasingly interconnected with existing energy grids, advancements in grid integration and smart grid technologies are essential for optimizing energy production, storage, and distribution. Demand response is one of the smart grid options, energy storage systems (e.g., batteries, pumped hydro), and real-time monitoring and control enable seamless integrating intermittent renewable energy sources and improving the dependability and stability of the system.

Data Analytics and Predictive Maintenance The advent of big data analytics and predictive maintenance techniques allows operators to monitor and optimize the performance of airborne renewable energy systems in real time. By analyzing vast amounts of operational data, machine learning algorithms can identify patterns, detect anomalies, and predict maintenance needs, optimizing energy production, reducing downtime, and extending equipment lifespan.

Carbon-Neutral Technologies Development in response to the necessity of lowering greenhouse gas emissions and the rising awareness of climate change. These are innovations designed to eliminate or offset the carbon dioxide emissions generated by human activities, thereby achieving a net-zero carbon footprint. Originating from the environmental awareness and energy crises of the twentieth century, these technologies encompass renewable energy sources like solar, wind, and hydropower, which replace fossil fuels with clean energy. Bioenergy from organic materials and advanced energy storage systems, such as lithium-ion batteries and pumped hydro storage, address the variability of renewable energy (Li et al. 2023). Carbon capture, utilization, and storage (CCUS) technologies aim to remove CO_2 from the atmosphere or industrial emissions, preventing it from contributing to global warming. Green hydrogen, produced via electrolysis using renewable energy, serves as a clean fuel for various applications. Energy-efficient buildings and smart grid technologies enhance energy conservation, while sustainable transportation, including electric vehicles and public transit, reduces emissions from the transportation sector. These technologies have gained significant momentum due to international climate agreements like the Kyoto Protocol and the Paris Agreement, which have fostered global commitments to reducing carbon emissions. Collectively, carbon-neutral technologies are essential for transitioning to a sustainable future and mitigating the impacts of climate change (Wang et al. 2021).

Swift expansion and development of aerial renewable energy are being propelled by these technical breakthroughs and advancements, which are opening the door to a future in energy that is more robust and sustainable. We can increase the pace of the shift to a low-carbon economy and open up new sources of clean, renewable energy by using cutting-edge technologies to harness the power of the air.

15.4 Integration and Grid Compatibility

These are crucial aspects of incorporating airborne renewable energy into the current energy infrastructure. As these innovative technologies become more prevalent, ensuring seamless integration with the grid is essential for maximizing their benefits and maintaining grid stability. Here's how integration and grid compatibility are addressed (Calvillo et al. 2016).

Intermittency Management Renewable energy sources that are carried aloft, including solar and wind power, are inherently intermittent, as they depend on weather conditions and diurnal cycles. Effective integration requires strategies for managing this variability and balancing supply and demand on the grid. This may involve deploying energy storage systems, implementing demand response programs, and optimizing the operation of flexible generation assets to smooth fluctuations in renewable energy output (Al Essa 2020).

Grid Stability and Reliability Integrating airborne renewable energy into the grid requires careful consideration of grid stability and reliability. Variations in wind speed and solar irradiance can affect grid frequency and voltage levels, potentially leading to disruptions in power supply. Grid-compatible inverters, voltage control devices, and advanced grid monitoring systems help maintain grid stability by dynamically adjusting power output and voltage levels in response to changing conditions (Al-Shetwi et al. 2020).

Transmission Infrastructure Airborne renewable energy systems are often located in remote or offshore areas with abundant wind or sunlight, necessitating the development of new transmission infrastructure to connect them to population centers and load centers. High-voltage transmission lines, underwater cables, and grid interconnections enable the efficient transfer of renewable energy from generation sites to end-users, reducing transmission losses and maximizing resource utilization (Li et al. 2010).

Grid Planning and Expansion Integrating airborne renewable energy requires comprehensive grid planning and expansion to accommodate the growing share of renewable generation. This includes identifying suitable locations for new wind farms and solar installations, assessing grid capacity and congestion, and upgrading existing infrastructure to support higher penetrations of renewable energy (Resch et al. 2021). Collaborative planning between utilities, regulators, and stakeholders is essential to ensure a coordinated and efficient transition to a renewable energy-dominated grid.

Grid Services and Ancillary Support Airborne renewable energy systems can provide valuable grid services and ancillary support functions, such as frequency regulation, voltage control, and reactive power compensation. Grid operators can leverage the flexibility and responsiveness of these systems to enhance grid stability and reliability, reducing the need for conventional generation reserves and enhancing overall system efficiency (Ramakrishnan et al. 2020).

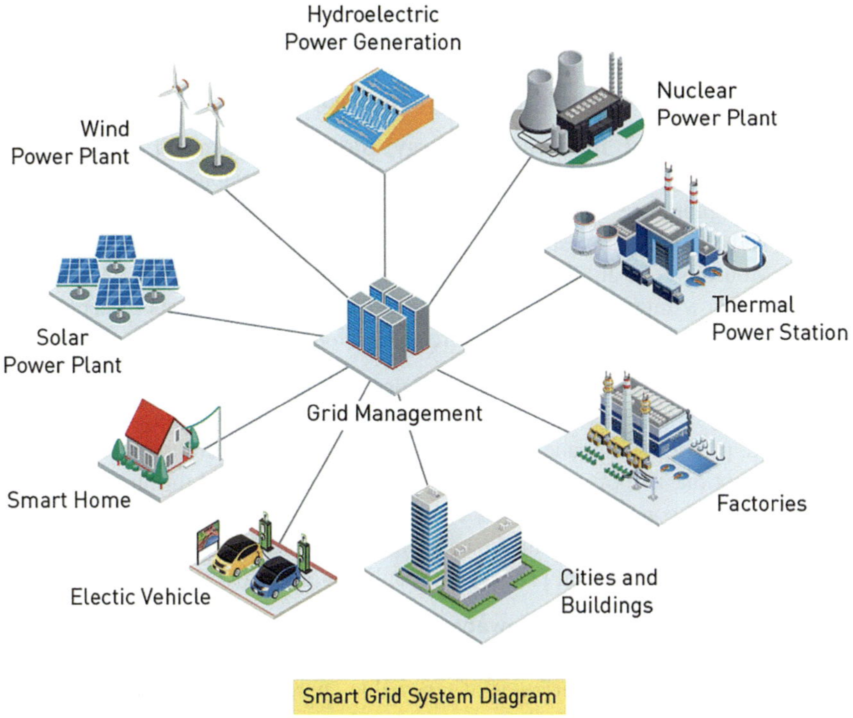

Fig. 15.2 Grid management and distribution

Effective integration and grid compatibility are essential for unlocking the full potential of airborne renewable energy and transitioning toward a cleaner, more sustainable energy future. By addressing technical, regulatory, and market challenges, we can build a resilient grid that harnesses the power of the air to meet the energy needs of society while minimizing environmental impact, Fig. 15.2 shows the grid management techniques.

15.5 Policy and Regulatory Frameworks

It is essential in determining how the deployment is made, integration, and adoption of airborne renewable energy technologies. By establishing supportive policies, governments can incentivize investment, remove barriers to entry, and create a conducive environment for innovation and market growth. Here's an overview of key policy and regulatory considerations (Raza et al. 2022).

Renewable Energy Targets and Mandates Establishing aggressive renewable energy goals and regulations sends a strong message to the market and increases demand for renewable energy technology used in aircraft. These goals can be national renewable energy targets, which seek to achieve a specified proportion of renewable energy in the total energy mix, which mandate utilities to get a set amount of their power from renewable sources (Hafner and Tagliapietra 2020).

Feed-in Tariffs and Renewable Energy Incentives Financial assistance for renewable energy producers is given via FITs and renewable energy incentives, which also guarantee fixed payments for electricity produced from renewable sources and encourage investment in renewable energy projects that include aircraft (Milad Mousavian et al. 2020). Other incentives may include tax credits, grants, rebates, and low-interest loans, which help reduce the upfront costs of deployment and improve the economic viability of renewable energy investments.

Net Metering and Grid Connection Policies Net metering policies allow renewable energy system owners to offset their electricity bills by exporting excess energy to the grid, providing a financial incentive for distributed generation and encouraging the adoption of rooftop solar panels and small-scale wind turbines. Clear and transparent grid connection policies ensure timely and cost-effective interconnection of renewable energy systems to the grid, minimizing administrative hurdles and streamlining the deployment process (Iliopoulos et al. 2020).

Market Mechanisms and Renewable Energy Certificates Market mechanisms such as renewable energy certificates (RECs) create tradable commodities that represent the environmental attributes of renewable energy generation, allowing utilities, businesses, and individuals to buy and sell renewable energy credits to meet regulatory requirements and demonstrate environmental stewardship (Adamczyk and Graczyk 2020). Renewable energy auctions, competitive bidding processes, and power purchase agreements (PPAs) provide additional mechanisms for procuring renewable energy at competitive prices and driving down costs through market competition (Adamczyk and Graczyk 2020).

Grid Integration and System Flexibility Regulatory frameworks should facilitate the integration of airborne renewable energy into the grid by promoting grid flexibility, enhancing transmission planning and coordination, and incentivizing investments in grid modernization and smart grid technologies (Inês et al. 2020). Clear technical standards, grid codes, and interconnection procedures ensure compatibility between renewable energy systems and existing grid infrastructure, maintaining grid stability and reliability as renewable energy penetration increases (Cambini et al. 2020).

Environmental Regulations and Permitting Processes Environmental regulations and permitting processes ensure that airborne projects utilizing renewable energy are built and operated in compliance with environmental laws and regulations, safeguarding ecosystems, habitats, and natural resources (Díaz-Rodríguez et al. 2023). Environmental impact assessments, habitat mitigation plans, and stakeholder engagement processes help identify and address potential social and ecological effects of the growth of renewable energy, encouraging the design and administration of sustainable projects (Deme et al. 2022).

Research, Development, and Demonstration (RD&D) Funding Government funding for RD&D initiatives supports innovation and technology advancement in airborne renewable energy, driving down costs, improving performance, and accelerating commercialization (Kassouri et al. 2022). Public-private partnerships, university-industry collaborations, and innovation grants facilitate knowledge transfer, technology transfer, and capacity building, fostering a vibrant research and innovation ecosystem in renewable energy (Bioret et al. 2023).

A comprehensive policy and regulatory framework is essential for creating an enabling environment for airborne renewable energy deployment, driving market growth, and achieving sustainability and energy security objectives. By aligning policies with broader energy and climate goals, governments can harness the power of airborne renewable energy to transition toward a cleaner, more resilient, and sustainable energy future.

15.6 Conclusion

A revolutionary step toward a greener, more resilient energy future is represented by the quest to harness the power of the air for a sustainable society. We have seen throughout this investigation the enormous potential of aerial renewable energy technology to address urgent environmental issues, reduce global warming, and spur economic prosperity. From wind turbines towering over landscapes to solar-powered drones soaring high above, the sky has emerged as a boundless source of renewable energy waiting to be harnessed. As it navigates the complexities of transitioning toward a sustainable energy system, it is clear that no single solution can meet the diverse needs and challenges we face. Airborne renewable energy complements existing renewable energy sources, offering unique advantages such as high energy density, scalability, and flexibility. It can reduce our dependency on fossil fuels and lessen our impact on the environment while increasing new potential for renewable energy generation through portfolio diversification and innovation. To fully use aerial renewable energy, governments, industry players, and communities throughout the world must work together. Encouraging policies, legal frameworks, and market processes are necessary to boost investment, lower entry barriers, and expand markets. Collaboration between public and private sectors, technology innovation, and knowledge sharing are critical for overcoming technical challenges, driving down costs, and accelerating the deployment of airborne renewable energy solutions. The transition to a sustainable energy future must be inclusive, equitable, and just, ensuring that benefits are shared equitably and vulnerable communities are not left behind. Community engagement, capacity building, and empowerment play a central role in fostering a sense of ownership and participation in the energy transition, empowering individuals and communities to shape their energy future. Climate change, environmental degradation, and energy insecurity, the vision of harnessing the power of the air for a sustainable tomorrow offers hope and inspiration. By embracing the boundless potential of airborne renewable energy, we can create a world where energy is abundant, clean, and accessible to all, where the air we breathe fuels not only our progress but also the preservation of our planet for generations to come. Together, let us rise to the challenge and seize the opportunity to shape a brighter, more sustainable future for ourselves and future generations.

15.7 Future Recommendation

To harness the power of the air for a sustainable tomorrow, it is crucial to expand and innovate within the field of wind energy. This entails not only increasing the deployment of onshore and offshore wind turbines but also advancing technology to enhance efficiency and reduce costs. Investing in research and development is essential to create more efficient blade designs, optimize turbine placement using sophisticated meteorological models, and integrate smart technologies for real-time performance monitoring and predictive maintenance (Bioret et al. 2023). Additionally, promoting the development of airborne wind energy systems, such as high-altitude kites and drones, can tap into stronger and more consistent winds at higher altitudes, potentially revolutionizing the wind energy sector. Governments should implement supportive policies, including subsidies, tax incentives, and streamlined permitting processes, to accelerate the adoption of wind energy (Hafez et al. 2023). Moreover, enhancing grid infrastructure to accommodate the variability of wind power and advancing energy storage solutions, such as large-scale batteries and pumped hydro storage, will ensure a stable and reliable energy supply. Collaboration between public and private sectors, alongside international partnerships, will be vital to share best practices, foster innovation, and finance large-scale projects. By prioritizing these strategies, we can significantly increase wind energy capacity, reduce reliance on fossil fuels, and move closer to achieving a sustainable and carbon-neutral future.

Acknowledgments The authors thank NIT Uttarakhand for supporting this work.

References

Adamczyk J, Graczyk M (2020) Green certificates as an instrument to support renewable energy in Poland—strengths and weaknesses. Environ Sci Pollut Res 27:6577–6588. https://doi.org/10.1007/s11356-019-07452-5

Ahmad N, Beheiry S (2023) Sustainability in the data age: harnessing the power of data for a greener future. Computer (Long Beach Calif) 56:106–110. https://doi.org/10.1109/MC.2023.3314221

Ajiere SI, Diagi BE, Edokpa D, Onyejekwe IM, Onyeananam UL (2024) Perceived impact of air quality and its implications for a sustainable environment in Rivers State. Int J Innov Sci Res Technol 9:197–206

Al Essa MJM (2020) Power management of grid-integrated energy storage batteries with intermittent renewables. J Energy Storage 31:101762. https://doi.org/10.1016/j.est.2020.101762

Al-Shetwi AQ, Hannan MA, Jern KP, Mansur M, Mahlia TMI (2020) Grid-connected renewable energy sources: review of the recent integration requirements and control methods. J Clean Prod 253:119831. https://doi.org/10.1016/j.jclepro.2019.119831

Anil LR (2018) Feasibility analysis of high altitude wind power system. In: National conference on research as a motivational source for a better tomorrow in making genius (RMSBTMG-2018)

Anjum A, Chauhan J (2023) Land use potential and related effect of power generation technologies in India: a sustainability perspective. Scope 13:449–458

Awewomom J, Dzeble F, Takyi YD, Ashie WB, Ettey ENYO, Afua PE, Sackey LNA, Opoku F, Akoto O (2024) Addressing global environmental pollution using environmental control techniques: a focus on environmental policy and preventive environmental management. Discov Environ 2. https://doi.org/10.1007/s44274-024-00033-5

Bioret L, Zhu Y, Krupnick A, Bergman A (2023) Tracking and evaluation of demonstration programs at the US Department of Energy. Resources for the Future

Calvillo CF, Sánchez-Miralles A, Villar J (2016) Energy management and planning in smart cities. Renew Sust Energ Rev 55:273–287. https://doi.org/10.1016/j.rser.2015.10.133

Cambini C, Congiu R, Jamasb T, Llorca M, Soroush G (2020) Energy systems integration: implications for public policy. Energy Policy 143:111609. https://doi.org/10.1016/j.enpol.2020.111609

Chen X, Eder MA, Shihavuddin ASM, Zheng D (2021) A human-cyber-physical system toward intelligent wind turbine operation and maintenance. Sustain For 13:1–10. https://doi.org/10.3390/su13020561

Cui Q, He L, Han G, Chen H, Cao J (2020) Review on climate and water resource implications of reducing renewable power curtailment in China: a nexus perspective. Appl Energy 267:115114. https://doi.org/10.1016/j.apenergy.2020.115114

Deme GG, Ewusi-Mensah D, Olagbaju OA, Okeke ES, Okoye CO, Odii EC, Ejeromedoghene O, Igun E, Onyekwere JO, Oderinde OK, Sanganyado E (2022) Macro problems from microplastics: toward a sustainable policy framework for managing microplastic waste in Africa. Sci Total Environ 804:150170. https://doi.org/10.1016/j.scitotenv.2021.150170

Díaz-Rodríguez N, Del Ser J, Coeckelbergh M, López de Prado M, Herrera-Viedma E, Herrera F (2023) Connecting the dots in trustworthy artificial intelligence: from AI principles, ethics, and key requirements to responsible AI systems and regulation. Inf Fusion 99:101896. https://doi.org/10.1016/j.inffus.2023.101896

Hafez FS, Sa'di B, Safa-Gamal M, Taufiq-Yap YH, Alrifaey M, Seyedmahmoudian M, Stojcevski A, Horan B, Mekhilef S (2023) Energy efficiency in sustainable buildings: a systematic review with taxonomy, challenges, motivations, methodological aspects, recommendations, and pathways for future research. Energy Strateg Rev 45:101013. https://doi.org/10.1016/j.esr.2022.101013

Hafner M, Tagliapietra S (2020) The geopolitics of the global energy transition. Lecture notes in energy. Springer, Cham

Huang Y, Chen Z, Wu B, Chen L, Mao W, Zhao F, Wu J, Wu J, Yu B (2015) Estimating roof solar energy potential in the downtown area using a GPU-accelerated solar radiation model and airborne LiDAR data. Remote Sens 7:17212–17233. https://doi.org/10.3390/rs71215877

Huang Y, Yu B, Hu Z, Wu J, Wu B (2012) Locating suitable roofs for utilization of solar energy in downtown area using airborne LiDAR data and object-based method: a case study of the Lujiazui region, Shanghai. In: Proc. 2nd Int. Work Earth Obs Remote Sens Appl EORSA 2012. IEEE, pp 322–326. https://doi.org/10.1109/EORSA.2012.6261192

Iliopoulos TG, Fermeglia M, Vanheusden B (2020) The EU's 2030 climate and energy policy framework: how net metering slips through its net. Rev Eur Comp Int Environ Law 29:245–256. https://doi.org/10.1111/reel.12339

Inês C, Guilherme PL, Esther MG, Swantje G, Stephen H, Lars H (2020) Regulatory challenges and opportunities for collective renewable energy prosumers in the EU. Energy Policy 138:111212. https://doi.org/10.1016/j.enpol.2019.111212

Karthik MV (2024) Harmonizing sustainable transportation: a comprehensive hybrid vehicle integrating CNG, in-wheel motors, and rooftop solar panels for environmental excellence. Int J Progress Res Eng Manag Sci 4(2):561–565

Kassouri Y, Bilgili F, Peter Majok Garang A (2022) Are government energy technology research, development, and demonstration budgets converging or diverging? Insights from OECD countries. Technol Anal Strateg Manag 34:563–577. https://doi.org/10.1080/09537325.2021.1914330

Li F, Qiao W, Sun H, Wan H, Wang J, Xia Y, Xu Z, Zhang P (2010) Smart transmission grid: vision and framework. IEEE Trans Smart Grid 1:168–177. https://doi.org/10.1109/TSG.2010.2053726

Li H, Hao Y, Xie C, Han Y, Wang ZR (2023) Emerging technologies and policies for carbon–neutral transportation. Int J Transp Sci Technol 12:329–334. https://doi.org/10.1016/j.ijtst.2022.09.002

Milad Mousavian H, Hamed Shakouri G, Mashayekhi AN, Kazemi A (2020) Does the short-term boost of renewable energies guarantee their stable long-term growth? Assessment of the dynamics of feed-in tariff policy. Renew Energy 159:1252–1268. https://doi.org/10.1016/j.renene.2020.06.068

Osman AI, Chen L, Yang M, Msigwa G, Farghali M, Fawzy S, Rooney DW, Yap PS (2023) Cost, environmental impact, and resilience of renewable energy under a changing climate: a review. Environ Chem Lett 21:741–764. https://doi.org/10.1007/s10311-022-01532-8

Pajek L, Košir M (2021) Exploring climate-change impacts on energy efficiency and overheating vulnerability of bioclimatic residential buildings under central european climate. Sustain For 13:6791. https://doi.org/10.3390/su13126791

Platzer MF, Sarigul-Klijn N (2021) Sustainable aviation. In: The green energy ship concept. SpringerBriefs in applied sciences and technology. Springer, Cham. https://doi.org/10.1007/978-3-030-58244-9_11

Ramakrishnan J, Hashemi S, Traholt C (2020) Assessment of energy storage systems for multiple grid service provision. In: Proc. - 2020 IEEE 14th international conference on compatibility, power electronics and power engineering (CPE-POWERENG), pp 333–339. https://doi.org/10.1109/CPE-POWERENG48600.2020.9161622

Rasugu Ayub H, Jakanyango Ambusso W, Muriuki Manene F, Mongeri Nyaanga D (2021) A review of cooking systems and energy efficiencies. Am J Energy Eng 9:1. https://doi.org/10.11648/j.ajee.20210901.11

Raza MA, Khatri KL, Ul Haque MI, Shahid M, Rafique K, Waseer TA (2022) Holistic and scientific approach to the development of sustainable energy policy framework for energy security in Pakistan. Energy Rep 8:4282–4302. https://doi.org/10.1016/j.egyr.2022.03.044

Resch M, Buhler J, Schachler B, Sumper A (2021) Techno-economic assessment of flexibility options versus grid expansion in distribution grids. IEEE Trans Power Syst 36:3830–3839. https://doi.org/10.1109/TPWRS.2021.3055457

Sacchi R, Bauer C, Cox B, Mutel C (2022) When, where and how can the electrification of passenger cars reduce greenhouse gas emissions? Renew Sust Energ Rev 162:112475. https://doi.org/10.1016/j.rser.2022.112475

Shirin S, Jamal A, Emmanouil C, Singh VP, Yadav AK (2023) Assessment and characterization of waste material used as backfilling in an abandoned mine. Int J Coal Prep Util 43(8):1402–1410. https://doi.org/10.1080/19392699.2022.2118259

Shirin S, Yadav AK (2014) Physico-chemical analysis of municipal wastewater discharge in Ganga river, Haridwar district of Uttarakhand, India. Curr World Environ 9(2):536–543. https://doi.org/10.12944/CWE.9.2.39

Stambouli AB, Khiat Z, Flazi S, Kitamura Y (2012) A review on the renewable energy development in Algeria: current perspective, energy scenario and sustainability issues. Renew Sust Energ Rev 16:4445–4460. https://doi.org/10.1016/j.rser.2012.04.031

Wang F, Harindintwali JD, Yuan Z, Wang M, Wang F, Li S, Yin Z, Huang L, Fu Y, Li L, Chang SX, Zhang L, Rinklebe J, Yuan Z, Zhu Q, Xiang L, Tsang DCW, Xu L, Jiang X, Liu J, Wei N, Kästner M, Zou Y, Ok YS, Shen J, Peng D, Zhang W, Barceló D, Zhou Y, Bai Z, Li B, Zhang B, Wei K, Cao H, Tan Z, Zhao L, He X, Zheng J, Bolan N, Liu X, Huang C, Dietmann S, Luo

M, Sun N, Gong J, Gong Y, Brahushi F, Zhang T, Xiao C, Li X, Chen W, Jiao N, Lehmann J, Zhu YG, Jin H, Schäffer A, Tiedje JM, Chen JM (2021) Technologies and perspectives for achieving carbon neutrality. Innovations 2:100180. https://doi.org/10.1016/j.xinn.2021.100180

Wang R, Ding C (2023) Bilateral impact of digital economy on air pollution: emissions increase and reduction effects. Front Environ Sci 11:1–18. https://doi.org/10.3389/fenvs.2023.1134334

Yadav AK, Shirin S, Emmanouil C, Jamal A (2022) Effect of seasonal and meteorological variability of air pollution in Singrauli coalfield. Aerosol Sci Eng 6(1):61–70. https://doi.org/10.1007/s41810-021-00124-3

Yan Z, Jiang L, Huang X, Zhang L, Zhou X (2023) Intelligent urbanism with artificial intelligence in shaping tomorrow's smart cities: current developments, trends, and future directions. J Cloud Comput 12. https://doi.org/10.1186/s13677-023-00569-6

Chapter 16
Electric Transportation Paving the Way for Eco-Friendly Mobility

Vivekanand Kumar

Abstract The transportation industry is becoming a vital front in the fight against climate change and the world's growing reliance on fossil fuels. This chapter explores the transformative potential of electric transportation to pave the way for eco-friendly mobility. Electric vehicles (EVs) offer a promising solution that leverages advance in battery technology, infrastructure development, and policy support to address environmental concerns and improve energy security. This chapter discusses the state of electric mobility today and looks at the main factors, such as environmental advantages, new technology, and government regulations, that contribute to the increasing use of electric cars. Explore the challenges facing widespread adoption of electric vehicles, such as range anxiety, limitations in the charging infrastructure, and cost barriers, while highlighting ongoing efforts to overcome these obstacles.

Furthermore, this chapter discusses the broader implications of electric transportation beyond environmental sustainability, including its potential to revolutionize urban mobility, foster innovation in the energy sector, and create new economic opportunities. Concerns about how electric vehicles affect the environment over their whole lifecycle are also covered, considering factors such as manufacturing processes, battery disposal, and energy sourcing. Drawing on insights from academic research, industry reports, and policy analyses, this chapter offers a thorough summary of how electric vehicles contribute to the advancement of environmentally responsible mobility. Examining the benefits and drawbacks of adopting electric vehicles can help educate decision-makers, business leaders, and the public about the necessity of making the switch to more environmentally friendly modes of transportation. Ultimately, making the transition to electric cars is crucial to creating a future that is both environmentally sustainable and more robust for coming generations.

V. Kumar (✉)
Department of Mechanical Engineering, Sandip University, Madhubani, India
e-mail: vivekanand.kumar@sandipuniversity.edu.in

A. K. Yadav (ed.), *Powering Future: Clean Energy Solutions for a Sustainable Planet*,
Clean Energy Production Technologies, https://doi.org/10.1007/978-981-96-5055-2_16

Keywords EV · Sustainability · Mobility · Electric transportation

16.1 Introduction

The need to stop climate change and reduce greenhouse gas emissions has caused a paradigm shift in transportation in recent years. The use of electric vehicles has demonstrated potential in mitigating the adverse environmental impacts associated with traditional fossil fuel-fueled automobiles. This chapter delves into the developments, obstacles, and potential of electric transportation, clarifying its essential function in promoting eco-friendly mobility.

The pressing need to combat climate change and cut greenhouse gas emissions is causing changes in the transportation sector. The use of electric vehicles has become apparent as a viable means of achieving environmentally sustainable mobility. We explore the world of electric vehicles (EVs) in this chapter, exploring their history, environmental impact, technological advances, challenges, and the transformative potential they hold for sustainable mobility.

The need for mobility has increased along with the rate of urbanization around the world, which has increased desire for automobiles that can transport passengers be divided into three categories: airborne (airplanes and helicopters), aquatic (ships and ferries), and surface-based (cars, buses, and trains). Because all of those vehicles burn traditional fossil fuels, the transportation sector makes a contribution to the world's greenhouse gas (GHG) emissions. (Razi and Dincer 2022). In addition to being bad for the environment, the continuous use of hydrocarbon fuels depletes these finite natural resources. Therefore, for sustainable development, a cleaner energy source is essential. EVs are a viable answer to reduce the emissions that come from the transportation industry in the urban transportation domain. An EV-based transportation sector is becoming more likely due to elements including financial sustainability, technological advancement, government backing, and large outlays from original equipment manufacturers (OEMs).

While reducing emissions, the use of electric vehicles will maintain current power systems. Electric vehicles can assist the power system by returning generated electricity to the grid during peak hours. Subject to significant expenditures and sustained financial support for research and development, the outlook for the electric car industry is quite positive. But a variety of issues make EVs' public acceptability a significant barrier to their wider implementation. The main challenge here is the installation of a significant quantity of charging stations, akin to those seen at gas stations. There is a growing movement to investigate the enormous potential of electric vehicles (EVs) due to the pressing need to lessen the significant environmental damage that internal combustion engine cars do (Schulz and Rode 2022; Tarei et al. 2021; Zhou et al. 2023). In recent times, there has been a rise in the market shares of electric vehicles due to government initiatives such as the establishment of London's Ultra-Low Emission Zone (ULEZ) and the European Union's stricter CO_2 tailpipe regulations for the automotive sector (Henderson et al. 2022; Iwan et al. 2021). The need for vehicles that are efficient, clean, and environmentally friendly is driving up the adoption of electric vehicles, which is being fueled by significant investments made by the global automotive industry in this technology (Altenburg et al. 2016; Fevang et al. 2021; Secinaro et al. 2022). The need to switch from gasoline and diesel to a more environmentally friendly energy source is what is driving this demand (Kar et al. 2022; Liao and Correia 2022). But there are several drawbacks to this transition to electric vehicles, including their short range and high battery costs, as well as issues with policymakers, manufacturers, and consumer attitudes (Durate et al. 2021; Li et al. 2020; Liao and Correia 2022; Patil et al. 2023). Therefore, in order to provide the best possible understanding and assistance for this shift, it is essential that these issues be carefully assessed. Research on electric vehicles (EVs) has advanced significantly, with notable contribution (Patil et al. 2023; Xiaoli et al. 2019; Secinaro et al. 2022). These developments have not, however, completely eliminated some restrictions, leaving gaps in our knowledge of the wider EV landscape. Aims to bridge the knowledge gap by first examining the attitudes, desires, economic viability, and infrastructure challenges around electric vehicles (EVs) before recommending energy-sourcing strategies (Greene et al. 2020). The drawbacks of this reliance on survey data were also emphasized, reiterating the concerns, and a significant gap continues to exist in the EV research now in publication (Lieven et al. 2011), brought up regarding possible bias in EV adoption projections (Al-Alawi and Bradley 2013) and underlined the necessity of looking for data sources outside surveys. Furthermore, a regional disparity is apparent, with the majority of studies conducted in the areas of technology, finance, and policy (Fevang et al. 2021; Schulz and Rode 2022). Reviews by Liao and Correia (2022) and Patil et al. (2023) revealed that no African nation was mentioned in any of the analyzed articles. This disparity makes it difficult to comprehend the particular opportunities and constraints emerging nations like Nigeria face when implementing electric car adoption (Rajper and Albrecht 2020; Tarei et al. 2021) and acknowledged the emphasis on the United States and Europe at the moment and emphasized the need for future study to take emerging markets into account. Singh et al. (2020) asked for more research in emerging nations to improve comprehension of customer behavior (Farinloye et al. 2024). The transportation industry is a strong candidate to be among the main sources of CO_2 emissions worldwide (Brückmann and Bernauer 2020). It is hardly unexpected that the deployment of electric vehicles has aroused scholarly interest given this indisputable reality. Our current research contributes to the ongoing conversation among academics who are extensively exploring different aspects of this shift, ranging from thorough literature studies (Patil et al. 2022; Liao and Correia 2022) to scrutinizing acceptance by consumers behaviors (Secinaro et al. 2022; Tarei et al. 2021), and examining perceptions of alternative energy sources for mobility (Li et al. 2020; Kar et al. 2022). These investigations have shed light on a number of tactics that encourage the public to embrace EVs, such as subsidies and improved infrastructure for charging EVs. At the same time, they have investigated push factors, such as carbon taxes and rules meant to phase out cars that run on fossil fuels (Brückmann and Bernauer 2020; Farinloye et al. 2024).

16.2 Rise of Electric Vehicles

When inventors began experimenting with electric propulsion technologies in the early nineteenth century, the idea of electric vehicles was born. However, electric vehicles did not become widely accepted as competitive alternatives to automobiles with until the late nineteenth and early twentieth centuries, internal combustion engines. The invention of lithium-ion batteries revolutionized electric car technology by enabling longer charging times and increased range. Government incentives, mounting environmental concerns, and the increasing popularity of electric vehicles can be attributed to developments in battery technology. With their cutting-edge designs and enhanced performance, automakers such as Tesla, Nissan, and Chevrolet have led the EV market and raised the bar for electric vehicles relative to their competition for internal combustion engines.

There has been a renewed push for electric mobility (EM) since 2005. This time, attempts to electrify transportation are driven by concerns about climate change rather than urban pollution, and peak oil also playing a part. Following Hurricane Katrina in 2005, the public's knowledge of the detrimental effects of climate change had increased and worldwide awareness was further heightened by Al Gore's Inconvenient Truth documentary, which aired in May 2006. When taken as a whole, these incidents encouraged legislators to create market mechanisms and regulatory frameworks to reduce carbon emissions (Dijk et al. 2013). The use of electric vehicles (EVs) has increased dramatically globally in the twenty-first century, primarily due to environmental concerns and the requirement for environmentally friendly modes of transportation. Scholarly investigations have been mindful of these revolutionary patterns, augmenting an expanding corpus of work that enhances our comprehension of several aspects of the electric vehicle environment. The study by Liao and Correia (2022) study is noteworthy because it delves into demand dynamics, usage trends, and potential impacts of micromobility and electric vehicle sharing. This research sheds light on how urban mobility is progressing. Liao and Correia (2022) examined consumer preferences for electric cars and offered insights on the dynamics of the market. Xiaoli et al. (2019) studied the developments in EV technology and consumer behavior was mapped. Rajper and Albrecht (2020) examined the opportunities in underdeveloped nations. According to Haghani et al. (2023), there is a deficiency in the literature on EV research, especially in poor nations such as Africa. This underscores the necessity of conducting detailed studies in a variety of international contexts. When combined, these studies look at a wide range of topics related to the adoption of electric vehicles, including technology, consumer preferences, and policy consequences. This knowledge helps us comprehend these vehicles better. Haghani et al. (2023) observed a conspicuous void in the scholarly literature on the study of electric vehicles (EVs) from numerous developing nations, with Africa serving as a prominent instance. Subsequent research centered on developing nations will enhance global comprehension of the electric car landscape, promoting ecologically conscious and sustainable transportation options around the world. Even though developed countries were the first to introduce electric vehicles into transportation networks, developing countries are looking into doing the same more and more. This underscores the need for research that takes into account a variety of global contexts. These comprise the higher expenses of owning electric cars, the cost of building infrastructure to charge them, and the requirement of sufficient electrical storage facilities. If these issues are not resolved, it may be difficult for electric cars to become widely used in these areas.

Tarei et al. (2021) studied adoption hurdles for electric vehicles with an emphasis on India. Their study brought to light a number of obstacles, such as those related to technology, infrastructure, finances, behavior, and outside difficulties. These obstacles were classified and prioritized, giving decision-makers a framework for efficiently allocating resources. Critical obstacles that impede the widespread adoption of electric vehicles include low consumer awareness, inadequate charging infrastructure, and poor vehicle performance. Jaiswal et al. (2022) conducted research to shed light on the factors that influence intentions of Indian consumers to adopt electric automobiles. Their findings demonstrated that, with the moderation of financial incentive programs, attitudes, perceived value, perceived ease of use, and perceived danger all had a direct and indirect impact on adoption intentions. These studies highlight the many benefits and drawbacks of using electric vehicles, underscoring the necessity for tailored strategies to address the many problems surrounding their adoption in developing nations (Farinloye et al. 2024).

16.3 Categories of Electric Cars

There are four categories of electric vehicles available:

16.3.1 Battery Electric Vehicles (BEVs)

Another name for BEVs is All-Electric Vehicles (AEV). Electric drivetrains fueled by batteries power the whole vehicle in BEV-equipped automobiles. A large battery pack that can be charged by connecting to the grid stores the vehicle's electricity. The charged battery pack then powers one or more electric motors, which drive the electric vehicle. The key components of a battery-electric car are the drive train, inverter, battery, control module, and electric motor. A DC battery that is converted to AC electricity powers the electric motor in a battery-electric vehicle (BEV). As soon as the accelerator is depressed, a signal is sent to the controller. The controller can adjust the vehicle's speed by modifying the frequency of the AC power that the inverter supplies to the motor. The motor then attaches and uses a gear to turn the wheels. An electric automobile's motor becomes an alternator when the brakes are applied or the car slows down. The Mahindra ZS, TATA Nexon, TATA Tigor, Mahindra E20 plus, Hyundai Kona, and Mahindra Verito are then powered by this energy that is then restored to the battery (Fig. 16.1).

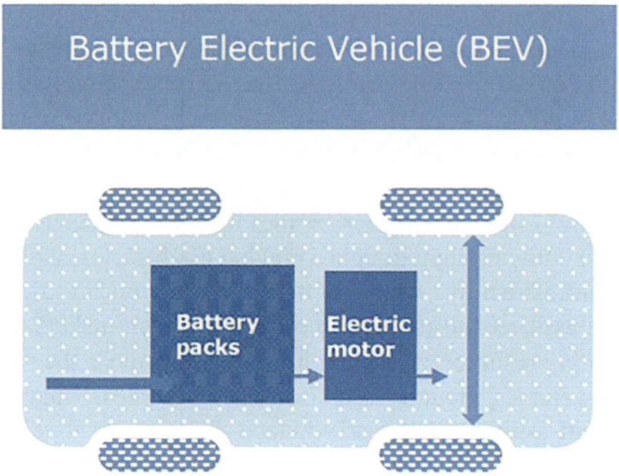

Fig. 16.1 Electric vehicle (https://e-amrit.niti.gov.in/types-of-electric-vehicles {Online Resource})

16.3.2 Hybrid Electric Vehicle (HEV)

HEVs also go by the titles parallel and series hybrids (Fig. 16.2). A hybrid electric vehicle (HEV) has an engine and an electric motor. The motor is powered by batteries, while the engine is run by fuel. Simultaneous rotation of the gearbox occurs due to the engine and electric motor. After that, this drives the wheels. In a hybrid electric car, the most important components are the engine, electric motor, fuel tank, control module, battery pack with controller and inverter. HEVs function similarly to a regular car in that the engine receives energy from the gasoline tank. The battery is powered by an electric motor. The gearbox can be turned simultaneously by the engine, electric motor, gasoline tank, control module, battery pack with controller and inverter.

16.3.3 Plug-in Hybrid Electric Vehicle (PHEV)

PHEVs are sometimes known as series hybrids (Fig. 16.3). They are all equipped with engines and motors. Either conventional fuel, like gasoline, or alternative fuel, like biodiesel, is your options. Moreover, it can be powered by a rechargeable battery pack. Outside charging of the battery is possible.

PHEVs Can Run in At Least Two Modes The car's battery and motor are its only sources of energy when it is in all-electric mode. Diesel and electricity are both used in hybrid mode. The key components of a plug-in hybrid electric car are the engine, battery, gasoline tank, control module, inverter, electric motor, and if it's an onboard model, the battery charger. When a plug-in hybrid electric vehicle (PHEV) starts up,

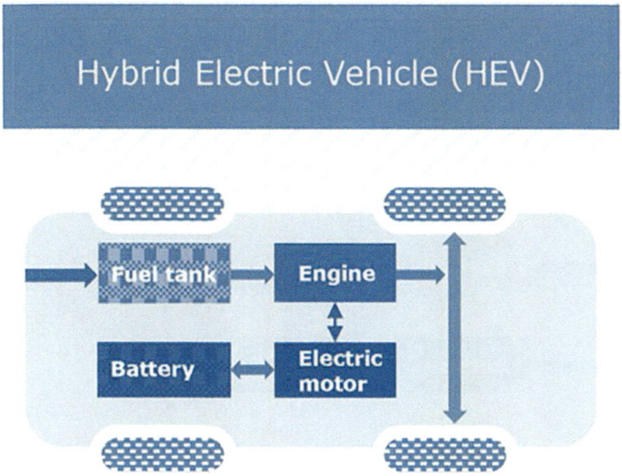

Fig. 16.2 Hybrid electric vehicle (https://e-amrit.niti.gov.in/types-of-electric-vehicles {Online Resource})

it uses electricity exclusively until its battery runs out. The engine takes over and the vehicle runs like a standard plug-in hybrid when the battery runs out. PHEVs can be charged by regenerative braking, starting an engine, or plugging in an external electric power source. When the brakes are engaged, the electric motor uses its capacity as a generator to charge the battery. Smaller engines can be used since the electric motor increases engine power, increasing fuel efficiency without compromising performance. Several vehicles are hybrids: Audi A3 Etron, BMW i8, BMW X5 xdrive40e, Fiat 500e, Hyundai Sonata, Fiat 330e, Porsche Panamera S E hybrid, Fiat 500e, Hyundai Sonata, Kia Optima, Ford C-Max Energi, Mini Cooper SE Countryman, Ford Fusion Energi, Mercedes C350e, Mercedes S550e, Mercedes GLE550e, and Volvo XC90 T8.

16.3.4 Fuel Cell Electric Vehicle

Zero-emission vehicle is another term for FCEVs. They employ "fuel cell technology" to generate the electricity required to run the vehicle. The chemical energy of the fuel is instantly converted into electric energy. The key components of a fuel cell electric vehicle are the hydrogen storage tank, electric motor, fuel cell stack, battery with converter, and controller. The basic tenet of FCEVs is that they can generate the electricity required to power vehicles on the vehicle itself, as demonstrated by the Hyundai Nexo, Honda Clarity Fuel Cell, Hyundai Tucson FCEV, Toyota Mirai, and Riversimple Rasa (Fig. 16.4).

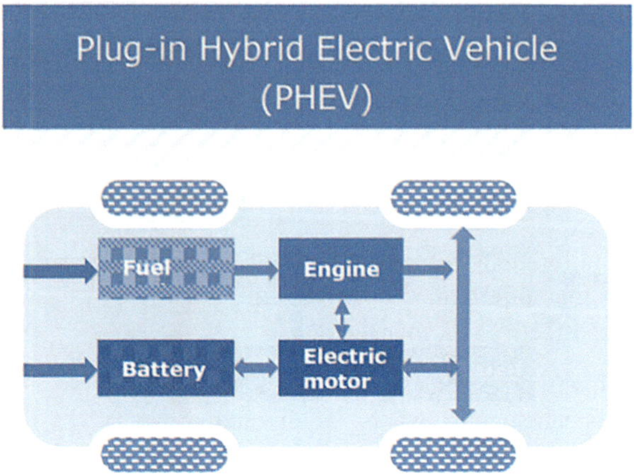

Fig. 16.3 Plug-in hybrid electric vehicle (https://e-amrit.niti.gov.in/types-of-electric-vehicles {Online Resource})

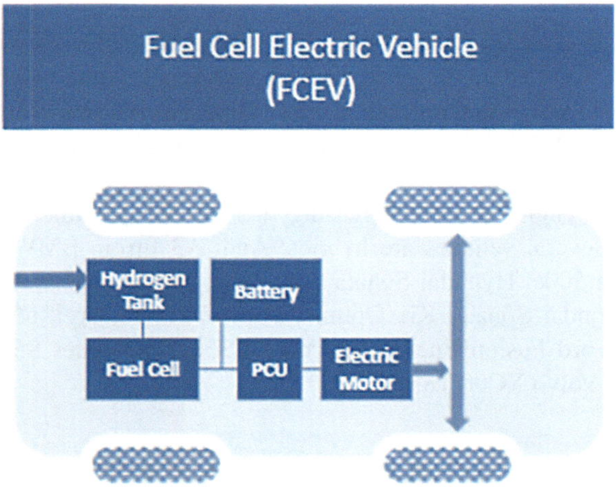

Fig. 16.4 Fuel cell electric vehicle (https://e-amrit.niti.gov.in/types-of-electric-vehicles {Online Resource})

16.4 Environmental Benefits

Comparing traditional internal combustion engine vehicles (ICEVs) with electric vehicles (EVs) reveals significant advantages for the environment.

16.4.1 Reduced Greenhouse Gas Emissions

Electric cars don't emit any emissions at the exhaust, which helps lower air pollution and greenhouse gas emissions. Even after deducting emissions from the electricity generation process, electric vehicles operating on sustainable energy sources such as solar or wind tend to generate lower total emissions than internal combustion engines (ICEVs).

16.4.2 Improved Air Quality

Electric cars enhance air quality, especially in cities where air pollution is a big concern, because they don't release pollutants from their tailpipes including nitrogen oxides, particulate matter, and volatile organic compounds.

16.4.3 Reduced Noise Pollution

Urban regions see less noise pollution because electric motors are quieter than internal combustion engines. Particularly in densely populated places, this may result in benefits to public health and quality of life.

16.4.4 Energy Efficiency

EVs are often more energy-efficient than ICEVs because electric motors are more efficient than internal combustion engines. This reduced energy waste is achieved by converting a larger proportion of the battery energy into electricity to move the car.

16.4.5 Conservation of Resources

Compared to ICEVs, electric vehicles often utilize fewer natural resources. Both types of cars require materials to be manufactured, but because electric cars often require less maintenance and have fewer moving components, resulting in a lower resource requirement over their lifetime.

16.4.6 Potential for Grid Integration and Energy Storage

EV batteries have the potential to be used as energy storage devices, which might help stabilize the electrical grid and store extra renewable energy produced when demand is low. This could improve overall energy efficiency and the ability to integrate renewable energy sources.

16.4.7 Reduced Dependence on Fossil Fuels

Reducing the use of fossil fuels, which are finite resources with significant environmental consequences associated with their extraction, transportation, and burning, is possible with the help of electric vehicles. This is because they require less gasoline and diesel fuel.

16.4.8 Promotion of Renewable Energy

The use of electric vehicles can encourage the growth and development of renewable energy sources, such wind and solar energy, to assist meet the world's growing demand for electricity. There is potential to significantly minimize greenhouse gas emissions and environmental harm by generating power using renewable energy sources rather than fossil fuels. Broadly speaking, the adoption of electric vehicles has great promise for reducing global warming, enhancing air quality, and promoting eco-friendly modes of transportation (Yadav et al. 2022a, b).

16.5 Challenges and Barriers

Even if they are better for the environment, electric cars still have a number of issues that prevent their wide use. Three main obstacles still face consumers: expensive upfront prices, inadequate charging infrastructure, and range anxiety. Lithium-ion battery production and disposal also raise issues with resource depletion and environmental sustainability. Governments, corporations, and research organizations must collaborate to provide charging infrastructure, advance battery technology, and promote the usage of electric vehicles in order to solve these challenges.

Despite the environmental and technological advantages of electric vehicles, there are several barriers impeding their broad implementation. Consumers fear running out of battery power or experiencing range anxiety more than anything else, especially in places with inadequate charging infrastructure. Furthermore, even though an electric car may have a lower lifetime cost of ownership due to reduced maintenance and fuel costs, its initial cost is still higher than that of a conventional vehicle. A complex strategy that includes infrastructure investment, consumer education, legislative support, and technological innovation is needed to address these issues. Governments, industry players, and academic institutions need to work together to accelerate the shift to electric vehicles and create an environment in which everyone can afford, use, and access electric vehicles.

16.6 Technological Innovations

The advancement of electric vehicles, their designs, and the infrastructure necessary for charging them are what are propelling this trend. Innovations such as wireless charging, rapid charging, and solid-state batteries have the potential to raise the effectiveness, cost, and convenience of electric cars. Furthermore, the future of mobility is changing due to advances in autonomous driving and vehicle-to-grid connectivity, bringing in a period of connected and sustainable transportation systems.

Due to their high energy density and quick recharge times, lithium-ion batteries are now frequently seen in electric cars. However, there is hope for more cost savings and performance enhancements through research into other battery chemistries, including lithium sulfur and solid-state batteries. In an effort to improve aerodynamics, efficiency, and safety, automakers are also investigating cutting-edge materials and designs. Furthermore, as wireless charging networks, vehicle-to-grid integration, and fast charging networks advance, the functionality and convenience increase, making it more feasible and accessible for users.

Battery-electric vehicles (BEVs) saw a resurgence in popularity in the United States in the 1960s and 1970s, primarily as a result of the detrimental effects of air pollution and rising oil prices, following their early debut and collapse in the late nineteenth and early twentieth centuries. Less than 4000 BEVs had been sold globally by the end of the 1970s. Following a period of low activity, public interest in BEVs increased once more in the latter part of the 1980s and the early 1990s, giving environmentalists fresh optimism that BEVs would eventually reach the mainstream market. This was mostly brought about by California's recent regulatory push and, to a lesser extent, by environmental laws and initiatives supported in Europe.

Plug-in hybrid electric automobiles (PHEVs), battery-electric vehicles (BEVs), hybrid electric vehicles (HEVs), and extended-range battery-electric vehicles (E-REVs) are some of the current models featuring a variety of unique technology. Despite not having gasoline engines, these cars offer longer electric driving ranges than plug-in hybrid electric cars (PHEVs) and emit no pollutants from their exhaust. An electric motor that is stored in batteries powers the internal combustion engine of a plug-in hybrid electric vehicle (PHEV). The engine can run on conventional or alternative fuels. To charge the batteries in these cars, all they have to do is plug them into a power outlet. Certain car types have a range of more than 70 miles on just gasoline or electricity. However, the main components of a hybrid electric vehicle (HEV) are an electric motor that runs on battery power and an internal combustion engine that can run on conventional or alternative fuel. There is no requirement to plug it in to a power source because the internal combustion engine and regenerative braking charge the battery (Biresselioglu et al. 2018).

Commercial vehicles, both light and heavy duty are significant sources of air pollution, noise, traffic, and CO_2 emissions. Although they make up only 13% of all vehicles in Europe, these cars are responsible for about a third of all CO_2 emissions from road transportation (EEA 2016). Therefore, switching to electric mobility for freight transportation offers a great opportunity to reduce its influence on the environment. In addition, because of its high rate of vehicle purchases, extensive use, and awareness of lifetime vehicle expenses, the freight transportation sector is probably going to be an early user of electric mobility (Sierzchula et al. 2014; Klauenberg et al. 2016). In other words, businesses may understand that the cheap operating expenses more than compensate for the large purchase price (Sierzchula et al. 2014; Wikström et al. 2015). Organizations also strive to reduce their negative effects on the environment and improve public perception (Sierzchula et al. 2014). Conversely, in the EU-28, passenger automobiles account for 83.4% of inland passenger transport and nearly two thirds of total road traffic (EEA 2016).

16.7 Policy and Regulation

Government rules and policies are very important in determining how electric vehicles develop. Consumers are urged to switch to electric automobiles by incentives, including tax credits, rebates, and subsidies, which also drive investment in sustainable energy infrastructure. Automobile manufacturers are also compelled to expedite the shift toward electric mobility due to regulations requiring zero-emission vehicle sales and emission targets. The long-term viability of electric vehicles is supported by policies that promote the use of renewable energy standards and energy efficiency measures. Policymakers can help move to a sustainable transportation environment by implementing supportive laws into place and encouraging cooperation between the public and commercial sectors.

16.7.1 Socioeconomic Impact of Electric Transportation

The switch to electric vehicles will have a significant socioeconomic impact, changing sectors, generating new employment opportunities, and improving energy security. Transportation electrification improves energy independence and reduces dependence on foreign oil, thus reducing geopolitical risks. Furthermore, the move to electric vehicles encourages entrepreneurship and innovation, which stimulates economic growth and competitiveness. There will be more opportunities for employment in a range of industries as the demand for specialized labor in the manufacturing, maintenance, and charging infrastructure of electric vehicles grows. Furthermore, since electric vehicles require less fuel and maintenance than conventional vehicles, the electrification of transportation might result in financial savings for both businesses and consumers. Generally speaking, the adoption of electric vehicles has the potential to spur a resilient, equitable, and prosperous economic change.

The concept of sociotechnical transition plays a crucial role in elucidating the relationship between radical innovations and system change and dynamic stability and gradual change. It uses terms like "niches," which are safe havens where potentially revolutionary ideas can surface, and "sociotechnical landscape," which refers to outside advancements that have the ability to put pressure on established systems, or more accurately, "regimes," to describe change. The idea of a sociotechnical regime is crucial in explaining stability and how car mobility is restricted to internal combustion engines because society has adjusted to their use in terms of expected power and speed, information and maintenance networks, regulations (such as those pertaining to safety and maximum speed), cultural acceptance, etc. Because it emphasizes coevolution, circular causality, hype-disappointment cycles, hype-disappointment cycles, and mutually reinforcing advancements and (often unexpected) alignments, the transition approach departs from simple drivers and linear cause-and-effect linkages.

Prior studies have mostly concentrated on the individual level, with little attention paid to the formal and collective levels, respectively. As a result, more elements were examined individually than on a group one. As a result, the factors influencing decision-making at the individual level are better understood. Here, socioeconomic and demographic factors, as well as customer attitudes and views regarding electric vehicles, are the main subjects of research. But according to Anable (2005), it is also important to pay attention to a person's drive, adaptability, and capacity for change, in addition to factors that could influence preferences and decisions. This is particularly crucial as using an electric vehicle forces drivers to adjust their routines and way of life and to schedule their trips in advance. As a result, it is necessary to broaden the scope of the research to look into the variables influencing a person's willingness and capacity for change. At the formal social unit level, purchasing decisions are influenced by a number of critical elements, including the economic performance of the electric vehicle, its operational qualities like range and charging time, and the charging infrastructure. Regulations, policy practices, incentives, and environmental variables (such as noise, CO_2 emissions) appear to be significant influencing elements in addition to these characteristics. However, despite the current extensive focus of study on the formal social unit, it is notable that important issues such as the availability and security of raw materials have not been considered as factors that influence decisions. Therefore, security and raw resource availability must be taken into account while making decisions about EVs at this level. Finally, but just as importantly, attitudes regarding electric vehicles appear to have a big impact on group decision-making. At this level, the characteristics of electric vehicle use and charging infrastructure and solutions are deemed pertinent. It is interesting to observe that the collaborative decision-making level has gotten a lot less research attention than other levels. Because there is a dearth of knowledge regarding significant elements that could affect decision-making at this level, it would be fascinating to investigate how group decision-making effects intentions to acquire electric vehicles. In particular, a number of limitations and obstacles influence the way choices are made at this level. In the context of EV purchasing desires, it would be fascinating to investigate the degree to which group decision-making facilitates or obstructs EV purchase decisions. Regretfully, there are considerably fewer drivers than obstacles standing in the way of the successful commercialization of electric automobiles. The analysis of motivators and impediments at each of the three formal social unit levels makes this clear. The low rate of electric car adoption and dispersion is indicative of this discrepancy. Given this discrepancy, governments, cities, and automakers need to step up and invest in laws and technologies that make driving electric cars more advantageous and, in turn, more appealing. The actors involved in decision-making are nearly in agreement regarding the obstacles that hinder the industry's growth for electric vehicles, the predominant concern being the absence of adequate infrastructure for charging EVs. To overcome this obstacle, decision-makers and energy suppliers must make sure that enough infrastructure is built for charging, through international cooperation across the European Union, so that a charging grid is available throughout the continent. Most of research found that the main obstacle to EV adoption is their high purchase cost. While these studies concur on this issue, further investigation is required to completely grasp the total cost of ownership, or long-term cost. The argument that the total cost of ownership (TCO) is less than for traditional cars is made by fewer academics (Haddadian et al. 2015; Bubeck et al. 2016). Tax rises on fossil fuels and conventional vehicles themselves could improve the TCO. Da Silva and Moura (2016) claim that this increase is one of the best strategies for spreading the use of electric vehicles. Therefore, future studies are required to examine the total cost of ownership (TCO) of electric vehicles and to give prospective EV users a better grasp of the calculation of the TCO, as there have been few studies that have looked at the TCO and there is still a lack of awareness. Furthermore, the actors in the three decision-making processes pointed out that distrust exists, particularly when it comes to the advantages for the environment. It is crucial to remember that in order to foster a genuine comprehension of benefits, policymakers, energy providers, and automakers should all offer clear, accessible, and intelligible public information. Particular obstacles that impact the formal decision-making stage are raw material and electrical availability. These are important challenges that must be met since they have an impact on the overall viability of electric mobility. To do this, governments and energy suppliers from all over the world should work together to develop policies that ensure a stable supply of electricity and remove any remaining obstacles to the market acceptability of EVs. Official social class, Unit of group decision-making unit on a personal basis, Lack of infrastructure to facilitate charging, Absence of infrastructure for charging, Absence of infrastructure for charging, Financial constraints (such as the cost of buying an EV), Economic constraints (such as the cost of buying an electric vehicle, a limited capacity for loading, and fluctuating oil and energy prices), Financial constraints (such as the expense of buying an electric vehicle or replacing its batteries), Technical limitations (range of electric batteries, battery technology, charging time), Operational limitations (big fleet problems and restricted vehicle availability on the grid), Technological limitations and technological security (such as battery life, driving range, inadequate acceleration, and noise), Lack of faith in the advantages of the environment, Lack of knowledge, information, and trust, Lack of faith in the advantages of the environment, Provision of electricity, Specific factors (e.g., ignorance about the causes and effects of climate change, perception of climate change as a distant threat), Supply of raw materials, Additional considerations/practicability (such as a change in driving style or lifestyle), and Limited supply of models are available. Obstacles that mainly impact the individual level unit include things like the necessity to alter one's lifestyle or climate change, which is a far-off threat. These obstacles can be overcome by educating the public about the ongoing and growing harm posed by CO_2 emissions, as well as by offering clear and reliable information on how each person can help create an environment with less CO_2 emissions. Environmental protection is the primary driver behind the use of electric mobility at all levels of decision-making. It should be noted that most prospective owners of electric vehicles assert that they lack sufficient confidence in the environmental performance of these vehicles. Thus, it is essential to make the environmental advantages of electric vehicles clear and to make this information more widely available. Therefore, policymakers have a duty to disseminate information about the environmental performance of electric vehicles, especially how they help reduce the dependence on fossil fuels and result in significant reductions in energy and CO_2 emissions. Furthermore, financial considerations such as lower fuel prices are consistently cited as motivators. A deeper comprehension of the whole expenses of ownership can be attained by starting with this perception. In other words, even if an electric vehicle costs more to buy than a traditional automobile, its lower running costs, including fuel, have a beneficial impact on the total cost of ownership. Specific motivators such as taxes, rewards, and regulations are what largely influence the formal decision-making level. These drivers may have an impact on the spread of electric mobility as a whole. As a result, encouraging advantageous taxation policies and financial incentives, such as lowering annual and purchase taxes, circulation fees, and company car tax concessions, is becoming increasingly crucial.

At the collective decision-making level, technical criteria and public acceptability emerged as the key motivators due to the ongoing superior technical performance and greater range of electric vehicles, as well as the public's favorable attraction of battery leasing programs. Functional features like the installation of smart charging systems, which enable EVS charging at home, and the vehicle's reduced noise level in comparison to traditional cars were identified as particular motivators at the individual decision-making level. As indicated previously, barriers work better than motivators. As a result, there is comparatively little market acceptance of electric mobility. As a result, as previously explained, there is an urgent need for governments, practitioners, and researchers to work together to remove obstacles and increase motivators. Therefore, it is unquestionably necessary to adopt appropriate policies at the local, regional, and European levels. These include enhancing the grid's infrastructure, better dispersing the infrastructure for charging, enacting subsidies like circulation taxes and registering vehicles with distinct CO_2 emissions, offering incentives such as tax breaks and exemptions, funding R&D, providing funding for in-depth scholarly research, and continuously and regularly updating the public with high-quality information (Biresselioglu et al. 2018).

16.8 Future of Electric Transportation

Electric mobility has an impact on economic growth, energy security, and urban resilience in addition to environmental sustainability as it develops. Opportunities to reduce transportation-related emissions and increase urban responsibility are presented by electrification of delivery fleets, public transportation, and ride-sharing services. The combination of energy storage technologies with renewable energy sources creates a more resilient and decentralized energy system and facilitates collaboration between the transportation and power sectors. Although there will be difficulties and unknowns, electric transportation has a promising future. The performance, accessibility, and affordability of electric vehicles will continue to increase, thanks to technological developments, increasing their competitiveness with conventional automobiles. Urban liability will be enhanced and transportation-related emissions will be further reduced by using electricity in ride-sharing services, delivery vehicles, and public transportation. In addition, synergies between the energy and transportation sectors will be made possible by integrating electric vehicles with renewable energy sources and smart grid technology, opening possibilities for demand response, energy storage, and system optimization. But governments, businesses, and communities everywhere must continue to work together if electric transportation is to reach its full potential. Together, we can clear the path for fair, efficient, and clean mobility in the future, leaving a lasting legacy for future generations. Reviewing the progress of BEVs over the past 40 years revealed that they have had periods of exhilaration and disappointment, but they have not been able to expand beyond their narrow market niche. Internal combustion engine vehicles were too powerful for both normal manufacturers' adapted traditional automobiles to be BEVs and newly introduced market entrants' specifically designed BEVs. Most BEV sales have come from unusual markets, such as fleet users committed to environmental causes and demonstration projects, where incentives are provided. The sales of the Toyota Prius, a hybrid electric vehicle (HEV), which were sold in excess of three million units worldwide between December 1997 and March 2011 (Toyota Motor Corporation 2011), stand in stark contrast to the underwhelming experiences with BEVs (Dijk et al. 2013).

16.9 Global Data on Electric Vehicles

Using electric vehicles is a major step in decarbonizing transportation. Compared to gasoline or diesel vehicles, electric cars usually have a lesser carbon footprint over the course of their lifetime. Despite having a larger carbon footprint during production, a vehicle's "carbon debt" usually disappears quickly after it is put to use. Cleaner energy-mix countries save more carbon, and this trend will continue as long as countries continue to decarbonize their electrical networks (Ritchie 2024).

16.9.1 Sales of New Cars

Electric vehicle sales began slowly but are now rapidly increasing in several areas. Approximately one in every four new cars sold globally in 2023 was electric. In China, the percentage was over 90%, while in Norway, it was about 40%. We can examine these global patterns in the chart that follows (Figs. 16.5 and 16.6). Plug-in hybrids and fully battery-electric cars are included in this definition of "electric cars."

16.9.2 Fully Electric Cars vs. Plug-in Hybrids

Batteries-powered vehicles and plug-in hybrids are both categorized as "electric cars." The difference is that whereas fully battery-electric vehicles lack an internal combustion engine, plug-in hybrids have an electric motor, rechargeable battery, and an external combustion engine that runs on gasoline. This implies that if the owner neglected to charge the battery, a plug-in hybrid may be operated like a regular gasoline vehicle. When driving longer distances, plug-in hybrids' smaller and shorter-range batteries cause the vehicle to switch to gasoline operation once the battery runs out. Plug-in hybrids typically emit more carbon dioxide than battery-electric vehicles since they frequently run on gasoline. They typically emit fewer pollution than cars that run on gasoline or diesel though sales of electric cars are split down by these two technologies in the first graphic (Fig. 16.7). This is expressed as a percentage of annual sales of new autos. The percentage of newly sold electric vehicles that run exclusively on batteries is shown Fig. 16.8.

Fig. 16.5 Share of new cars sold that are electric 2010 to 2023 (Ritchie 2024)

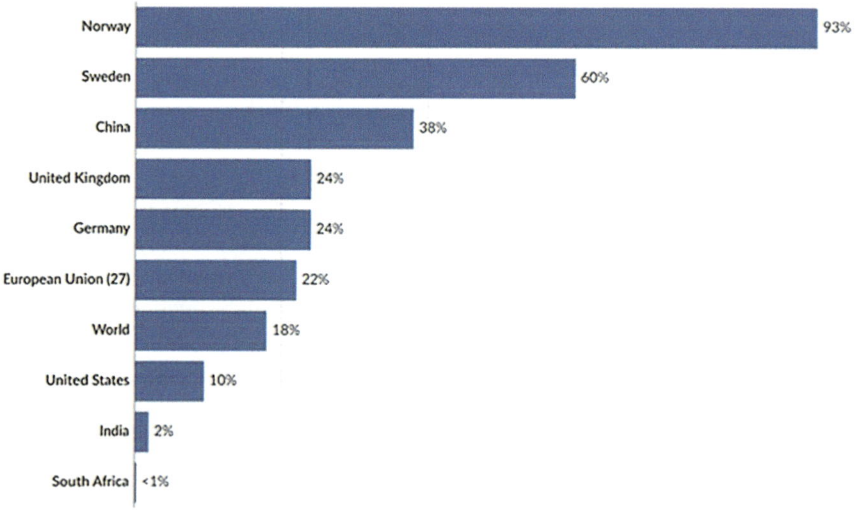

Fig. 16.6 Share of new cars sold that are electric 2023 (Ritchie 2024)

16.9.3 New Electric Cars are Sold Each Year

The total number of new electric automobiles sold is depicted in the graph below. Once more, plug-in hybrids and fully battery-electric vehicles are included (Fig. 16.9).

With information from the IEA on sales of electric vehicles and EVs as a percentage of all new cars, we can calculate the exact number of new cars of each type

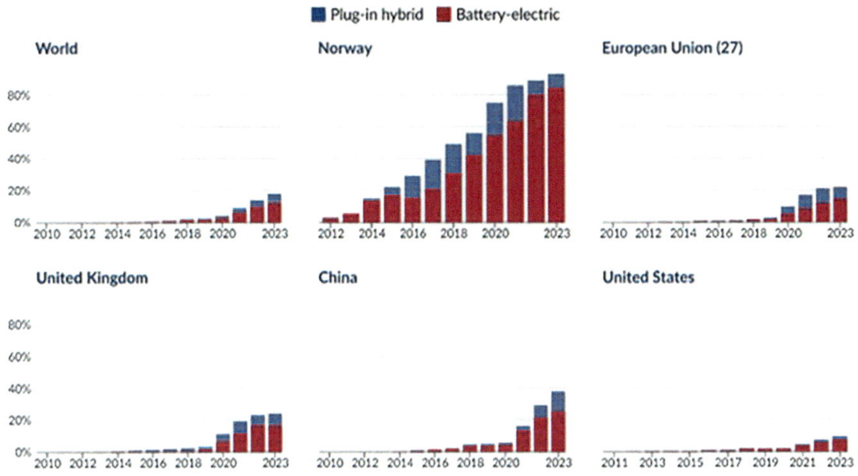

Fig. 16.7 Share of new cars sold that are battery-electric and plug-in hybrid, 2010–2023 (Ritchie 2024)

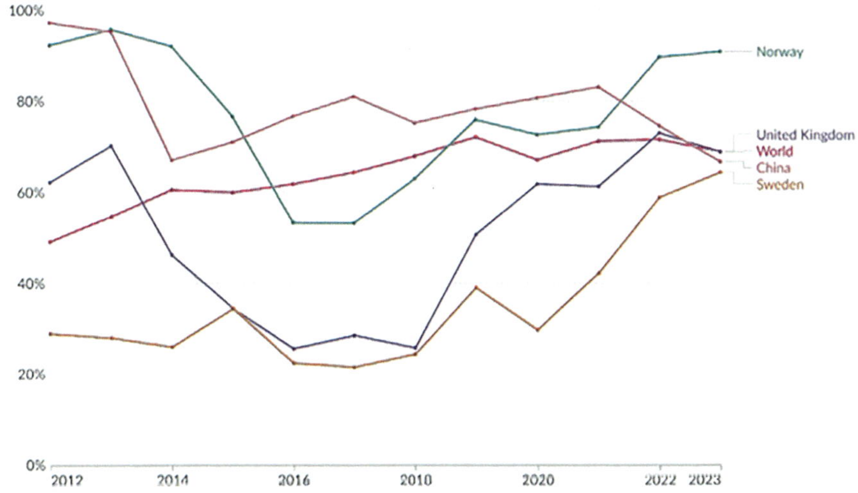

Fig. 16.8 Share of new electric cars that are fully battery-electric 2012–2023 (Ritchie 2024)

sold each year. According to these figures, 2018 was the biggest year ever for sales of nonelectric cars worldwide. This matches previous public estimates; Bloomberg New Energy Finance, for example, said that sales reached their peak in 2017 (Fig. 16.10).

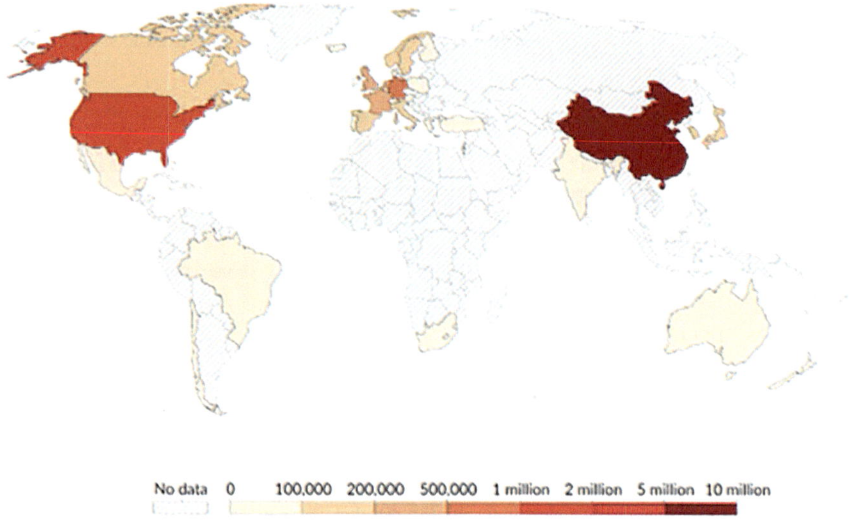

Fig. 16.9 Number of new cars sold in 2023 (Ritchie 2024)

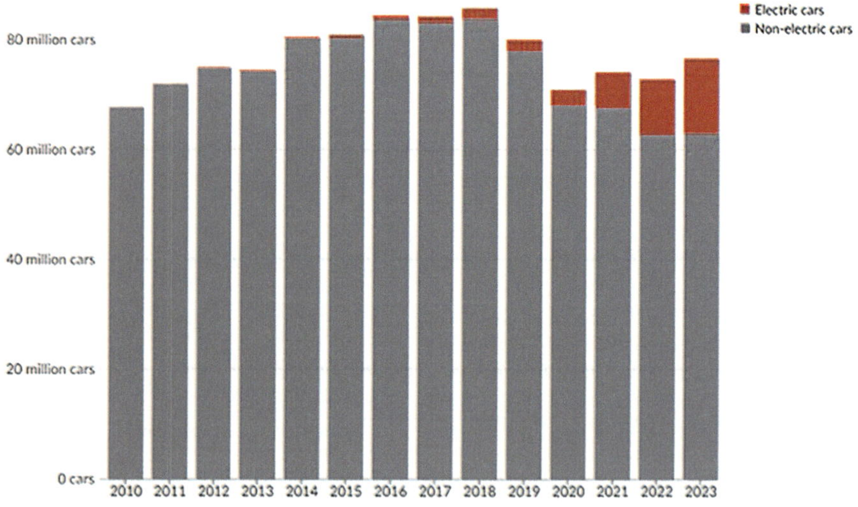

Fig. 16.10 Number of new cars sold, by type, world (Ritchie 2024)

16.9.4 Stocks of Cars in Use

The vehicles now in use are a representation of sales statistics spanning more than 10 years. This can be represented by the "stocks" metric. It is an estimate of the total number of automobiles in use and shows the remaining balance between the cars

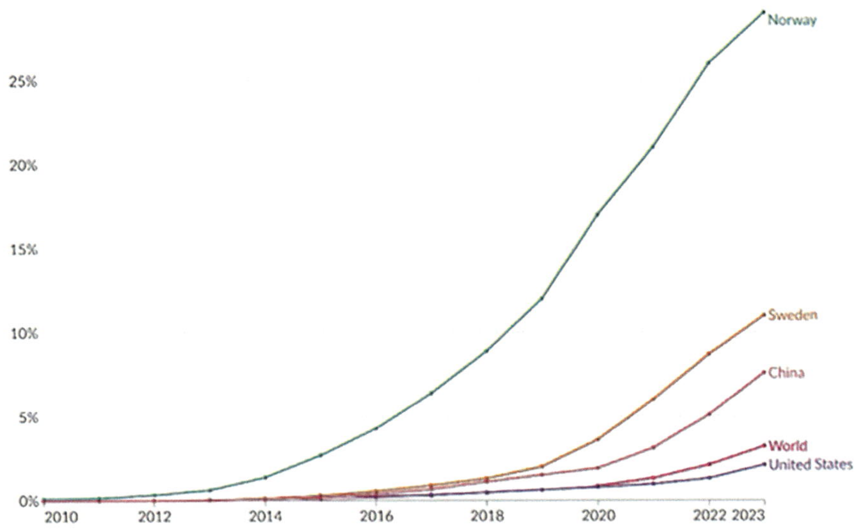

Fig. 16.11 Share of electric cars currently in use during 2010–2023 (Ritchie 2024)

that are being added and the cars that are being retired. People use their cars for extended periods of time, thus it takes time for fresh sales to noticeably impact auto stocks. Thus, compared to the share of new sales, the percentage of electric vehicles on the road is substantially lower. The chart below displays the percentage of electric vehicles in use (Fig. 16.11).

16.9.5 Electric Cars are on the Road

The number of electric vehicles in use at any one time is the total number of vehicles sold over the course of a year, less any vehicles that have been removed from service. Below is a list of all the stocks of electric vehicles. Over 40 million electric vehicles are currently in use globally, and this number is rising swiftly. This number was only 26 million in 2022 (Fig. 16.12).

16.10 Conclusion

Electric transportation is a sustainable substitute for traditional fossil fuel-powered cars, marking a paradigm shift in mobility. Electric vehicles are leading the way in environmentally friendly mobility, thanks to their zero tailpipe emissions, minimal environmental impact, and innovative technology. However, achieving the full potential of electric transportation requires concerted efforts to overcome

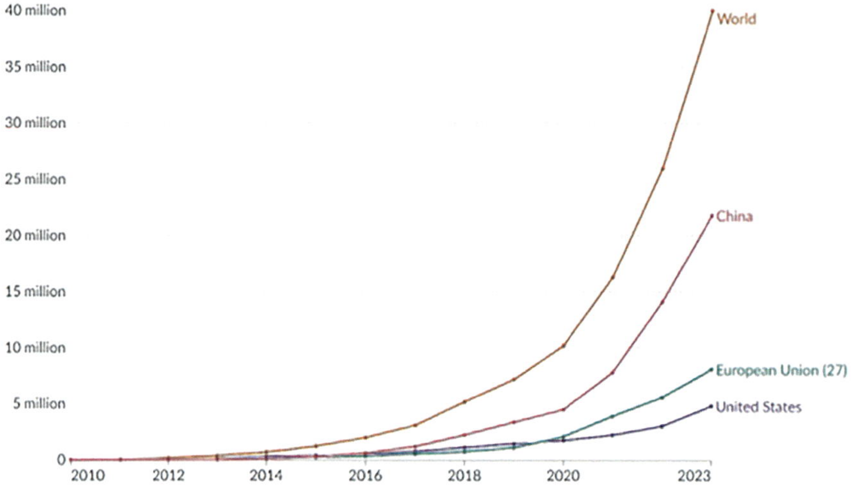

Fig. 16.12 Electric car stocks during 2010–2023 (Ritchie 2024)

challenges, promote policy support, and foster technological innovation. By embracing electric mobility, we can accelerate the transition toward a cleaner, greener future, where transportation is not only sustainable, but also equitable and accessible to all. As we embark on this journey, let us take advantage of the opportunities presented by electric transportation to create a better world for ourselves and future generations.

Acknowledgment I want to sincerely thank everyone who helped to complete this book chapter on electrical vehicles and for their support. My sincere thanks go to my mentor and advisor, Dr. Akhilesh Kumar Yadav, whose guidance, insight, and encouragement have been invaluable throughout. Your expertise and constructive feedback have greatly enriched the content of this chapter. Lastly, I would want to express my sincere gratitude to my family and friends for their unwavering support and tolerance during the lengthy hours I spent writing and conducting research. Your encouragement and strength have been a source of strength.

References

Anable J and Gatersleben B (2005) All work and no play? The role of instrumental and affective factors in work and leisure journeys by different travel modes. Transportation Research Part A: Policy and Practice 39(2-3 SPEC. ISS.):163–181

Al-Alawi BM, Bradley TH (2013) Review of hybrid, plug-in hybrid, and electric vehicle market modeling studies. Renew Sust Energ Rev 21:190

Altenburg T, Schamp EW, Chaudhary A (2016) The emergence of electromobility: comparing technological pathways in France, Germany, China and India. Sci Public Policy 43(4):464–475

Biresselioglu ME, Demirbag Kaplan M, Yilmaz BK (2018) Electric mobility in Europe: a comprehensive review of motivators and barriers in decision making processes. Transp Res Part A Policy Pract 109(October 2017):1–13

Brückmann G and Bernauer T (2020) What drives public support for policies to enhance electric vehicle adoption? Environmental Research Letters 15(9)

Bubeck S, Tomaschek J and Fahl U (2016) Perspectives of electric mobility: Total cost of ownership of electric vehicles in Germany. Transport Policy 50:63–77.

Da Silva ARCB, De Moura FABF and Dias WS (2016) Wave packet dynamics under effect of a pulsed electric field. Solid State Communications 236:12–16

Dijk M, Orsato RJ, Kemp R (2013) The emergence of an electric mobility trajectory. Energy Policy 52:135–145

Durate G, Silva A, Baptista P (2021) Assessment of wireless charging impacts based on Rea-world driving patterns: case study in Lisbon, Portugal. Sustain Cities Soc 71:102952

EEA (2016) Regulations 33 and 41 of the Immigration (European Economic Area) Regulations 2016.

Farinloye T, Oluwatobi O, Ugboma O et al (2024) Driving the electric vehicle agenda in Nigeria: The challenges, prospects and opportunities. Transp Res Part D Trans Environ 130(September 2023):104182

Fevang E, Figenbaum E, Fridstrøm L et al (2021) Who goes electric? The anatomy of electric car ownership in Norway. Transp Res Part D: Transp Environ 92(February):1–20

Greene DL, Kontou E, Borlaug B et al (2020) Public charging infrastructure for plug-in electric vehicles: what is it worth? Transp Res Part D: Transp Environ 78:102182

Haddadian G, Khodayar M and Shahidehpour M (2015) Accelerating the Global Adoption of electric vehicles: Barriers and drivers. Electricity Journal 28(10):53–68. Elsevier Inc.

Haghani M, Coughlan M, Crabb B, et al. (2023) A roadmap for the future of crowd safety research and practice: Introducing the Swiss Cheese Model of Crowd Safety and the imperative of a Vision Zero target. Safety Science 168:106292. Elsevier Ltd.

Henderson A, Cao M, Liu Q (2022) Access-based consumption, behaviour change and future mobility: insights from visions of car sharing in Greater London. Future Transp 2(1):216–236

Iwan S, Nürnberg M, Jedliński M et al (2021) Efficiency of light electric vehicles in last mile deliveries – Szczecin case study. Sustain Cities Soc 74(July):103167

Jaiswal A, Arun CJ, Varma A (2022) Rebooting employees: upskilling for artificial intelligence in multinational corporations. The International Journal of Human Resource Management 33(6):1179–1208. https://doi.org/10.1080/09585192.2021.1891114

Kar SK, Bansal R, Harichandan S (2022) An empirical study on intention to use hydrogen fuel cell vehicles in India. Int J Hydrog Energy 47(46):19999–20015

Klauenberg J, Rudolph C and Zajicek J (2016) Potential Users of Electric Mobility in Commercial Transport-Identification and Recommendations. Transportation Research Procedia 16:202–216

Li L, Wang Z, Chen L et al (2020) Consumer preferences for battery electric vehicles: a choice experimental survey in China. Transp Res Part D Transp Environ 7(8):102185

Liao F, Correia G (2022) Electric carsharing and micromobility: a literature review on their usage pattern, demand, and potential impacts. Int J Sustain Transp 16(3):269–286

Lieven T, Mühlmeier S, Henkel S et al (2011) Who will buy electric cars? An empirical study in Germany. Transp Res Part D: Transp Environ 16:236

Patil PV, Khan MK, Korulla M, et al. (2022) Design optimization of an AUV for performing depth control maneuver. Ocean Engineering 266(P5):112929.

Patil P, Kazemzadeh K, Bansal P (2023) Integration of charging behavior into infrastructure planning and management of electric vehicles: a systematic review and framework. Sustain Cities Soc 88(June):104265

Rajper SZ, Albrecht J (2020) Prospects of electric vehicles in the developing countries: a literature review. Sustainability (Switzerland) 12(5):1906

Razi F, Dincer I (2022) A review of the current state, challenges, opportunities and future directions for implementation of sustainable electric vehicle infrastructure in Canada. J Energy Storage 56(PB):106048

Ritchie H (2024) Tracking global data on electric vehicles. Our World in Data. https://ourworldindata.org/electric-car-sales
Schulz F, Rode J (2022) Public charging infrastructure and electric vehicles in Norway. Energy Policy 160(October 2021):112660
Sierzchula W (2014) Factors influencing fleet manager adoption of electric vehicles. Transportation Research Part D: Transport and Environment 31:126–134
Secinaro S, Calandra D, Lanzalonga F et al (2022) Electric vehicles' consumer behaviours: mapping the field and providing a research agenda. J Bus Res 150:399–416
Singh V, Singh V, Vaibhav S (2020) A review and simple meta-analysis of of factors influencing adoption of electric vehicles. Transp Res Part D Transp Environ 86:102436
Tarei PK, Chand P, Gupta H (2021) Barriers to the adoption of electric vehicles: evidence from India. J Clean Prod 291:125847
Toyota Motor Corporation (2012) Annual report 2011. EPPO Bulletin 42(3):606–616
Wikström M, Hansson L, Alvfors P (2015) An end has a start-investigating the usage of electric vehicles in commercial fleets. Energy Procedia 75(0):1932–1937
Xiaoli S, Zhengguo L, Xiaolin W et al (2019) Technology development of electric vehicles: a review. Energies 13(90):1–29
Yadav AK, Sahoo SK, Patra AC et al (2022a) Enrichment factor and chemical composition of size separated airborne particulate matter of Singrauli coalfield, India. Aerosol Sci Eng 6(4):414–436. https://doi.org/10.1007/s41810-022-00155-4
Yadav AK, Shirin S, Emmanouil C, Jamal A (2022b) Effect of seasonal and meteorological variability of air pollution in Singrauli coalfield. Aerosol Sci Eng 6(1):61–70. https://doi.org/10.1007/s41810-021-00124-3
Zhou W, Cleaver CJ, Dunant CF et al (2023) Cost, range anxiety and future electricity supply: A review of how today's technology trends may influence the future uptake of BEVs. Renew Sustain Energy Rev 173(June 2022):113074

Chapter 17
Unlocking Grid Resilience Through Customer-Centric Behavior Analysis with Deep Learning Techniques

V. R. Revathy, M. B. Yashoda, and Hari S. Prabhu

Abstract In the current era of rapidly evolving energy systems and increasing frequency of grid disruptions, grid resilience has become more critical than ever before. This study explores how deep learning methods play a crucial role in examining customer behavior to support grid modernization efforts that enhance resilience in energy systems. Deep learning is valuable for extracting complex patterns from large datasets, providing insights into consumption patterns, preferences, and socio-economic factors. Long Short-Term Memory (LSTM) networks and Convolutional Neural Networks (CNNs) enable accurate demand forecasting, anomaly detection, and personalized service recommendations. Furthermore, deep learning supports the development of predictive models for demand response, load forecasting, and energy efficiency optimization, empowering utilities to improve grid stability and resilience.

Nevertheless, integrating deep learning for customer behavior analysis in grid modernization projects presents challenges such as data privacy concerns, computational complexity, and interpretability issues. Overcoming these challenges requires interdisciplinary collaboration, robust governance frameworks, and efforts to enhance model transparency and accountability. The study highlights the transformative potential of deep learning in understanding consumer behavior dynamics for grid modernization. By leveraging advanced analytics and machine learning techniques, utilities can create a resilient energy future that meets evolving consumer needs and societal trends. The paper advocates for continued research and collaboration to fully utilize deep learning in shaping the energy systems of the future.

V. R. Revathy (✉)
Department of Computer Science (UG), Kristu Jayanti College (Autonomous), Bengaluru, India

M. B. Yashoda
Department of Computer Science and Engineering, Nitte Meenakshi Institute of Technology, Bangalore, India

H. S. Prabhu
FormLabs, Bangalore, India

A. K. Yadav (ed.), *Powering Future: Clean Energy Solutions for a Sustainable Planet*, Clean Energy Production Technologies, https://doi.org/10.1007/978-981-96-5055-2_17

Keywords Deep learning · Energy grid modernization · Energy management strategies · Energy conservation efforts

17.1 Introduction

A fundamental component of the modern energy transformation is the idea of the smart grid. Recent technologies harnessing energy from sustainable resources are opening new opportunities to meet the growing demand, characterized by a significant increase in component diversification tailored for microgrid compatibility (Ghorashi Khalil Abadi et al. 2022; Mutarraf et al. 2022). These systems use tiny power sources like solar panels, batteries, and other generators to handle small electrical demand sections. Many scholars have looked into making these grids better in areas where dispersion is more constrained. In addition to offering financial advantages by decreasing the frequency of supply disruptions and the associated energy costs, these technologies also improve energy conservation and lower outage rates. The Smart Grid offers many opportunities for consumers to save energy and for utilities to operate the grid in a more efficient, effective, and reliable way. The dynamic energy environment and cutting-edge innovations impact consumer behavior in smart grids. By giving users access to contemporaneous information on energy costs and utilization, smart grids enable users to become more aware of and adaptable to usage patterns. Customers can use this insight to optimize their energy usage depending on changing charges or to make informed decisions about when to consume less during peak times. Furthermore, smart grids facilitate the inclusion of electric automobiles and clean energy sources, further influencing consumer behavior in energy efficiency and conservation. Moreover, the transition to automation and intelligent meters improves customer convenience by allowing for remote energy consumption tracking and management. Finally, by encouraging a more active and engaged approach to monitoring energy consumption, these developments in smart grid technologies are changing consumer behavior and paving the way for a safer and cleaner future. Recent strides in deep learning and neural network algorithms have sparked a surge of interest in harnessing these innovations for smart grid stability analysis. Deep learning, a specialized branch of machine learning, excels at uncovering hierarchical data representations using intricate neural networks. Unlike conventional methods, deep learning models autonomously distill complex features from raw data. With their multi-layered neural architectures, these models adeptly capture intricate patterns, nonlinear relationships, and sophisticated abstractions from vast, high-dimensional datasets (Liu and Lang 2019). When it comes to grid failure scenarios, ML approaches are helpful. ML methods additionally offer a perspective on the gathered data. The authors (Tiwari et al. 2022) trained and tested various machine learning algorithms employing the SG dataset publicly available on Kaggle, to forecast grid reliability. Using the classifier assessment measures, the SVM-based model yielded the best results.

This research proposes an approach utilizing deep learning techniques for the prediction and optimization of grid activity to discover an intelligent system that effectively controls the power needs of a grid. They combine the best features of each method, forecast system consumption on a regression problem where these algorithms excel, and optimize potential actions to be taken in the upcoming time frame, ensuring that you meet the intended usage requirements—whether they need to be increased or decreased (Gao et al. 2021; Lin et al. 2022). In particular, we will employ neural networks to forecast energy usage and to carry out the necessary activities to manage an ecosystem of batteries, electric vehicle charging stations, and buildings (Y. Zhang et al. 2023; L. Zhang et al. 2022). The rest of the article is as follows: In Sect. 17.2, the related studies on the usage of ML for smart grid resilience are reviewed. Section 17.3 introduces the architecture of the proposed system. The ML models used for customer behavior analysis are discussed in Sect. 17.4, followed by a conclusion and future enhancement in Sect. 17.5.

17.2 Review

In recent years, Smart Grid (SG) stability assessment using machine learning (ML) and deep learning (DL) algorithms has become a research field. SGs hold great promise as a substitute for traditional electrical networks. A thorough investigation of the use of ML in smart grids was conducted by Hossain et al. (2019). They suggested that integrating an SG system with the Internet of Things may provide approaches for cost-effective estimation of load and data collection. Surplus data are produced by SGs, and ML is the more effective approach for evaluating stability with suitable decision-making than traditional methods. Zheng et al. (2019) also created various machine learning models to investigate the effects of consumer-scheduled loads and everyday energy usage. K-Nearest Neighbor and Gaussian process regression demonstrated consistent results with greater precision among the different machine learning techniques.

The possibilities of applying machine learning (ML) algorithms for feature extraction and large-scale data analysis for the development of contemporary industrial structures, such as smart grids, were covered (Gunel and Ekti 2019). Researchers noted that adding a knowledge layer to the conventional power network for data collecting, storage, and analysis can help address the challenges conventional methodologies face when analyzing massive amounts of data (Kumar and Yadav 2024; Kulwant and Yadav 2024). Utilizing the data collected by the New York ISO, Alquthami et al. (2022) designed a unique enhanced decision tree classifier (EDTC) model for SG stability analysis. An evaluation of EDTC's efficacy to the other machine learning models—SVM, KNN, ANN, linear regression, and DT—was conducted. The results verified that EDTC was better than other methods with a 99.07% success rate. A balanced dataset outperformed an unbalanced dataset in the effectiveness of classifiers for grid stability estimation, as demonstrated by Ibrar et al. (2022). Further, the XGBoost technique beat the other machine learning techniques in the experiment, with an accuracy score of 94.7%; however, when XGBoost and randomly sampled data combined, the level of precision rose to 96.8%. The authors (Candanedo et al. 2017) investigated the development of predictive models for estimating the energy consumption of household appliances. The authors developed predictive models for energy consumption in a low-energy house and their work highlighted the potential for such models to enhance energy efficiency and conservation efforts in residential environments (Candanedo et al. 2017). Also their work was primarily centered on the development of predictive models and does not delve into the application of these models within a smart grid context. By exploring the recent works of literature, we have concluded that there is a gap that needs to connect smart grid technology and customer behavior analysis through the application of AI models in a more efficient way. In the present work, we have used state-of-the-art machine learning algorithms to explore the need for smart grid technologies and predictive analytics to understand customer behavior patterns, explore the possibilities to enhance the data collection, identification of the irregularities in the network, and so on.

17.3 Methodology

Through analyzing customer behavior patterns, smart grids can incorporate smart technologies in traditional power grids that upgrade and enhance the power distribution by monitoring, optimizing, and managing it more efficiently. For example, the customer usage pattern based on time, day of the week; peak demand for power; anticipation of the high consumption, and many more. While the study of Candanedo et al. (2017) does not explicitly involve smart grid integration. In this work, we have presented the findings and developed models that have potential applications within a smart grid framework. The accurate predictive modeling using the deep learning algorithms could enhance smart grid operations, as previously discussed, by providing valuable insights for real-time energy management and optimization.

17.3.1 System Architecture

The generic proposed architecture used for this research work is shown in Fig. 17.1. It illustrates a machine learning pipeline that generates outcomes as predictions based on an input dataset. The ML pipeline process used in the present work begins with the identification of a suitable dataset. The raw data in the dataset contains various features required to build ML models. The dataset that is used for this research work is adapted from the work of Candanedo et al. (2017) and the authors have collected the data using a wireless sensor network that stores data from a low-energy house. This raw dataset undergoes a feature engineering phase that involves transforming or extracting the necessary features that enhance the models' performance

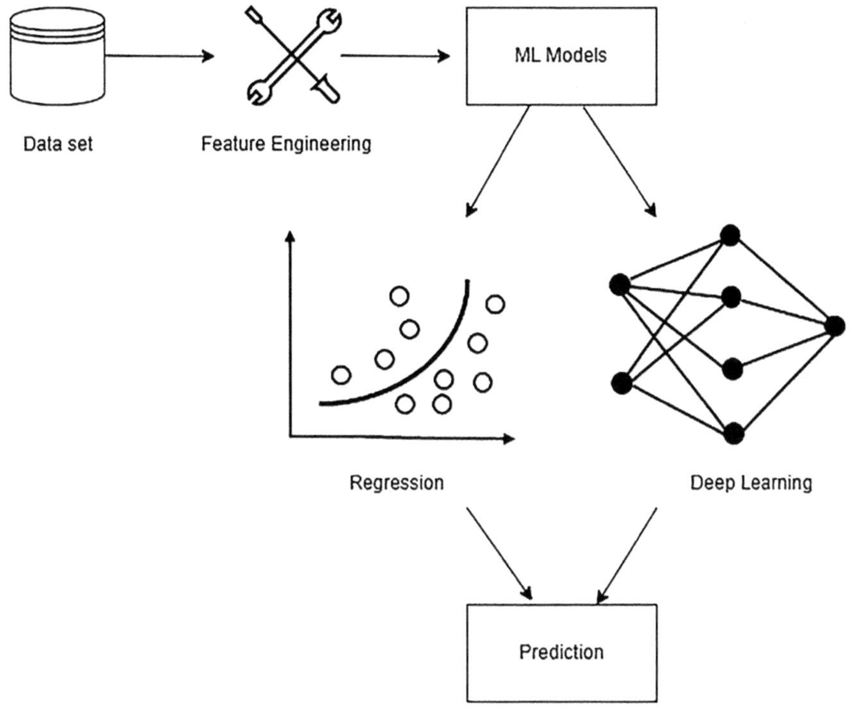

Fig. 17.1 System architecture

and can help in better predictive results. Regression and deep learning models were built to predict the total energy consumption in the house.

17.3.2 Dataset

The dataset contains detailed information on energy usage and environmental conditions, recorded with high temporal resolution in a low-energy house situated in Belgium (Candanedo et al. 2017). It includes features such as appliance energy consumption (in Wh), light fixture energy usage (in Wh), and temperatures (in Celsius) and humidities (%) across various indoor and outdoor locations: kitchen, living room, laundry room, office, bathroom, outside (north side of the building), ironing room, teenager room 2, and parents' room. Furthermore, the Chievres weather station provides external meteorological data like outside temperature (in Celsius), pressure (in mmHg), humidity (in %), wind speed (in m/s), visibility (in km), and dew point (in °C). This comprehensive dataset, spanning hourly measurements, incorporates interpolated data from the nearest airport weather station over 4.5 months, obtained with permission from Reliable Prognosis.

17.3.3 Data Visualization: Exploratory Data Analysis (EDA)

The Exploratory Data Analysis conducted for the present dataset helps to identify the various trends and patterns in energy consumption in the various rooms in the low-energy house.

17.3.3.1 Feature Engineering

In order to do better predictive modeling, we have extracted a set of new features to identify the time-specific, interaction based, and short-term statistics of energy consumption. For time-based analysis, we have extracted information such as hour, day of the week, and month from the date column. For interaction-based analysis, this work considered the temperature and humidity readings to capture their combined effect.

Visualizing Per Day Consumption

The Heatmap in Fig. 17.2 shows the hourly consumption of day power consumption of the appliances that are installed in various rooms explained in the dataset section. Considering all 7 days a week, the energy usage is less from 1:00 am to 6:00 am. It increases gradually from 7:00 am to 5:00 pm, and the usage is maximum at 6:00 pm to 7:00 pm daily. The analysis shows that the usage of appliances is maximum on Fridays, Saturdays, and Sundays of the week.

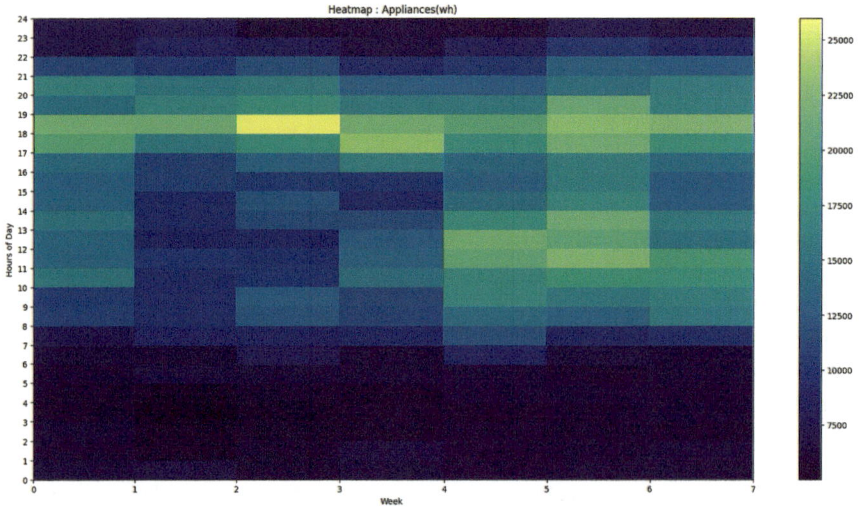

Fig. 17.2 Heatmap for total appliances energy consumption vs days of the week

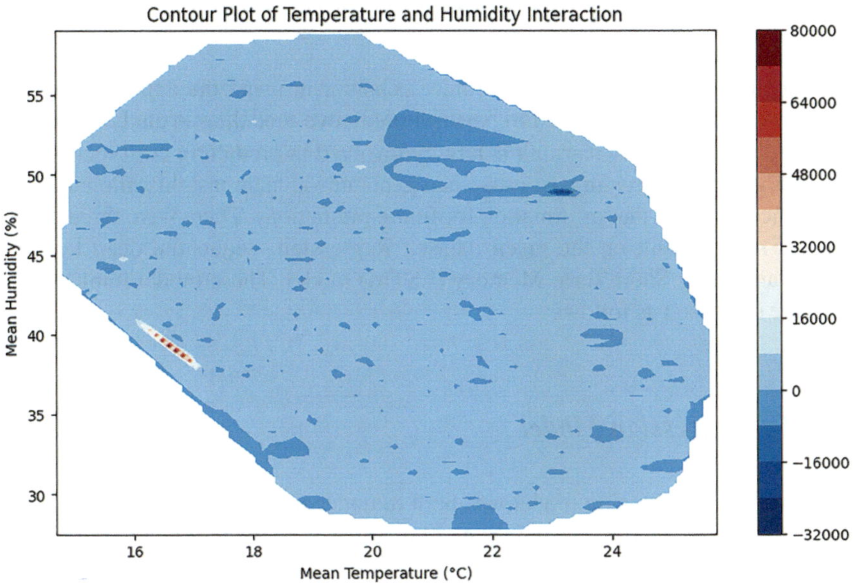

Fig. 17.3 Temperature and humidity interaction

Visualizing Interaction Features

Figure 17.3 illustrates a contour plot of temperature and humidity interaction on energy consumption. From a view of color interpretation, the color bar placed on the right side ranges from blue which represents lower values and migrates towards red (higher values). In the plot, it is visible that the color shades of blue appear the most. This simply indicates lower energy consumption. There is a very small region in the plot with red and orange shades that points to the area of higher energy consumption. This area maintains 16–18 degrees as mean temperature and 40–45% mean humidity. The implication of the above findings is that within the given range of humidity and temperature, there shows significantly higher energy consumption for appliances.

Since most of the portion of the graph is uniform light blue, the environmental factors combination that is humidity-temperature combination does not drastically affect the energy consumption except for the identified hotspot in red. The high consumption spot in the plot which is a relatively small section indicates specific conditions under which energy consumption spikes. This can infer overuse of heating or cooling appliances to keep comfortable indoor conditions at the specific temperature-humidity range.

17.4 Machine Learning Models

By visualizing and anticipating customer behavior in using the appliances of the present, we have a further need to perform predictive modeling to analyze the data more in detail. Machine learning is largely involved in predicting customer behavior. Analysis of patterns in past usage can generate valuable insights into customer preferences and use that information for future predictions. There were three different ML models built on the given dataset: Regression, Sequential deep learning model, and Long-Short Term Memory (LSTM) model. These model implementations are discussed as follows:

17.4.1 Regression Model

A simple linear regression model was built to understand and analyze the relationship between the energy consumption column that is appliances and various other columns such as temperature, humidity, and time-based features. As most of the features in the dataset hold continuous values and exhibit linear relationships, linear regression is a suitable choice for capturing these patterns. A simple interpretation of coefficients shows how each independent feature contributes to predicting energy usage. The Actual vs. Predicted Appliance Energy Consumption scatter plot shown in Fig. 17.4 shows the performance of the linear regression model. Each point on the

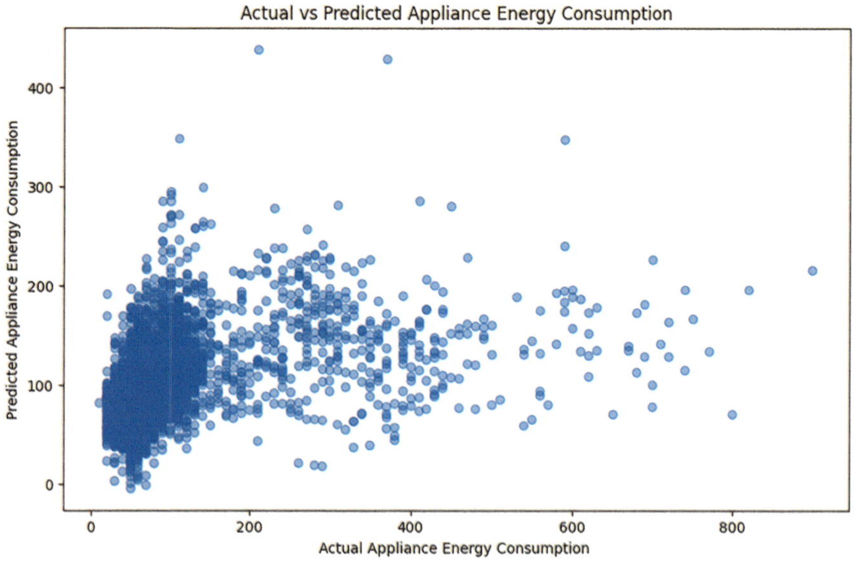

Fig. 17.4 Actual and predicted energy consumption using a simple linear regression model

plot represents a single observation from the test dataset, where the x-axis shows the actual energy consumption and the y-axis shows the predicted energy consumption by the model. The plots show the data points near the range that corresponds to the lower range of energy consumption. The scatter points are clustered mostly towards the lower left area of the plot. There is a significant scattering of points away from the 45-degree diagonal line, predicting higher consumption, and indicating inaccurate results of energy usage. Model accuracy is declining when the actual energy consumption is increasing. The model was evaluated using the following evaluation metrics Mean Absolute Error (MAE), Mean Squared Error (MSE), and R^2 score, and their respective values are 52.55, 8312.85, and 0.169.

17.4.2 Deep Learning (DL) Models

In the realm of smart homes, the quality of human behavior detection has been further enhanced, making the recognition process more intelligent. Artificial intelligence and deep learning may also increase sensing accuracy and find possible patterns in data.

The deep learning models are found to be suitable for predictive modeling in energy consumption because of their ability to deal with complex data, datasets with nonlinear relationships, and also the huge volumes of energy consumption data which get generated tremendously using IoT devices, wireless sensor networks, etc. Deep learning models which are discussed in the literature survey have predicted various energy demands by considering various factors that influence consumption like technology and socio-economic factors. The predictive capability of the deep learning model is vital for analyzing the context of the present dataset of this work.

Also, DL models can forecast customer consumption patterns ensuring reliability which we consider as crucial and have the capability to enhance energy usage and distribution to build efficient smart grid systems. Through developing such smart grid systems there exists the wide possibility of integration of renewable energy sources that leads to sustainable energy consumption and overall reduction of energy consumption for our living styles. The two different DL models used in this work are the sequential deep learning model and the Long-Short-Term Memory (LSTM) model.

17.4.2.1 Sequential Deep Learning Regression Model

The implementation of the DL regression model was done using Google Colab with Python—TensorFlow and Keras libraries. The model forecasts the energy consumption of the appliances considering the internal room and environmental conditions. All predictor features except the date column were selected as input features for the DL model. The target column for the model is appliances. The input features were standardized and this is to ensure that each input feature can contribute evenly during the training process.

A proportion of data was taken to perform training and testing. The DL model was constructed using the Sequential Keras library and is formed of multiple dense (fully connected) layers. The energy prediction DL architecture has three hidden layers, which get activated using ReLU activation functions. The dropout layers and early stopping prevent overfitting and the model is compiled with the Adam optimizer. The model learns the training data of energy consumption with a validation split, and the loss and validation loss over epochs are noted. The performance is evaluated using mean squared error (MSE), mean absolute error (MAE), and R-squared (R^2) metrics and their respective results are 7272.65, 46.80, and 0.27. These results show the DL model's accuracy and predictions are comparatively higher than the results of the simple linear regression model.

The visualization of the training and validation loss over epochs is shown in Fig. 17.5 which shows the model's learning curve and a scatter plot (Fig. 17.6) of predicted versus actual appliance energy consumption visually evaluates the model's performance. In Fig. 17.5, the blue line (training loss) and the orange line (validation loss) decrease over the epochs. This implies that the model is learning the energy consumption data and is improving the predictions on the validation dataset. Initially, the training loss shows sudden spikes than the validation but rapidly decreases. The inference from the visualization is that the DL model fits suitably over the train data. Overall, the model showed good learning behavior with no evidence of overfitting or underfitting, and it shows to be well-tuned for the energy consumption dataset.

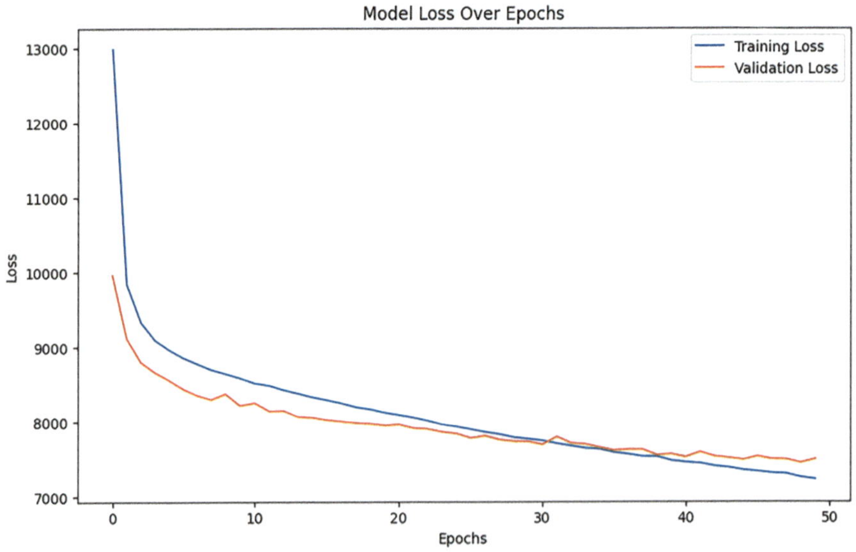

Fig. 17.5 Analysis of validation loss over epochs

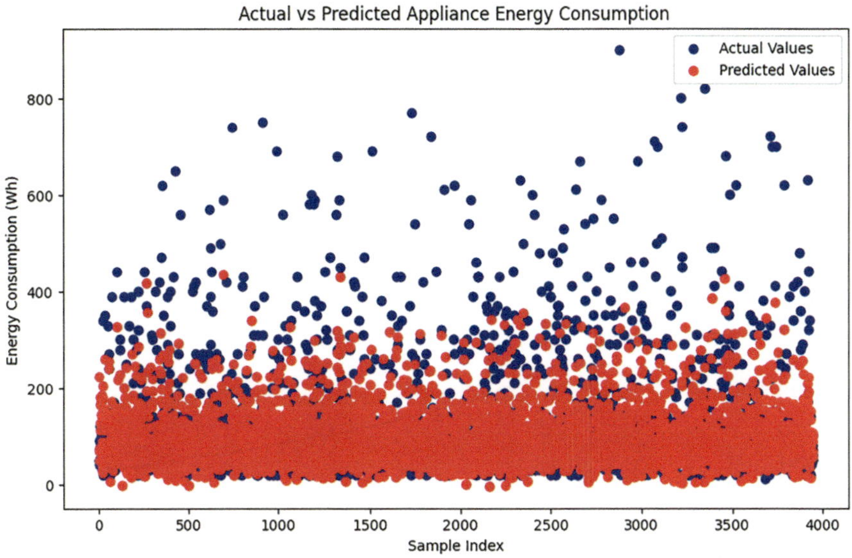

Fig. 17.6 Actual and predicted appliance energy consumption using sequential DL model

17.4.2.2 Long Short-Term Memory (LSTM)

Long Short-Term Memory (LSTM) network, a type of recurrent neural network (RNN) is used to predict the energy consumption of appliances based on various conditions discussed in the previous section. The dataset is preprocessing, feature engineering is already done for the sequential model, and the LSTM model is built using the Sequential API from TensorFlow and Keras. The proposed architecture consists of two LSTM layers, dropout layers to prevent overfitting, a dense layer with ReLU activation, and a final output layer that predicts energy consumption. The LSTM model is compiled using the Adam optimizer, which adapts the learning rate dynamically, and the mean squared error (MSE) as the loss function. The MSE, MAE and R^2 metrics evaluation for the LSTM model are 6471.64, 40.714, and 0.35, respectively.

The graph (Fig. 17.7) illustrates the performance of the LSTM model over 50 epochs. The lines representing both the training and validation loss are declining over time indicating a good learning process on the train-validation data. Towards the end of 50 epochs, the training loss continues to decrease slightly while the validation loss level is steady. This indicates the model is giving its optimal performance and the minimum divergence between training and validation loss implies the present model is showing a significant generalization. Thus, no clear sign of overfitting or underfitting is present in the graph and the model fits similarly over train-validation data splits.

The scatter plot in Fig. 17.8 depicts the actual predicted appliance energy consumption given by the LSTM model. The x-axis plots the sample index of values

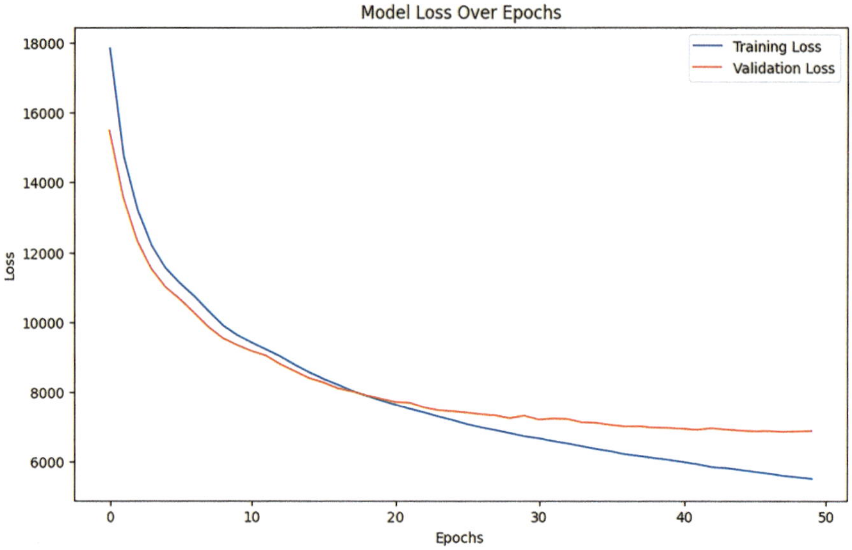

Fig. 17.7 LSTM Model training-validation loss over epochs

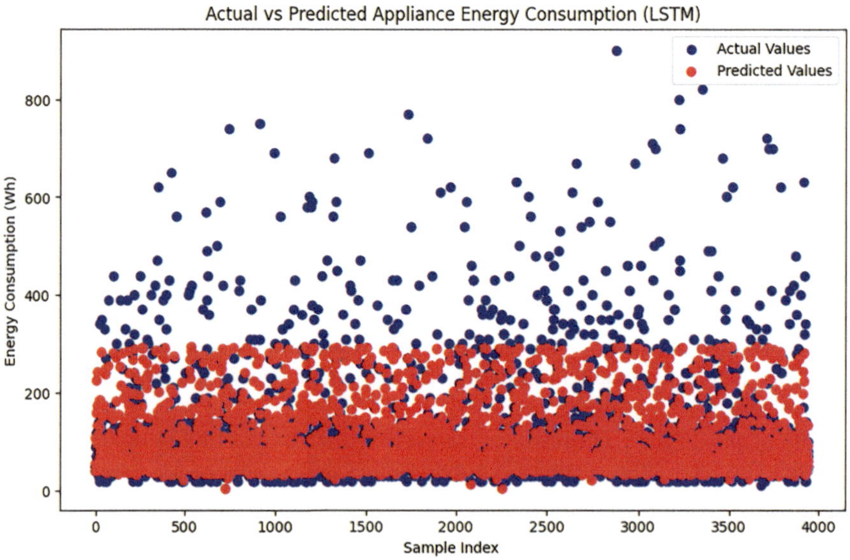

Fig. 17.8 Actual and predicted appliance energy consumption using LSTM

starting with 0, 500, 100, and so on. The y-axis displays the energy consumption in Wh (Watt-hour). The red dots show the values forecasted by the LSTM model, and the blue dots represent actual energy consumption values. Some blue dots representing the actual values are significantly higher than the main cluster of red dots.

This scenario further indicates the outlier instances that exhibited higher actual energy consumption than the predicted values. These outliers indicate that some patterns are not suitably learnt by the model. The overall scattering pattern in the plot indicates the model is reasonably performing well for most of the consumption levels even though there is difficulty in generalizing the extremely high consumption levels. The reason behind this can be there are comparatively very less instances of very high consumption in the dataset. Also, this seems to be a very common challenge for regression algorithms when the target variable like appliances is skewed.

17.5 Discussion

In this work, we have analyzed various findings where the smart grid technology integration can enhance and help manage the energy consumption of appliances. The demand prediction can be based on an hourly basis, daily patterns, and seasonal trends. By analyzing the hourly and daily patterns of energy consumption, the smart grid can understand the demand peaks and can accordingly optimize the energy distribution. We have visualized the weekday and weekend patterns that show various demands of energy consumption. The smart grid can perform real-time management by analyzing the hourly patterns that help in load balancing. This will also help to implement strategies that alert the customers to reduce their use during peak times. Time-based predictive modeling helps the smart grids create schedules for operating high-energy consumption home appliances by incentivizing the users to use them during off-peak hours. Thus, smart grid can achieve overall energy efficiency and can automate the control over heating, lighting, and cooling appliances according to the users' typical usage behavior. A time-based smart grid can also integrate renewable energy sources and align the energy demand with an additional supply. Notifying the customers of their time-based usage patterns will alert them about their energy use.

17.6 Conclusion and Future Work

To create a more sustainable energy system and lessen the effects of climate change, renewable energy resources must be integrated into networks. To forecast energy consumption and maximize the usage of renewable energy sources, machine learning hybrid models can be helpful. As consumer demand and power supply fluctuate, machine learning becomes an increasingly important tool for decision-making, supporting the stability and dependability of smart grids. The study's conclusions highlight the potential of deep learning and machine learning methods for smart grid stability assessments. This study emphasizes how crucial deep learning methods—in particular, DNNs—are to improving the robustness and efficiency of smart grids. It opens the door for clever and effective management/control systems that can handle the difficulties of contemporary power grids and bring sustainability through energy conservation. This research work can be further extended by adapting the dataset(s) which contain data about the seasonal trends of energy consumption, country-wise energy diversity, different energy sources, etc. Also, this work can be enhanced by applying different other deep learning algorithms.

Acknowledgments We are thankful to Dr. Akhilesh Kumar Yadav for allowing us to be a part of this project. Also, we express our gratitude to him for his support and assistance throughout the creation of this chapter.

References

Alquthami T, Zulfiqar M, Kamran M, Milyani AH, Rasheed MB (2022) A performance comparison of machine learning algorithms for load forecasting in smart grid. IEEE Access 10:48419–48433. https://doi.org/10.1109/ACCESS.2022.3171270

Candanedo LM, Feldheim V, Deramaix D (2017) Data driven prediction models of energy use of appliances in a low-energy house. Energ Buildings 140:81–97

Gao L, Liu T, Cao T, Hwang Y, Radermacher R (2021) Comparing deep learning models for multi energy vectors prediction on multiple types of building. Appl Energy 301:117486. https://doi.org/10.1016/j.apenergy.2021.117486

Ghorashi Khalil Abadi SA, Habibi SI, Khalili T, Bidram A (2022) A model predictive control strategy for performance improvement of hybrid energy storage systems in DC microgrids. IEEE Access 10:25400–25421. https://doi.org/10.1109/ACCESS.2022.3155668

Gunel K, Ekti AR (2019) Exploiting machine learning applications for smart grids. In: 2019 16th International multi-conference on systems, signals & devices (SSD). IEEE, pp 679–685. https://doi.org/10.1109/SSD.2019.8893252

Hossain E, Khan I, Un-Noor F, Sikander SS, Md. Samiul Haque Sunny. (2019) Application of big data and machine learning in smart grid, and associated security concerns: a review. IEEE Access 7:13960–13988. https://doi.org/10.1109/ACCESS.2019.2894819

Ibrar M, Hassan MA, Shaukat K, Alam TM, Khurshid KS, Hameed IA, Aljuaid H, Luo S (2022) A machine learning-based model for stability prediction of decentralized power grid linked with renewable energy resources. Wirel Commun Mob Comput 2022:1–15. https://doi.org/10.1155/2022/2697303

Kulwant M, Yadav AK (2024) Enhancing water conservation and management in India: a holistic approach. In: Yadav AK, Yadav K, Singh VP (eds) Integrated management of water resources in India: a computational approach. Springer Nature, Cham, pp 121–137. https://doi.org/10.1007/978-3-031-62079-9_7

Kumar A, Yadav AK (2024) Challenges and opportunities for water quality monitoring and management in India. In: Yadav AK, Yadav K, Singh VP (eds) Integrated management of water resources in India: a computational approach. Springer Nature, Cham, pp 493–511. https://doi.org/10.1007/978-3-031-62079-9_28

Lin L, Gao L, Kedzierski MA, Hwang Y (2022) A general model for flow boiling heat transfer in microfin tubes based on a new neural network architecture. Energy AI 8:100151. https://doi.org/10.1016/j.egyai.2022.100151

Liu H, Lang B (2019) Machine learning and deep learning methods for intrusion detection systems: a survey. Appl Sci 9:4396. https://doi.org/10.3390/app9204396

Mutarraf MU, Guan Y, Terriche Y, Chun-Lien S, Nasir M, Vasquez JC, Guerrero JM (2022) Adaptive power management of hierarchical controlled hybrid shipboard microgrids. IEEE Access 10:21397–21411. https://doi.org/10.1109/ACCESS.2022.3153109

Tiwari S, Jain A, Ahmed NMOS, Charu, Alkwai LM, Dafhalla AKY, Hamad SAS (2022) Machine learning-based model for prediction of power consumption in smart grid- smart way towards smart city. Expert Syst 39. https://doi.org/10.1111/exsy.12832

Zhang L, Gao Y, Zhu H, Tao L (2022) A distributed real-time pricing strategy based on reinforcement learning approach for smart grid. Expert Syst Appl 191:116285. https://doi.org/10.1016/j.eswa.2021.116285

Zhang Y, Yang Q, An D, Li D, Zongze W (2023) Multistep multiagent reinforcement learning for optimal energy schedule strategy of charging stations in smart grid. IEEE Trans Cybern 53:4292–4305. https://doi.org/10.1109/TCYB.2022.3165074

Zheng Y, Suryanarayanan S, Maciejewski AA, Siegel HJ, Hansen TM, Celik B (2019) An application of machine learning for a smart grid resource allocation problem. In: 2019 IEEE Milan PowerTech. IEEE, pp 1–6. https://doi.org/10.1109/PTC.2019.8810875

Chapter 18
Empowering Clean Energy Transition Through Advanced Battery Technologies

M. B. Yashoda, V. R. Revathy, Vrinda Shivashetty, and Hari S. Prabhu

Abstract The rapid growth in global energy demand has prompted a shift towards clean energy sources to mitigate the environmental impacts of traditional fossil fuels. Advanced battery technologies have emerged as a key enabler in this transition, offering grid stability, energy storage, and increased integration of renewable energy sources. This research explores the impact of advanced battery technologies on the empowerment of the clean energy transition. The study examines the current state of advanced battery technologies, including lithium-ion, flow, and solid-state batteries, and their role in promoting clean energy solutions. The research aims to identify key drivers and barriers to the widespread deployment of these technologies by analyzing data on market trends, policy frameworks, and technological advances.

Furthermore, the research investigates the economic, environmental, and social implications of advanced battery technologies in the context of the clean energy transition. By conducting case studies and data analysis, the study aims to quantify the potential benefits of integrating advanced batteries into the energy system, including cost savings, emissions reductions, and job creation. Furthermore, research assesses the challenges and opportunities associated with deploying advanced battery technologies, such as resource constraints, recycling and disposal issues, and regulatory hurdles. By engaging with industry experts, policymakers, and stakeholders, the study seeks to develop recommendations to promote a

M. B. Yashoda (✉)
Department of Computer Science and Engineering, Nitte Meenakshi Institute of Technology, Bengaluru, India

V. R. Revathy
Department of Computer Science, Kristu Jayanti College (Autonomous), Bengaluru, India

V. Shivashetty
Department of Information Science and Engineering, Sai Vidya Institute of Technology, Bengaluru, India

H. S. Prabhu
FormLabs, Bangalore, India

A. K. Yadav (ed.), *Powering Future: Clean Energy Solutions for a Sustainable Planet*, Clean Energy Production Technologies, https://doi.org/10.1007/978-981-96-5055-2_18

supportive environment to accelerate the adoption of these technologies. Overall, this research contributes to the growing body of literature on clean energy transition by shedding light on the role of advanced battery technologies in driving sustainable energy solutions. By exploring the impact of these technologies from multiple perspectives, the study aims to inform decision-makers and stakeholders about the opportunities and challenges in transitioning towards a cleaner and more resilient energy future.

Keywords Clean energy transition · Battery technology · Sustainable energy systems · Energy storage systems · Lithium-ion · Electric vehicles · Smart energy · Renewable energy

18.1 Introduction

The global landscape of energy demand has undergone profound changes in recent decades, driven by a combination of population growth, industrialization, and urbanization. With burgeoning economies and expanding infrastructures, energy demand has surged, placing unprecedented strain on traditional fossil fuel resources such as coal, oil, and natural gas. However, this growth in energy consumption has come at a significant cost to the environment, as the combustion of fossil fuels releases harmful pollutants and greenhouse gases, exacerbating climate change and posing serious health risks to communities around the world. Recognizing the urgent need to address these environmental challenges, there has been a growing consensus among policymakers, industry leaders, and civil society advocates to transition to cleaner and more sustainable energy sources.

In response to increasing environmental concerns, there has been a concerted global effort to accelerate the shift towards clean energy sources, characterized by renewables such as solar, wind, hydroelectric and geothermal power. Unlike fossil fuels, renewable energy technologies harness natural resources that are abundant and inexhaustible, and emit minimal or no greenhouse gases during electricity generation. In addition, advances in renewable energy technologies have led to significant cost reductions, making them increasingly competitive with conventional fossil fuels in many parts of the world. This transition to clean energy sources represents not only a critical imperative to mitigate the impacts of climate change but also an economic opportunity to foster innovation, job creation, and sustainable development.

However, despite the momentum behind the clean energy transition, significant challenges remain in achieving a rapid and widespread deployment of renewable energy technologies. Advanced battery technologies play a pivotal role in enabling the transition to clean energy by addressing key challenges associated with the integration of renewable energy sources into the grid. One of the primary hurdles in scaling up renewables such as solar and wind power is their inherent intermittency, as energy production fluctuates depending on weather conditions and time of day.

Advanced batteries offer a solution to this problem by providing efficient energy storage capabilities, allowing excess energy generated during periods of high production to be stored for use during times of low production. This not only helps to balance supply and demand on the grid, but also improves the stability and reliability (Moghimian et al. 2023) reducing the need for backup fossil fuel-based power plants and allowing a higher penetration of renewables.

Self-driving vehicles will be in operation by 2030. They are going to be shared and powered by electricity. As stated by the SAE in 2010, some commercially used vehicles had automation level 1 installed. Modern automobile engineering is at level 3, with some even having level 4 automation. They can use them for fresh transport services because they have systems for communication integrated with machine learning elements. The widespread application of Internet of Things (IoT) technology in modern life makes it easy to inspect, control, and keep track of a wide range of objects. Additionally, this will facilitate seamless cooperation and simplify data exchange (Alam et al. 2017; Munoz et al. 2018; Frustaci et al. 2018).

Artificial intelligence (AI), especially its productive subset known as machine learning (ML), emerges as a promising approach with the potential to revolutionize battery research and development (R&D). For example, AI techniques require less expertise and require shorter development times for designing complex battery systems compared to conventional model-based frameworks (Hossain Lipu et al. 2022a, 2022b). Furthermore, they demonstrate high efficacy when there are ample data available and excel in managing uncertainties such as noise, temperature variations, and aging effects. Additionally, they have self-learning capabilities for parameterization and rapid online execution (Gabbar et al. 2021).

18.2 Review

Developing green energy for neighborhoods requires incorporating energy from natural resources such as water, sunlight, and the wind. Renewable power sources have a considerably smaller effect on global warming than traditional energy sources such as fossil fuels, natural gas, or oil, even after considering the periodic implications of the fluctuation in a dynamic power system (Blaabjerg et al. 2006). There are issues with reliability of power, security, and islanding safety due to the growing number of distributed energy production units connected to the grid infrastructure. Remote operating system control is implemented to ensure compliance with grid connectivity standards (Messagie et al. 2014; Van Mierlo 2018). Utilization of green energy is increasing. In this case, we would have to rely on renewable energy sources or support funding for energy storage technologies. The capacity influences the efficiency of electric vehicles. Through "smart charge management," automobile batteries can store electricity generated from solar or wind sources. When additional power is required, the system can release the stored energy to replenish itself. The research findings attribute the prevalence of lithium-ion batteries in electric vehicles to their advantageous characteristics (Li et al. 2018). Because it controls vehicle vitality, lifespan, electrical performance, and protection, such a battery is essential for electric vehicles. Several novel technological advances in battery science, structure, and production have reduced the total price of electric vehicles and increased their reliability (Philippot et al. 2019; Schmuch et al. 2018). The authors (Armand and Tarascon 2008) offer an in-depth examination of various cutting-edge battery technologies such as lithium-ion, sodium-ion and lithium-sulfur batteries. It delves into their operational principles and the materials used and outlines challenges and prospects for future advancement. The analysis thoroughly examines the environmental consequences and sustainability aspects related to advanced battery technologies (Larcher and Tarascon 2015). It delves into various approaches aimed at enhancing the sustainability of battery materials and manufacturing techniques, while also considering the potential for recycling and repurposing (Luz et al. 2024) examines the main hurdles and factors pertinent to battery management in electric vehicles (EVs). It addresses issues such as battery aging, thermal regulation, and safety concerns, providing perspectives on current advances and potential avenues for future research.

The study (Luz et al. 2024) examines recent progress in electric vehicle (EV) battery technology, examining advanced lithium-ion alternatives and alternatives. The review discusses enhancements in lithium-ion batteries, including increased energy density, faster charging, and longer lifespan, as well as advances in electrode materials and manufacturing techniques. Furthermore, it explores alternative technologies such as solid-state batteries, highlighting their transformative potential in energy storage. Collaboration among industry, academia, and government is crucial for innovation, while ongoing research is urged to improve battery performance, lower costs, and improve environmental sustainability in EV battery manufacturing and recycling.

Renewable energy presents a promising avenue for tackling greenhouse gas emissions and promoting environmental conservation. Conventional electrical grids demonstrate inefficiency, experiencing notable energy losses during transmission, and maintaining excess capacity to meet peak demand. In contrast, grids provide regulated networks that efficiently handle renewable resources, streamline energy consumption, and enhance security and privacy measures. Integration of technologies with renewable energy requires the establishment of resilient communication networks, effective demand management strategies, and the development of appropriate regulatory frameworks, as underscored in the examination of Indian energy legislation and subsequent policy proposals. Based on the inclusion of clean energy sources and battery energy storage devices in the grid examines the challenges and opportunities associated with this technology change (Kataray et al. 2023; Kumar et al. 2022; Patel et al. 2022). Batteries are classified according to their kind and purpose in the battery technology marketplace (Fig. 18.1).

Battery types are classified within the battery technology market, spanning lead-acid batteries, lithium-ion batteries, lithium-metal batteries, nickel-cadmium batteries, nickel-metal batteries, and various other variants. By 2022, the lithium-ion segment will have secured a substantial 49.7% market share and is expected to maintain its leadership position in the forecast period. Although lithium-ion batteries have dominated, ongoing research is actively exploring alternative technologies.

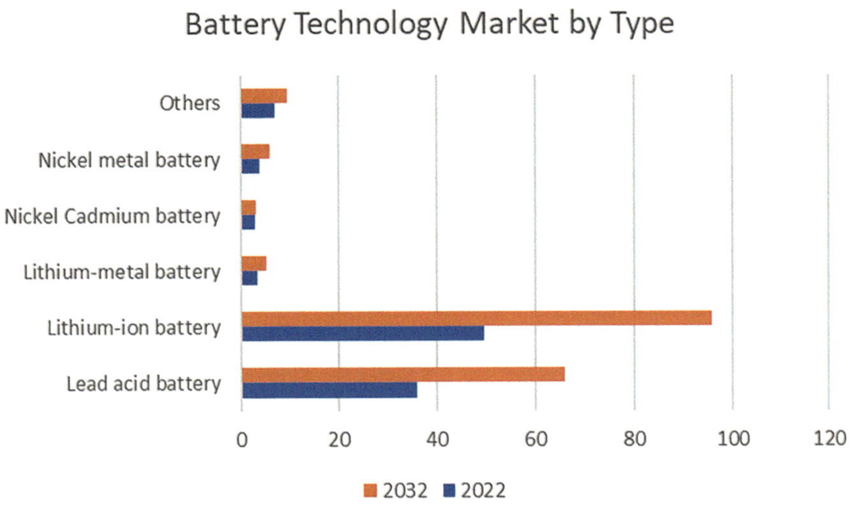

Fig. 18.1 Battery technology market by type

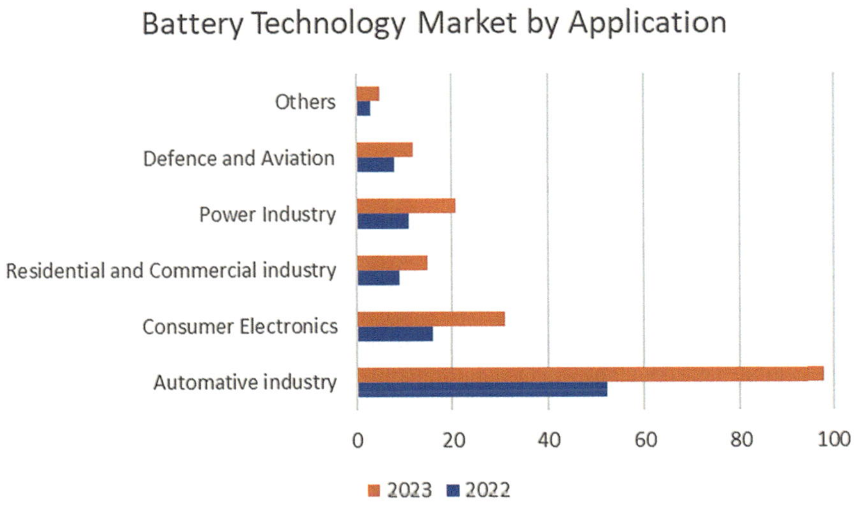

Fig. 18.2 Market for battery technology by application

There is a growing imperative for environmentally friendly and sustainable battery solutions, leading the industry to focus on battery recycling and repurposing initiatives to mitigate environmental impact. In 2022, the automotive segment claimed a majority share of 52.1% in the battery technology market to defend its supremacy in the forthcoming period. The automotive industry is experiencing a notable shift towards electric vehicles, intended to mitigate carbon emissions and lessen dependence on fossil fuels. Within this transition, battery technology emerges as a pivotal element, fuelling electric drivetrains (Fig. 18.2).

Previous research highlights the crucial contribution of advanced batteries to clean energy solutions, bolstering grid stability through surplus renewable energy, and alleviating supply-demand imbalances. These batteries facilitate efficient energy storage, thus streamlining the integration of renewable energy by addressing intermittent generation challenges. Additionally, advanced batteries play a pivotal role in the adoption of renewable energy adoption, providing dependable and economical storage options that enhance energy self-sufficiency and environmental sustainability. Together, they significantly advance clean energy technologies while fostering a resilient, reliable, and environmentally sustainable energy infrastructure.

18.3 Methodology

The automotive sector is currently experiencing a significant shift from traditional internal combustion engine vehicles, including those powered by diesel and gasoline, to the emerging dominance of second-generation hybrid and electric vehicles. Within this transition, lithium-ion (li-ion) batteries have emerged as a focal point of interest for the automotive industry. A critical element in electric vehicles is the battery management system (BMS), which plays a pivotal role in overseeing and regulating the performance of the battery pack. Given that the battery pack constitutes the most substantial and costly component of an electric vehicle, meticulous monitoring and control are imperative.

The remaining useful life data set appears to be relevant to battery Remaining Useful Life (RUL) prediction, which is an important task in battery management systems (BMS) and predictive maintenance, which impacts sustainable energy. This data set provides a variety of variables that explain the behavior and characteristics of batteries during several charging and discharging cycles. The data set has each row representing one cycle of battery operation. There are multiple columns, each representing a particular aspect or measurement obtained during the battery's operation. The columns include the Cycle_Index field indicates the cycle's index or number. It is a sequential number that begins at 1 and increases with each successive cycle. This index is critical for assessing the battery performance over time. Discharge time (s): This feature counts the number of seconds it takes for the battery to discharge during a specific cycle. Discharge time can be an important measure of battery health because it can indicate capacity decline. Decrement 3.6–3.4 V (s): This column records the time in seconds that the battery voltage lowers from 3.6 V to 3.4 V during discharge. This precise voltage range can provide information about the battery's performance and degradation characteristics. Maximum discharge voltage (V): This attribute shows the maximum voltage observed during the cycle's discharge phase. Monitoring the maximum discharge voltage can aid in detecting any irregularities or symptoms of decline in battery performance. Minimum voltage charging (V): This column records the minimum voltage achieved during the cycle's charging phase. Similar to discharge voltage, this statistic can reflect the efficiency and health over time.

Fig. 18.3 System architecture

Time at 4.15 V (s): This feature counts the seconds that the battery spends at 4.15 V when charging. This exact voltage level may be essential to understand the battery's charging behavior and efficiency.

Time-constant current (s): This column specifies the number of seconds that the battery is charged at a constant current. Consistent charging times with steady current are critical for battery health and longevity.

Charging time (s): This feature displays the total time, in seconds, required to charge the battery for a given cycle.

18.3.1 System Architecture

The architecture of the system in Fig. 18.3 outlines the holistic arrangement of the system, which includes its components, their interplay, and the data exchange among them. Within this framework, we have pinpointed the following components:

Preprocessing of Data: Data preparation is typically the initial phase in the deep learning workflow to format raw data into a desired structure. Additionally, employ data preprocessing to improve specific features or mitigate artifacts that might introduce bias into the neural network.

Splitting the Data: In deep learning, this method divides the data into training and testing sets. The training set is used for model development and training, often to estimate various parameters or evaluate different model performances. Following training, use the testing set to verify that the final model functions as intended by comparing its performance against the training data.

18.3.2 Exploratory Data Analysis

The data set contains details about the battery's life cycle. The Cycle_Index keeps track of each cycle, while Discharge Time measures how long it can deliver power. The decline of 3.6–3.4 V shows how quickly its voltage drops. We also see its highest and lowest voltages with maximum voltage discharge and minimum voltage charged. Moments of stability are revealed by T time at 4.15 V. It charges up as demonstrated by additional parameters such as time at constant current and charging time. Lastly, the remaining useful life (RUL) projects its remaining life. When taken as a whole, these figures depict the battery's journey from beginning to end. There are no missing values in the data set.

The degree to which two variables are linearly connected is statistically expressed by the term correlation. Figure 18.4 show the correlation between the different characteristics of the data set that are relevant to the prediction of the target variable RUL and relationship between the discharge time in seconds and the RUL of the battery. Plot the discharge time along the x-axis and the battery RUL along the y-axis. The relationship between discharge time (s) and RUL (Remaining Useful Life) is crucial to understanding battery performance and predicting battery life. The 3D plot shows the data points for "Discharge Time (s)" on the x-axis and "Remaining Useful Life (RUL)" on the y-axis. There is no obvious strong correlation between the two variables in the plot (Fig. 18.4). The data points are widely scattered, which indicates a weak or no clear relationship. The correlation coefficient between discharge time (s) and RUL is approximately 0.012 and as it is close to zero, the relationship between Discharge Time (s) and RUL indicates a very weak positive correlation. This implies that discharge time (s) alone is not a strong predictor of the battery's remaining useful life, and other factors or a combination of multiple factors might be needed to accurately predict the RUL.

The data indicate a strong positive association between time at 4.15 V (s) and discharge time (s), indicating that an increase in 4.15 V time is generally accompanied by an increase in discharge time (Fig. 18.5). The following are the reasons for their relationship: a) Battery characteristics: Batteries exhibit a range of voltage levels throughout the discharge cycle. A slower discharge process may be indicated by a battery spending a considerable amount of time at a particular voltage (4.15 V in this example), which lengthens the discharge time overall. State of Charge (SoC): The battery's state of charge may be indicated by the duration at 4.15 V. Higher levels of charge are typically associated with higher voltages. If the battery is held

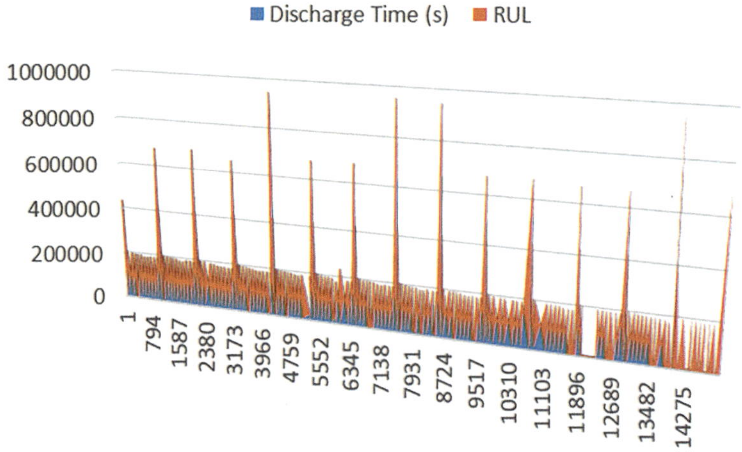

Fig. 18.4 3-D stacked plot for discharge time (s) and RUL (Remaining Useful Life) features

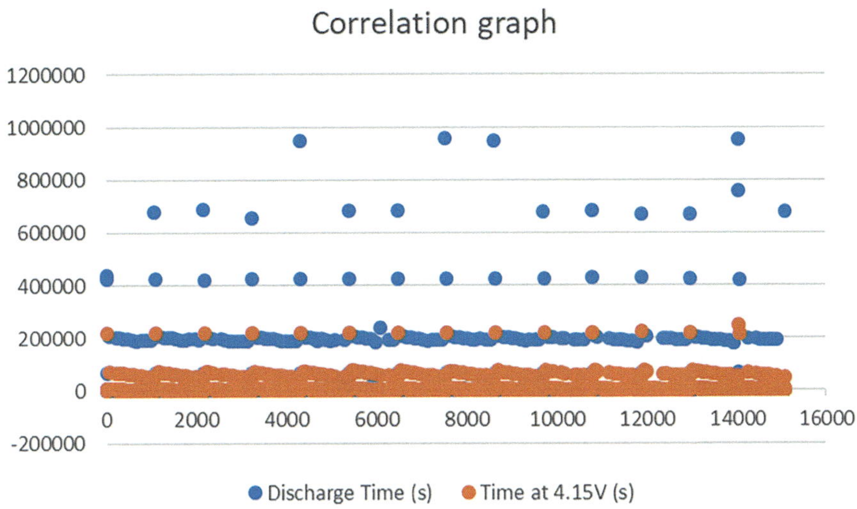

Fig. 18.5 Correlation between discharge time and Time at 4.15 V

at this higher voltage for an extended period of time, it may discharge more slowly, increasing the overall discharge time. Batteries deteriorate over time, and their voltage profiles change. If a degraded battery is forced to function at a higher voltage for a lengthy period of time (for internal resistance or other reasons), it will take longer to drain.

The scatter plot shows the data points for Max. Voltage discharge (V) on the x-axis and RUL on the y-axis (Fig. 18.6). The scatter plot would show data points that form an upward-sloping trend. As the maximum discharge voltage increases, so does the remaining useful life. The correlation coefficient would be 0.78. This number value implies a high positive correlation between the two variables. The high positive correlation indicates that changes in the maximum discharge voltage are closely related to changes in the remaining useful life.

Figure 18.7 shows the strong negative correlation between Min. Voltage Charge. (V) and Remaining Useful Life (RUL) in the data set. The points on the line plot show a clear downward trend. Batteries with greater minimum charging voltages would have a shorter usable life. Minimum voltage charging (V) is a reliable predictor of RUL. It can be used in predictive maintenance plans to properly estimate the remaining useful life of batteries. For battery management systems (BMS), this metric could be crucial. Monitoring Min. Voltage Charge. (V), BMS can provide timely alerts for maintenance or replacement, thereby preventing unexpected failures. To put this knowledge into practice, one could:

- Track and record data: Maintain a record in Min. Voltage Charge. (V) and RUL data to enhance prediction models' accuracy over time.
- Create predictive frameworks: Using machine learning techniques develops models that forecast RUL based on a range of characteristics. Because of its

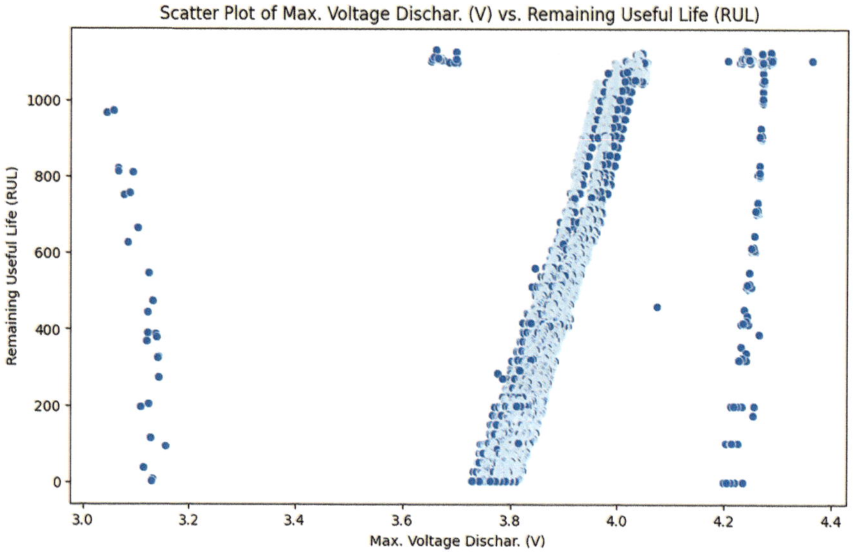

Fig. 18.6 Correlation between "Max. Voltage Dischar. (V)" and "RUL"

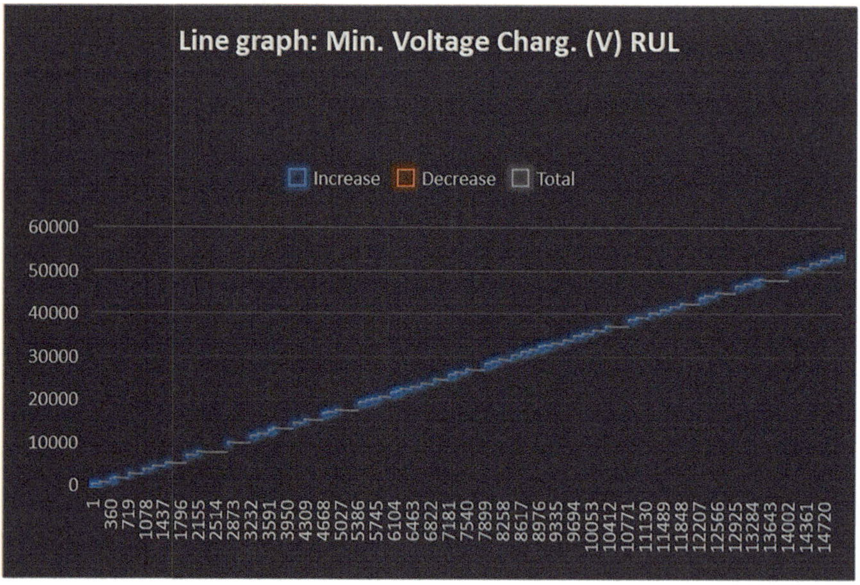

Fig. 18.7 Correlation between "Min. Voltage Charge. (V)" and "Remaining Useful Life (RUL)"

significant negative connection, the minimum voltage charge (V) is an important feature.

- Put alerts in place: To alert users or operators when the minimum voltage charge (V) reaches critical levels, indicating imminent battery degradation, and to set alarms in battery management systems. The endurance and dependability of battery-powered systems can be enhanced by utilizing the strong negative relationship between Min and Xiang et al. Voltage Charge. (V) and RUL to facilitate predictive maintenance and optimal battery management.

18.3.3 Gaussian Process Regression (GPR) and Its Applications

A potent class of statistical models known as Gaussian processes (GPs) enables us to specify a probability distribution over random function. We use our physical domain knowledge and/or accessible data to guide our modeling of the covariance between pairs of process samples rather than writing down an explicit mathematical formula for the function from which some observations are generated. GPs have several uses, ranging from high-dimensional smoothing and interpolation to modeling of stochastic physical processes, although this may sound abstract. In particular, throughout the past 10 years, the astronomical community has grown more and more in favor of Gaussian Process Regression (GPR). Time-domain data sets are becoming more accessible and important in astronomy, which is promoting their adoption. These are consistently made up of unknown or non-trivial random signals, whether instrumental or astrophysical, that must be modeled. These are typically nuisance signals that must be minimized to accurately identify or quantify other signals. There are moments when we wish to forecast the behavior or deduce the features of the stochastic behavior itself. GPR provides a convincing answer because it is mathematically straightforward, highly flexible, and based on statistical principles. It can also be easily integrated into a Bayesian inference framework. Nevertheless, GPR may appear a little mysterious to novice users due to the model's absence of a stated functional form. Moreover, the method's computing expense, which erroneously grows cubically with the size of the data collection, might be a challenge.

18.3.4 Applicability of GPR for This Research Work

The possible application of GPR should be tested by initially understanding the nature of the data. Gaussian Process Regression (GPR) can be used with continuous target variables, which means that the target can take any value within a range rather than being limited to discrete categories. The target feature for our data set is the

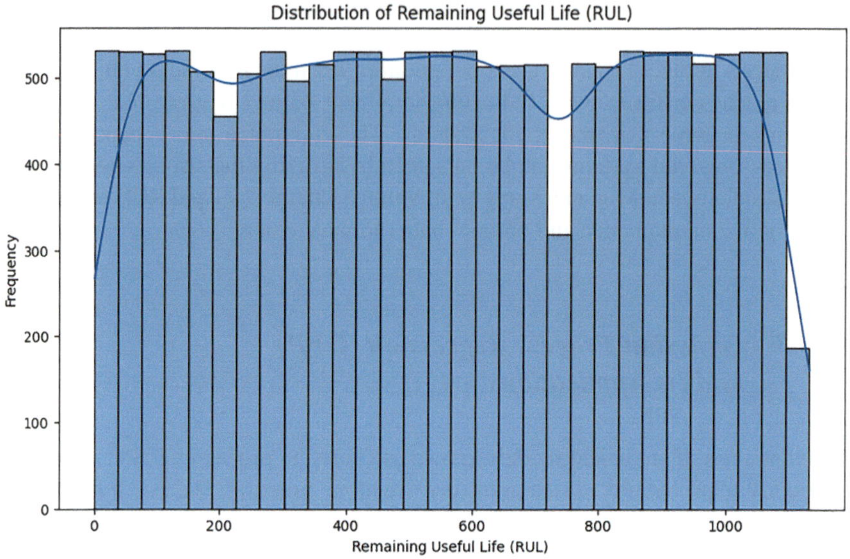

Fig. 18.8 Distribution of RUL

Remaining Useful Life (RUL), a continuous measure of the remaining useful life of a battery. In the data set, RUL refers to the number of cycles that remain before the battery is no longer usable. Figure 18.8 shows the distribution of RUL across the frequency. The values are not discrete categories, but rather numeric values that can take on a variety of forms, making them continuous. This makes GPR a suitable method to predict RUL because it can handle the continuous nature of the data. The histogram in Fig. 18.8 shows how the RUL values are spread across different ranges. It also shows a relatively even distribution of RUL values across different ranges. It is also clear that the summary statistics in Table 18.1 indicate that there is no extreme skewness or kurtosis, which concludes the finding that the class distribution is even, with similar numbers of instances in each class.

Any collection of observed data points can be modeled by a joint Gaussian distribution since GPR presupposes that the data can be represented by a Gaussian process. Because of this presumption, GPR can produce point forecasts as well as uncertainty estimates for those predictions, which can be useful in determining how reliable the model output is. The covariance function, which measures the degree of correlation between two points in the input space, is generally assumed to be stationary in the GPR. This indicates that rather than depending on the precise locations of the points in the input space, the covariance just depends on the distance between them. It is simpler to model a process using a Gaussian process when it is stationary, which indicates that the statistical characteristics of the process remain constant over time or space.

Table 18.1 Summary statistics of RUL

S. No.	Statistic	RUL
1	Count	15,064
2	Mean	554.194
3	Std. Deviation	322.435
4	Min	0.000
5	25th Percentile	277.000
6	50th Percentile	551.000
7	75th Percentile	839.000
8	Max	1133.000

18.4 Implementation

Before applying GPR, the preprocessing stage on the data. This involved checking for missing data, handling missing values, encoding categorical variables if present, and scaling the features. Standardizing or normalizing the features ensures that they are on a similar scale, which can improve the performance of the GPR model. Then the implementation was initiated with kernel selection. Kernel selection is performed by choosing a kernel function that measures the shape and smoothness of the predictions made by the GPR model. The rational quadratic kernel, the Matern kernel, and the radical basis function (RBF) kernel are popular options. Each kernel is unique and capable of capturing various kinds of relationships within the data. To determine which kernel best fits your data, you may need to experiment with a few different options, as the choice of kernel can greatly impact the model's performance.

18.4.1 Model Training and Validation

For performing the model training, initially the data set was divided into the training and validation sets. The battery residual useful life (RUL) training set is used to train the GPR model, while the validation set is used to evaluate the model's performance. Fit the GPR model to the training data during the training phase by fine-tuning the kernel's hyper parameters to best match the observed data points. The smoothness and flexibility of the predictions are affected by the selection of the kernel functions. Since it can represent smooth functions, the Radial Basis Function (RBF) kernel is a preferred option for many applications, and this work has also used it. During the validation, it is evaluated how well the model performs on the validation set to understand if it has a good generalization to new data. The R^2 score and the mean squared error (MSE) are two metrics that can be used to assess how reliable and accurate the predictions are.

18.5 Experimental Analysis

This chapter extensively examines various methodologies and approaches designed to precisely gauge the remaining useful life (RUL) Li-ion batteries using artificial intelligence (AI) techniques. Three leading machine learning methodologies, Gaussian Regression, Neural Networks, and Kernel Regression, are used to explore the complex intricacies of estimating a battery's RUL. The results of the experimental research are noteworthy in that they demonstrate the extraordinary efficiency and elevated precision of nonparameterized Gaussian Regression in discharge estimation. As an illustration, the Gaussian regression produces insignificant errors of 0.0031, 0.0014, and 0.0098 for mean and median absolute and root mean squared error (RMSE), respectively, and achieves an unsurpassed R2-score of 0.988. These outcomes underscore the effectiveness of artificial intelligence methodologies in accurately predicting the state of lithium-ion batteries, advocating their integration with battery management systems to enhance electric vehicle performance.

18.6 Conclusion

The remaining useful lifetime prediction has been carried out to plan the maintenance requirements of the automobile vehicle. Gaussian process regression can be a powerful tool for predicting the remaining useful life of batteries, provided that the data are appropriately pre-processed and the model is carefully tuned. Its ability to model complex relationships and provide uncertainty estimates makes it a valuable method in many regression tasks. However, due to its computational complexity, GPR is best suited for smaller data sets or scenarios where approximations can be applied effectively. By doing predictive maintenance, failures can be predicted and maintenance can be scheduled in advance. This reduces the cost and effort of performing maintenance. By predicting RUL, ML algorithms can assist in scheduling maintenance activities proactively. This can minimize downtime and reduce the overall maintenance costs associated with battery replacements or repairs. It can analyze various sensor data and operational parameters to optimize performance. This estimation helps to plan maintenance activities and prevent unexpected failures.

18.7 Future Research Directions

Future research should improve the energy density, durability, and safety of advanced battery technologies to increase their feasibility for widespread implementation in sustainable energy solutions. It is imperative to understand and address the environmental consequences of battery manufacturing and disposal to ensure that advanced battery technologies support sustainable energy solutions. Integrating advanced batteries with renewable energy sources such as solar and wind power could enhance energy generation and storage for a more sustainable energy grid. Researchers, industry collaborators, and policymakers must collaborate to advance advanced battery technologies for sustainable energy solutions.

Acknowledgment We are thankful to Dr. Akhilesh Kumar Yadav for allowing us to be a part of this project. Also, we express our gratitude to him for his support and assistance throughout the creation of this chapter. Special thanks to Mr. Krishnamoorthy A for his insightful feedback and constructive suggestions, which significantly enhanced the quality of this work and served as a constant source of motivation.

References

Alam F, Mehmood R, Katib I, Albogami NN, Albeshri A (2017) Data fusion and IoT for smart ubiquitous environments: a survey. IEEE Access 5:9533–9554. https://doi.org/10.1109/ACCESS.2017.2697839

Armand M, Tarascon JM (2008) Building better batteries. Nature 451:652. https://doi.org/10.1038/451652a

Blaabjerg F, Teodorescu R, Liserre M, Timbus AV (2006) Overview of control and grid synchronization for distributed power generation systems. IEEE Trans Ind Electron 53:1398–1409. https://doi.org/10.1109/TIE.2006.881997

Frustaci M, Pace P, Aloi G, Fortino G (2018) Evaluating critical security issues of the IoT world: present and future challenges. IEEE Internet Things J 5:2483–2495. https://doi.org/10.1109/JIOT.2017.2767291

Gabbar H, Othman A, Abdussami M (2021) Review of battery management systems (BMS) development and industrial standards. Technologies 9:28. https://doi.org/10.3390/technologies9020028

Hossain Lipu MS, Ansari S, Miah MS, Meraj ST, Hasan K, Shihavuddin ASM, Hannan MA, Muttaqi KM, Hussain A (2022a) Deep learning enabled state of charge, state of health and remaining useful life estimation for smart battery management system: methods, implementations, issues and prospects. J Energy Storage 55:105752. https://doi.org/10.1016/j.est.2022.105752

Hossain Lipu MS, Miah MS, Ansari S, Wali SB, Jamal T, Elavarasan RM, Kumar S et al (2022b) Smart battery management technology in electric vehicle applications: analytical and technical assessment toward emerging future directions. Batteries 8:219. https://doi.org/10.3390/batteries8110219

Kataray T, Nitesh B, Yarram B, Sinha S, Cuce E, Shaik S, Vigneshwaran P, Roy A (2023) Integration of smart grid with renewable energy sources: opportunities and challenges – a comprehensive review. Sustain Energy Technol Assess 58:103363. https://doi.org/10.1016/J.SETA.2023.103363

Kumar A, Shirin S, Ansari MI, Pandey G, Sharma SN, Yadav AK (2022) Fuzzy and neural network model-based environmental quality monitoring system: past, present, and future. In: Maurya SP, Yadav AK, Singh R (eds) Modeling and simulation of environmental systems: a computation approach. CRC Press. https://doi.org/10.1201/9781003203445-10

Larcher D, Tarascon JM (2015) Towards greener and more sustainable batteries for electrical energy storage. Nat Chem. https://doi.org/10.1038/nchem.2085

Li M, Jun L, Chen Z, Amine K (2018) 30 Years of lithium-ion batteries. Adv Mater 30. https://doi.org/10.1002/adma.201800561

Luz A, Kayode SO, Henry W, Sheriffdeen K (2024) Advancements in battery technology for electric vehicles: a comprehensive analysis of recent developments. Glob Mainstream J Innov Eng Emerg Technol 2(02):1–28

Messagie M, Mertens J, Oliveira L, Rangaraju S, Sanfelix J, Coosemans T, Van Mierlo J, Macharis C (2014) The hourly life cycle carbon footprint of electricity generation in Belgium, bringing a temporal resolution in life cycle assessment. Appl Energy 134:469–476. https://doi.org/10.1016/j.apenergy.2014.08.071

Moghimian S, Ouerdane H, Hoosh SM, Ouerdane H, Terzija V, Pozo D (2023) Assessing the value of energy storage systems for distribution grid applications. https://doi.org/10.48550/arXiv.2307.09380

Munoz R, Vilalta R, Yoshikane N, Casellas R, Martinez R, Tsuritani T, Morita I (2018) Integration of IoT, transport SDN, and edge/cloud computing for dynamic distribution of IoT analytics and efficient use of network resources. J Lightwave Technol 36:1420–1428. https://doi.org/10.1109/JLT.2018.2800660

Patel D, Kulwant M, Shirin S, Kumar A, Ansari MA, Yadav AK (2022) Artificial intelligence for air quality and control systems: status and future trends. In: Maurya SP, Yadav AK, Singh R (eds) Modeling and simulation of environmental systems: a computation approach. CRC Press. https://doi.org/10.1201/9781003203445-9

Philippot M, Alvarez G, Ayerbe E, Van Mierlo J, Messagie M (2019) Eco-efficiency of a lithium-ion battery for electric vehicles: influence of manufacturing country and commodity prices on GHG emissions and costs. Batteries 5:23. https://doi.org/10.3390/batteries5010023

Schmuch R, Wagner R, Hörpel G, Placke T, Winter M (2018) Performance and cost of materials for lithium-based rechargeable automotive batteries. Nat Energy 3:267–278. https://doi.org/10.1038/s41560-018-0107-2

Van Mierlo J (2018) The world electric vehicle journal, the open Access journal for the e-mobility scene. World Electr Veh J 9:1. https://doi.org/10.3390/wevj9010001

GPSR Compliance

The European Union's (EU) General Product Safety Regulation (GPSR) is a set of rules that requires consumer products to be safe and our obligations to ensure this.

If you have any concerns about our products, you can contact us on ProductSafety@springernature.com

In case Publisher is established outside the EU, the EU authorized representative is:

Springer Nature Customer Service Center GmbH
Europaplatz 3
69115 Heidelberg, Germany

Batch number: 10704307

Printed by Printforce, the Netherlands